AF734476

Recent Trends in Formal and Analytic Solutions of Diff. Equations

Contemporary Mathematics

782

Recent Trends in Formal and Analytic Solutions of Diff. Equations

Virtual Conference
Formal and Analytic Solutions of Diff. Equations
June 28–July 2, 2021
University of Alcalá, Alcalá de Henares, Spain

Galina Filipuk
Alberto Lastra
Sławomir Michalik
Editors

American Mathematical Society

2020 *Mathematics Subject Classification.* Primary 30D35, 34A26, 34C20, 34E05, 34M40, 35A02, 35C10, 35G10, 37J40, 38A13.

Library of Congress Cataloging-in-Publication Data

Names: Virtual Conference on Formal and Analytic Solutions of Diff. Equations (2021: Alcalá de Henares, Spain), author. | Filipuk, Galina, editor. | Lastra, Alberto, editor. | Michalik, Sławomir, editor.

Title: Recent trends in formal and analytic solutions of diff. equations: Virtual conference on formal and analytic solutions of diff. equations, June 28-July 2, 2021, University of Alcalá, Alcalá de Henares, Spain / Galina Filipuk, Alberto Lastra, Sławomir Michalik, editors.

Description: Providence, Rhode Island : American Mathematical Society, [2023] | Series: Contemporary mathematics, 0271-4132 ; volume 782 | Includes bibliographical references.

Identifiers: LCCN 2022032611 | ISBN 9781470466046 (paperback) | ISBN 9781470472092 (ebook)

Subjects: LCSH: Differential equations–Congresses. | Difference equations–Congresses. | AMS: Functions of a complex variable – Entire and meromorphic functions, and related topics – Distribution of values, Nevanlinna theory. | Ordinary differential equations – General theory – Geometric methods in differential equations. | Ordinary differential equations – Qualitative theory – Transformation and reduction of equations and systems, normal forms. | Ordinary differential equations – Asymptotic theory – Asymptotic expansions. | Ordinary differential equations – Differential equations in the complex domain – Stokes phenomena and connection problems (linear and nonlinear). | Partial differential equations – General topics – Uniqueness problems: global uniqueness, local uniqueness, non-uniqueness. | Partial differential equations – Representations of solutions – Series solutions. | Partial differential equations – General higher-order equations and systems – Initial value problems for linear higher-order equations. | Dynamical systems and ergodic theory – Finite-dimensional Hamiltonian, Lagrangian, contact, and nonholonomic systems – Perturbations, normal forms, small divisors, KAM theory, Arnold diffusion. | Difference and functional equations – Difference equations – Difference equations, scaling (q-differences).

Classification: LCC QA370 .V57 2021 | DDC 515/.35–dc23/eng20230106

LC record available at https://lccn.loc.gov/2022032611

Contemporary Mathematics ISSN: 0271-4132 (print); ISSN: 1098-3627 (online)

DOI: https://doi.org/10.1090/conm/782

∞ The paper used in this book is acid-free and falls within the guidelines
established to ensure permanence and durability.

Visit the AMS home page at `https://www.ams.org/`

10 9 8 7 6 5 4 3 2 1   23 22 21 20 19 18

Contents

Preface

The book presents the state of the art in the study of functional equations in the complex domain, including differential equations, difference equations, q-difference equations, moment-differential equations, partial differential equations, etc. For this reason, the title of the volume contains the word "diff." aiming to embrace all of them together. The volume consists of selected peer-reviewed contributions from the conference "Formal and Analytic Solutions of Diff. Equations on the Internet", which was held online from June 28th to July 2nd, 2021. The editors want to express sincere gratitude to all the participants of the conference who made it possible to exchange scientific ideas and results, to examine different directions for future research and to promote both new and existing collaborations of researchers in these topics, despite the world pandemic situation. The virtual conference venue was the School of Architecture of the University of Alcalá, located in the city center of Alcalá de Henares (Spain).

The topics of the conference were:

- Ordinary differential equations in the complex domain. Formal and analytic solutions. Stokes multipliers.
- Formal and analytic solutions of partial differential equations.
- Formal and analytic solutions of difference equations (including q-difference and differential-difference equations).
- Special functions (hypergeometric functions and others), orthogonal polynomials, continuous and discrete Painlevé equations.
- Integrable systems.
- Holomorphic vector fields. Normal forms.
- Asymptotic expansions, Borel summability.

Scientific Committee consisted of Galina Filipuk (University of Warsaw, Poland), Alberto Lastra (University of Alcalá, Spain), Stephane Malek (University of Lille, France), Jorge Mozo-Fernández (University of Valladolid, Spain), Yasunori Okada (Chiba University, Japan), Hidetoshi Tahara (Sophia University, Japan).

Organizing Committee consisted of Javier Jiménez-Garrido (University of Cantabria, Spain), Alberto Lastra (University of Alcalá, Spain), Ignacio Miguel (University of Valladolid, Spain), Andrew Pickering (Universidad Rey Juan Carlos, Spain).

This volume presents recent advances in the study of functional equations made by experts in the field. A. D. Bruno considers normal forms of a polynomial ordinary differential equation (ODE) and A. B. Batkhin studies the procedure of deriving homological equations of arbitrary order for the Hamiltonian normal form. Hamiltonian systems are also considered from the summability point of view in the contribution by M. Yoshino. E. Ciechanowicz deals with the so-called Painlevé sigma equations from the Nevanlinna theory point of view. G. Filipuk, A. Ligęza and A. Stokes focus on the third Painlevé equation and present relations between different Hamiltonian forms of this equation. In the last years, one finds a growing interest in the study of q-equations from the analytic and formal point of view. The contribution by S. Sasaki, S. Takagi and K. Takemura is an example of this direction. They consider the so-called q-Heun equation. The contribution by C. Zhang concerns the positive powers of the q-analogs of Euler series, which is of great interest and importance in the theory of summability of formal solutions to functional equations. The contribution by H. Ogawara deals with the differential transcendence of solutions for q-difference equation of Ramanujan function. Systems of difference equations coming from orthogonal polynomials are studied in the work by Y. Chen, G. Filipuk and M. N. Rebocho. Singular perturbed equations are also of interest in this volume. The work by T. Aoki and S. Uchida is an example of this. They search for the structure of the Voros coefficients of the generalized hypergeometric differential equation. T. Oshima studies Riemann-Liouville transform of linear differential equations on the Riemann sphere. M. Cafasso and S. Tarricone apply the Riemann-Hilbert problem technique to the Airy process. H. Tahara considers a family of nonlinear singular second order partial differential equations in the complex plane, determines uniqueness of solutions, and applies this result to the analytic continuation of solutions. As mentioned above, the theory of summability of formal solutions to functional equations is also considered in this volume. We refer to the contribution by M. Suwińska on the study of moment functional equations, and more precisely on moment integro-differential equations. The volume concludes with the state of the art regarding such emerging functional equations, in a work by A. Lastra, S. Michalik and M. Suwińska.

We hope that the present volume will be of interest not only for researchers in this field of knowledge but also for graduate students who aim to acquire new techniques and learn new results.

G. Filipuk
A. Lastra
S. Michalik

Contemporary Mathematics
Volume **782**, 2023
https://doi.org/10.1090/conm/782/15717

Normal forms of a polynomial ODE

Alexander D. Bruno

Abstract. We consider an ordinary differential equation (ODE) of order n, which is a polynomial of the independent variable x, the dependent variable y and all its derivatives up to the order n. To such equation we put in correspondence its Newton polygon. If the polygon has a vertex $(v, 1)$ and corresponding truncated equation has eigenvalues $\lambda_1, \ldots, \lambda_n$, then there exists such formal substitution $y = z + \varphi(x)$, where $\varphi(x)$ is a power series, that for $z = z' = \cdots = z^{(n)} = 0$ the transformed equation has only resonant terms $a_m x^m$, where $m = v + \lambda_k \in \mathbb{Z}$. It is true near each of two points: $x = 0$ and $x = \infty$.

1. Introduction

In the present paper we consider an ordinary differential equation

$$f\left(x, y, y', \ldots, y^{(n)}\right) = 0,$$

where f is a polynomial of its arguments, x is the independent variable and y is the dependent variable $x, y \in \mathbb{C}$.

We first recall some notions of power geometry [**1**], [**2**]. We set $X = (x, y)$. By a *differential monomial* $a(x, y)$ we mean the product of an ordinary monomial

$$cx^{r_1}y^{r_2} \stackrel{\text{def}}{=} cX^R, \tag{1.1}$$

where $c = \text{const} \in \mathbb{C}$, $R = (r_1, r_2) \in \mathbb{R}^2$, with finitely many derivatives of the form

$$d^l y/dx^l, \quad l \in \mathbb{N}. \tag{1.2}$$

A sum of differential monomials

$$f(X) = \sum a_i(X) \tag{1.3}$$

is called a *differential sum*.

To every differential monomial $a(X)$ one assigns its (vector) *exponent* $Q(a) = (q_1, q_2) \in \mathbb{R}^2$ by the following rules. For a monomial of the form (1.1) let

$$Q\left(cX^R\right) = R,$$

that is, $Q\left(cx^{r_1}y^{r_2}\right) = (r_1, r_2)$; for a derivative of the form (1.2) let

$$Q\left(d^l y/dx^l\right) = (-l, 1).$$

2020 *Mathematics Subject Classification.* Primary 34E05, 34A26; Secondary 34A25, 34C20.
Key words and phrases. ordinary differential equation, resonant term, normal form.

When differential monomials are multiplied, their exponents are summed as vectors:

$$Q(a_1 a_2) = Q(a_1) + Q(a_2).$$

The set $\mathbf{S}(f)$ of exponents $Q(a_i)$ of all the differential monomials $a_i(X)$ in a differential sum of the form (1.3) is called the *support* of the sum $f(X)$. Obviously, $\mathbf{S}(f) \in \mathbb{R}^2$. We denote by $f_Q(X)$ the sum of the monomials $a_i(X)$ in (1.3) for which $Q(a_i) = Q$. Then the differential sum (1.3) can be represented in the form

$$f(X) = \sum f_Q(X) \quad \text{over} \quad Q \in \mathbf{S}(f).$$

The closure $\Gamma(f)$ of the convex hull of the support $\mathbf{S}(f)$ is referred to as the *Newton polygon of the sum* $f(X)$. The boundary $\partial\Gamma(f)$ of the polygon $\Gamma(f)$ consists of vertices $\Gamma_j^{(0)}$ and edges $\Gamma_j^{(1)}$. These objects are called (generalized) *faces* $\Gamma_j^{(d)}$, where the superscript indicates the dimension of the face and the subscript is the index of the face. Corresponding to any face $\Gamma_j^{(d)}$ are the related *boundary subset* $\mathbf{S}_j^{(d)} = \mathbf{S}(f) \cap \Gamma_j^{(d)}$ of the set $\mathbf{S}$ and the *truncated sum*

$$\hat{f}_j^{(d)}(X) = \sum a_i(X) \quad \text{over} \quad Q(a_i) \in \mathbf{S}_j^{(d)}. \tag{1.4}$$

Let $\mathbb{R}_*^2$ be the plane conjugate to the plane $\mathbb{R}^2$ so that the inner product

$$\langle P, Q \rangle \stackrel{\text{def}}{=} p_1 q_1 + p_2 q_2$$

is defined for any $P = (p_1, p_2) \in \mathbb{R}_*^2$ and $Q = (q_1, q_2) \in \mathbb{R}^2$. Corresponding to any face $\Gamma_j^{(d)}$ in the plane $\mathbb{R}_*^2$ is its *normal cone*,

$$\mathbf{U}_j^{(d)} = \left\{ P : \begin{array}{ll} \langle P, Q \rangle = \langle P, Q' \rangle, & Q, Q' \in \mathbf{S}_j^{(d)} \\ \langle P, Q \rangle > \langle P, Q'' \rangle, & Q'' \in \mathbf{S}(f) \backslash \mathbf{S}_j^{(d)} \end{array} \right\}.$$

Let N_j be an outward normal vector to the edge $\Gamma_j^{(1)}$. The normal subspace $\mathbf{N}_j^{(1)}$ of an edge $\Gamma_j^{(1)}$ is the line passing through the origin $P = 0$ and the point $P = N_j$, and the normal cone is the ray of $\mathbf{N}_j^{(1)}$ that is directed outwards with respect to the edge $\Gamma_j^{(1)}$ of the polygon $\Gamma(f)$, that is, is spanned by the vector N_j. For a vertex $\Gamma_j^{(0)}$ the corresponding normal subspace is $\mathbf{N}_j^{(0)} = \mathbb{R}_*^2$, and the normal cone $\mathbf{U}_j^{(0)}$ is the open sector (angle) on $\mathbb{R}_*^2$ with vertex at $P = 0$ which is bounded by the rays that are the normal cones of the edges incident to $\Gamma_j^{(0)}$.

Thus, corresponding to any face $\Gamma_j^{(d)}$ are the normal cone $\mathbf{U}_j^{(d)}$ in $\mathbb{R}_*^2$ and the truncated equation

$$\hat{f}_j^{(d)}(X) = 0. \tag{1.5}$$

Corresponding to a vertex $\Gamma_j^{(0)} = \{Q\}$ is a truncated equation (1.5) with one-point support Q and with $d = 0$. We set $g(X) \stackrel{\text{def}}{=} X^{-Q} \hat{f}_j^{(0)}(X)$. Then the solution $y = cx^r$ of (1.5) satisfies the equation

$$g(X) = 0.$$

Substituting $y = cx^r$ into $g(X)$, we see that $g(x, cx^r)$ does not depend on x and c and so it is a polynomial in r, that is,

$$g(x, cx^r) \equiv \chi(r),$$

where $\chi(r)$ is the characteristic polynomial of the differential sum $\hat{f}_j^{(0)}(X)$. Hence, for a solution $y = cx^r$ of the equation (1.5) the exponent r is a root of the characteristic equation

$$\chi(r) \stackrel{\text{def}}{=} g(x, x^r) = 0, \tag{1.6}$$

and the coefficient c_r is arbitrary. Roots $r_j \stackrel{\text{def}}{=} \lambda_j$ of the equation (1.6) are called as *eigenvalues of the differential sum* $\hat{f}_j^{(0)}(X)$, corresponding to a vertex. If the differential sum $\hat{f}_j^{(0)}(X)$ has order l, then the characteristic equation (1.6) has l roots, $0 \le l \le n$.

THEOREM 1. *Let*

(1) *$f(X)$ be a polynomial in x, y, …, $y^{(n)}$;*
(2) *its Newton polygon $\Gamma(f)$ have a vertex $\Gamma_1^{(0)} = (v, 1)$ at the left side of its boundary $\partial\Gamma$;*
(3) *truncated differential sum $\hat{f}_1^{(0)}(X)$ have eigenvalues $\lambda_1, \ldots, \lambda_l$, $0 \le l \le n$;*
(4) *the most left point of the support $\mathbf{S}(f)$ in the axis $q_2 = 0$ be $(\alpha, 0)$. Evidently $\alpha \in \mathbb{Z}$.*

Then there exists such power series $\varphi(x)$ with integral increasing exponents, that after substitution

$$y = z + \varphi(x) \tag{1.7}$$

the transformed differential sum

$$g(x, z) = f(x, z + \varphi(x)) \tag{1.8}$$

for

$$z = z' = \cdots = z^{(n)} = 0 \tag{1.9}$$

has only resonant terms $b_m x^m$, where

$$m = v + \lambda_k \in \mathbb{Z} \tag{1.10}$$

and $m \ge \alpha$.

So here the eigenvalue λ_k is resonant if $\alpha - v \le \lambda_k \in \mathbb{Z}$. Proof of Theorem 1 is in Section 2.

THEOREM 2. *Let*

(1) *$f(x, y, y', \ldots, y^{(n)})$ be a polynomial in x, y, y', …, $y^{(n)}$;*
(2) *its Newton polygon $\Gamma(f)$ have a vertex $\Gamma_1^{(0)} = (v, 1)$ at the right side of its boundary $\partial\Gamma$;*
(3) *truncated differential sum $\hat{f}_j^{(0)}(X)$ have eigenvalues $\lambda_1, \ldots, \lambda_l$, $0 \le l \le n$;*
(4) *the most right point of the support $\mathbf{S}(f)$ in the axis $q_2 = 0$ be $(\beta, 0)$. Evidently $\beta \in \mathbb{Z}$.*

Then there exists such power series $\varphi(x)$ with integral decreasing exponents, that after substitution (1.7), the differential sum (1.8) for identities (1.9) has only resonant terms $b_m x^m$, where equality (1.10) is true, and $m \le \beta$.

So here the eigenvalue λ_k is resonant if $\beta - v \ge \lambda_k \in \mathbb{Z}$. Proof of Theorem 2 is in Section 3. Equations $g(x, z) = 0$ for (1.8) in situations of Theorems 1 and 2 we will call *normal forms*.

In Section 4 we describe a method for computation of eigenvalues λ_k for a simple power solution $y = cx^r$ of a truncated equation $\hat{f}_j^{(d)}(X) = 0$, where either $d \neq 0$ or $d = 0$, but its vertex $\Gamma_j^{(0)} = (n, w)$ has $w \neq 1$.

COROLLARY 1. *If the truncated sum $\hat{f}_j^{(0)}(X)$ has no integral eigenvalue $\lambda_k \geq \alpha - v$ (for Theorem 1) or $\lambda_k \leq \beta - v$ (for Theorem 2), then the initial equation $f(X) = 0$ has formal solution $y = \varphi(x)$. If the truncated sum $\hat{f}_j^{(0)}(X)$ contains the derivation $y^{(n)}$, then the series $\varphi(x)$ converges according to Theorem 3.4 in* [**2**].

REMARK 1. If the truncated sum $\hat{f}_j^{(0)}(X)$ has integral eigenvalue $\lambda_k \geq \alpha - v$ (for Theorem 1) or $\lambda_k \leq \beta - v$ (for Theorem 2), then the initial equation $f(X) = 0$ has formal power-logarithmic (or Dulac) solution

$$y = c_r x^r + \sum c_s x^s,$$

where c_r and c_s are polynomials in $\log x$. It converges under the same condition, that the truncated sum $\hat{f}_j^{(0)}(X)$ contains $y^{(n)}$ (see [**3**]).

2. Proof of Theorem 1

We will successively compute coeffitients $c_s = \text{const}$ of the series

$$\varphi(x) = \sum_{s=\alpha-v}^{+\infty} c_s x^s \tag{2.1}$$

according to increase the exponent s. For each integral $s \geq \alpha - v$, we obtain the following equation

$$\hat{f}_1^{(0)}(x, c_s x^s) + b_{s+v} x^{s+v} = 0, \tag{2.2}$$

where b_{s+v} is a constant obtained as a result of previous computation. But $\hat{f}_1^{(0)}(x, c_s x^s) = c_s x^{s+v} \chi(s)$, where $\chi(s)$ is the characteristic polynomial of the truncated sum $\hat{f}_1^{(0)}(X)$. So equation (2.2) has the form

$$c_s \chi(s) + b_{s+v} = 0.$$

If $\chi(s) \neq 0$, then we put $c_s = -b_{s+v}/\chi(s)$ and eliminate the term $b_{s+v}x^{s+v}$ from our equation. If $\chi(s) = 0$, then we cannot exclude the term $b_{s+v}x^{s+v}$ by such simple substitution. So the term $b_{s+v}x^{s+v}$ can be in the final equation. And $\chi(s) = 0$ only if s is an eigenvalue of the characteristic equation (1.6). The proof is finished.

3. Proof of Theorem 2

Now we will successively compute constant coefficients c_s of the series

$$\varphi(x) = \sum_{s=\beta-v}^{-\infty} c_s x^s \tag{3.1}$$

according to decrease exponents s. Then for each integral $s \leq \beta - v$, we obtain the equation (2.2), where constant coefficient b_{s+v} was obtained during the previous computation. Further proof coincides with the proof of Theorem 1 after the formula (2.2). The proof is finished.

4. Eigenvalues of a simple power truncated solution

Consider a truncated equation

$$\hat{f}_j^{(d)}(X) = 0, \tag{4.1}$$

which is different from equation $\hat{f}_1^{(0)}(X)$ considered in Theorems 1 and 2. Let it have the normal cone $\mathbf{U}_j^{(d)}$ and a power solution

$$y = c_r x^r, \quad c = \text{const} \in \mathbb{C} \tag{4.2}$$

with vector $\omega(1, r) \in \mathbf{U}_j^{(d)}$, where

$$\omega = \begin{cases} -1 & \text{for the case } x \to 0, \\ +1 & \text{for the case } x \to \infty. \end{cases}$$

Then the change of variables

$$y = c_r x^r + z \tag{4.3}$$

reduces the initial equation $f(X) = 0$ to the form

$$\tilde{f}(x, z) \stackrel{\text{def}}{=} f\left(x, cx^r + z\right) = 0, \tag{4.4}$$

where $\tilde{f}(x, z)$ is a differential sum such that all the points $Q = (q_1, q_2)$ of its support $\mathbf{S}(\tilde{f})$ have a non-negative integral coordinate q_2. One can apply to the equation (4.4) the above manipulations (that is, calculation of the support, the polygon, the truncation, and so on).

If the power solution (4.2) of the truncated equation (4.1) is simple, then the equation (4.4) is of the form

$$\tilde{f}(x, z) \stackrel{\text{def}}{=} \mathcal{L}(x) z + h(x, z) = 0, \tag{4.5}$$

where $\mathcal{L}(x)$ stands for a linear differential operator, the support $\mathbf{S}(\mathcal{L}z)$ consists of a single point $(v, 1)$ which is a vertex $\widetilde{\Gamma}_1^{(0)}$ of the polygon $\Gamma(\tilde{f})$, every point $Q = (q_1, q_2)$ of the support $\mathbf{S}(h)$ has a non-negative second coordinate $(q_2 \geq 0)$, and there is no point $Q = (v, 1)$ for which the normal cone of the vertex $\widetilde{\Gamma}_1^{(0)}$ contains a vector $P = (p_1, p_2)$ with $p_1 \omega > 0$.

By analogy with the Fréchet derivative [**4**] which is well known in functional analysis, we introduce for any differential sum $f(x, y)$ a formal *Fréchet derivative* (or, in other words, the *first variation*) $\delta f(x, y)/\delta y$, which has the following properties and is determined by them:

$$\begin{gathered} \delta\left(cx^{q_1} y^{q_2}\right)/\delta y = c q_2 x^{q_1} y^{q_2 - 1}, \\ \delta\left(d^l y/dx^l\right)/\delta y = d^l/dx^l, \\ \delta(f + g)/\delta y = \delta f/\delta y + \delta g/\delta y, \\ \delta(fg)/\delta y = (\delta f/\delta y) g + f(\delta g/\delta y). \end{gathered}$$

In the exposition below we omit the adjective 'formal' for brevity. By the second property, the Fréchet derivative is a linear differential operator, that is, it has the form

$$\sum_{k=0}^{l} g_k(x, y) \frac{d^k}{dx^k},$$

where $g_k(x, y)$ are differential sums.

THEOREM 3 ([**5**],[**6**],[**7**]). *Let the function* (4.2) *be a solution to the truncated equation* (4.1) *with* $\omega(1,r) \in \mathbf{U}_j^{(d)}$. *Then the operator* $\mathcal{L}(x)$ *in the equation* (4.5) *is*

$$\mathcal{L}(x) = \frac{\delta \hat{f}_j^{(d)}(x,y)}{\delta y} \quad on \quad y = c_r x^r. \tag{4.6}$$

That is, it is equal to the first variation calculated on the solution (4.2). *Moreover,* $\mathbf{S}(\mathcal{L}z) = (v,1)$, *where* $v = \langle Q_1, (1,r)\rangle - r$ *with* $Q_1 \in \Gamma_j^{(d)}$.

Hence, after the substitution (4.3) the equation $f(X) = 0$ acquires the form (4.5) if

$$\mathcal{L}(x) \not\equiv 0.$$

Let $\nu(k)$ be the characteristic polynomial of the differential sum $\mathcal{L}(x)z$, that is,

$$\nu(k) = x^{-v-k}\mathcal{L}(x)x^k.$$

If $\nu(k) \not\equiv 0$, then the roots $k_1, \ldots, k_s$, of the polynomial $\nu(k)$ are called *eigenvalues of the truncated solution* (4.2).

References

[1] A. D. Bruno, *Power Geometry in Algebraic and Differential Equations* (Russian), "Nauka", Moscow; English transl., Elsevier, Amsterdam, 2000., 1998. MR1734769

[2] A. D. Bruno, *Asymptotic behavior and expansions of solutions of an ordinary differential equation* (Russian, with Russian summary), Uspekhi Mat. Nauk **59** (2004), no. 3(357), 31–80, DOI 10.1070/RM2004v059n03ABEH000736; English transl., Russian Math. Surveys **59** (2004), no. 3, 429–480. MR2116535

[3] R. R. Gontsov and I. V. Goryuchkina, *On the convergence of formal Dulac series satisfying an algebraic ODE* (Russian, with Russian summary), Mat. Sb. **210** (2019), no. 9, 3–18, DOI 10.4213/sm9064; English transl., Sb. Math. **210** (2019), no. 9, 1207–1221. MR4017592

[4] V. M. Tikhomirov, *The Fréchet derivative*, Mathematical Encyclopaedia, vol. 5, Soviet Encyclopaedia, Moscow 1985, p. 666; English transl., Encyclopaedia of Mathematics, vol. 4, Kluwer, Dordrecht 1989, p. 93.

[5] A.D. Bruno, *Power geometry as a new calculus*, Analysis and applications—ISAAC 2001 (Berlin), Int. Soc. Anal. Appl. Comput., vol. 10, Kluwer Acad. Publ., Dordrecht, 2003, pp. 51–71, DOI 10.1007/978-1-4757-3741-7_4. MR2022739

[6] A. D. Bruno, *Power expansions of solutions of an algebraic or a differential equation* (Russian), Preprint no. 63, Keldysh Institute of Applied Mathematics, Moscow, 2000.

[7] A. D. Bruno, *Power expansions of solutions of an algebraic or a differential equation* (Russian), Dokl. Akad. Nauk **380** (2001), no. 2, 155–159. MR1873273

KELDYSH INSTITUTE OF APPLIED MATHEMATICS OF RAS, MIUSSKAYA SQ. 4, MOSCOW, 125047, RUSSIA

Email address: abruno@keldysh.ru

Contemporary Mathematics
Volume **782**, 2023
https://doi.org/10.1090/conm/782/15718

Computation of homological equations for Hamiltonian normal form

Alexander B. Batkhin

ABSTRACT. The procedure of deriving homological equations of arbitrary order, which solutions are used in iterative procedure of normalization of a Hamiltonian in a neighborhood of an equilibrium position, is considered. A formula for a homological equation of arbitrary order used in the method of normalization by means of the Lie series is proposed. The normalization procedure is applied to Hamiltonian of the Hill problem written in scaled regular variables. The resulting normal form of the Hill problem can be used to find sets of analyticity of the normalizing transformation.

1. Introduction

Normal form (NF) of a system of ordinary differential equations (ODE) computed near an invariant manifold (stationary point, periodic solution or invariant torus) is rather powerful technique for investigation of local dynamics of the phase flow in the vicinity of this invariant structure.

Even though the NF is a formal object it can be used for searching first integrals of the system, families of periodic solutions, for studying integrability, stability and bifurcations (see [**5, 7, 30**] for more details).

The special properties of Hamilton systems require specific algorithms for computation their NF. The goal of the presented work is to provide a procedure for constructing so called homological equation of any order, which is used in the modern normalization techniques.

Remark on notations.

- Boldface symbols like $\mathbf{x}, \mathbf{y}, \mathbf{u}, \mathbf{v}$ denote column-vectors in n-dimensional real $\mathbb{R}^n$ or complex $\mathbb{C}^n$ spaces.
- Boldface symbols like $\mathbf{p}, \mathbf{q}$ denote vectors in n-dimensional integer lattice $\mathbb{Z}^n$.
- $|\mathbf{p}| = \sum_{j=1}^{n} |p_j|$ denotes vector norm.
- For $\mathbf{x} = (x_1, \dots, x_n)^{\mathrm{T}}$ and $\mathbf{p} = (p_1, \dots, p_n)^{\mathrm{T}}$ we denote the multi-index power by $\mathbf{x}^{\mathbf{p}} \equiv \prod_{j=1}^{n} x_j^{p_j}$ and scalar product by $\langle \mathbf{p}, \mathbf{x} \rangle \equiv \sum_{j=1}^{n} p_j x_j$.

2020 *Mathematics Subject Classification.* Primary 34C20, 37J40.

Key words and phrases. Normal form, homological equation, Hamiltonian system, the Hill problem, set of analyticity.

- Let $\mathfrak{A}$ be a space of infinitely differentiable functions on $\mathbb{R}^{2n}$, then for a pair of functions $F, G \in \mathfrak{A}$ we denote a *Poisson bracket* by $\{F, G\} \equiv F * G$:
$$\{F, G\} \equiv \langle J \operatorname{grad} G, \operatorname{grad} F \rangle ,$$
where J is the symplectic unit.
- Notation $F * G^n$, $n > 1$ means multiple left-associated Poisson bracket with the following recurrent definition: $F * G^n = F * G^{n-1} * G$.

2. Hamiltonian normal form

We consider an analytic Hamiltonian system

$$\dot{\mathbf{x}} = \frac{\partial H}{\partial \mathbf{y}}, \quad \dot{\mathbf{y}} = -\frac{\partial H}{\partial \mathbf{x}} \tag{2.1}$$

with n degrees of freedom near its stationary point

$$\mathbf{x} = \mathbf{y} = 0.$$

The Hamiltonian function $H(\mathbf{x}, \mathbf{y})$ is expanded into convergent power series

$$H(\mathbf{x}, \mathbf{y}) = \sum H_{\mathbf{pq}} \mathbf{x}^{\mathbf{p}} \mathbf{y}^{\mathbf{q}}$$

with constant coefficients $H_{\mathbf{pq}}$, $\mathbf{p}, \mathbf{q} \geqslant 0$, $|\mathbf{p}| + |\mathbf{q}| \geqslant 2$.

Canonical transformations of coordinates $\mathbf{x}, \mathbf{y}$

$$\mathbf{x} = \mathbf{f}(\mathbf{u}, \mathbf{v}), \quad \mathbf{y} = \mathbf{g}(\mathbf{u}, \mathbf{v}), \tag{2.2}$$

preserve the Hamiltonian character of the initial system (2.1).

Let denote by $\mathbf{z} = (\mathbf{x}, \mathbf{y}) \in \mathbb{R}^{2n}$ the phase vector. Then the linear part of the system (2.1) can be written in the form

$$\dot{\mathbf{z}} = B\mathbf{z}, \quad B = \frac{1}{2} J \operatorname{Hess} H|_{\mathbf{z}=0}, \quad J = \begin{pmatrix} 0^n & E^n \\ -E^n & 0^n \end{pmatrix},$$

where J is symplectic unit matrix, E^n is identity matrix and $\operatorname{Hess} H$ is Hessian (second order partial derivatives) of function H.

Let $\lambda_1, \ldots, \lambda_{2n}$ be eigenvalues of the matrix B, which can be reordered in such a way that $\lambda_{j+n} = -\lambda_j$, $j = 1, \ldots, n$. Denote by $\boldsymbol{\lambda} = (\lambda_1, \ldots, \lambda_n)^{\mathrm{T}}$.

There exists [**3**, § 12, Theorem 12] a canonical formal transformation (2.2) in the form of power series, which reduces the initial system (2.1) into its *normal form*

$$\dot{\mathbf{u}} = \frac{\partial h}{\partial \mathbf{v}}, \quad \dot{\mathbf{v}} = -\frac{\partial h}{\partial \mathbf{u}},$$

defined by the normalized Hamiltonian $h(\mathbf{u}, \mathbf{v})$

$$h(\mathbf{u}, \mathbf{v}) = \sum_{j=1}^{n} \lambda_j u_j v_j + \sum h_{\mathbf{pq}} \mathbf{u}^{\mathbf{p}} \mathbf{v}^{\mathbf{q}},$$

containing only resonant terms $h_{\mathbf{pq}} \mathbf{u}^{\mathbf{p}} \mathbf{v}^{\mathbf{q}}$ with

$$\langle \mathbf{p} - \mathbf{q}, \boldsymbol{\lambda} \rangle = 0.$$

Here $0 \leqslant \mathbf{p}, \mathbf{q} \in \mathbb{Z}^n$, $|\mathbf{p}| + |\mathbf{q}| \geqslant 2$ and $h_{\mathbf{pq}}$ are constant coefficients.

If matrix B is semi-simple, i. e. geometrical and algebraic multiplicities of the eigenvalues $\boldsymbol{\lambda}$ are equal, then we have so called Birkhoff [**2**] or Cherry-Gustavson [**11**] normal form.

2.1. Methods of constructing canonical transformations. There exist several methods for constructing canonical transformations

- Method of Jacobi generating functions (so called the Birkhoff method) [**2**, **11**].
- Method of Lie series (so called the Deprit-Hori method) [**9**, **15**].
- Method of Poincare parametric generating function [**21**, **22**].
- Method of primitive functions [**12**].

Below we will describe in detail a variation of the Deprit-Hori method, called the method of *invariant normalization* [**29**]. Although this method is not universal in the sense that it applies only to nonzero simple (or semi-simple) eigenvalues, it is quite efficient and its software implementation is much easier than the Deprit-Hori method and, even more so, than the Birkhoff method.

The Deprit-Hori method uses an auxiliary Hamiltonian $G(\mathbf{X}, \mathbf{Y})$ called *generating Hamiltonian*, which define canonical system

$$\frac{\partial \mathbf{X}}{\partial \tau} = \frac{\partial G}{\partial \mathbf{Y}}, \quad \frac{\partial \mathbf{Y}}{\partial \tau} = -\frac{\partial G}{\partial \mathbf{X}} \tag{2.3}$$

with initial conditions $\mathbf{X}(0) = \mathbf{x}$, $\mathbf{Y}(0) = \mathbf{y}$. Generating function G is presented in the form of a series of a small parameter ε:

$$G(\mathbf{X}, \mathbf{Y}, \varepsilon) = \varepsilon G_1 + \varepsilon^2 G_2 + \cdots .$$

Let for $\tau = 1$ new variables $\mathbf{u}, \mathbf{v}$ are defined as the solution to the Cauchy problem (2.3): $\mathbf{u} = \mathbf{X}(1), \mathbf{v} = \mathbf{Y}(1)$. Transformation $(\mathbf{x}, \mathbf{y}) \to (\mathbf{u}, \mathbf{v})$ of the phase flow of the Hamiltonian system (2.1) is an univalent canonical transformation.

For small values of ε old variables $\mathbf{x}, \mathbf{y}$ can be written as Lie series of the new ones $\mathbf{u}, \mathbf{v}$:

$$\mathbf{x} = \mathbf{u} + \mathbf{u} * G + \frac{1}{2!}\mathbf{u} * G^2 + \frac{1}{3!}\mathbf{u} * G^3 + \cdots ,$$

$$\mathbf{y} = \mathbf{v} + \mathbf{v} * G + \frac{1}{2!}\mathbf{v} * G^2 + \frac{1}{3!}\mathbf{v} * G^3 + \cdots$$

New transformed Hamiltonian h is also related to the old one H by the following Lie series

$$h(\mathbf{u}, \mathbf{v}) = H(\mathbf{u}, \mathbf{v}) + H * G + \frac{1}{2!}H * G^2 + \frac{1}{3!}H * G^3 + \cdots$$

The procedure of nonlinear normalization of the real Hamiltonian H is more convenient to perform, when it is written in complex form. In [**5**, Subsect. 1G] were proposed such complex coordinates $\mathbf{z}$ that their complex conjugate values $\bar{\mathbf{z}}$ are not their canonically conjugate, but these coordinates make possible to write down universally quadratic NF for any sets of eigenvalues $\boldsymbol{\lambda}$. Here we will use another set of complex coordinates $\mathbf{z}$, which complex conjugate values are simultaneously canonically conjugate. Usually it is useful when the eigenvalues λ_j are either purely imagine or real.

There exists a formal transformation $\boldsymbol{\Phi} : (\mathbf{x}, \mathbf{y}) \to (\mathbf{z}, \bar{\mathbf{z}})$ which transforms the original Hamiltonian to the form

$$h(\mathbf{z}, \bar{\mathbf{z}}) = \sum h_{\mathbf{pq}} \mathbf{z}^{\mathbf{p}} \bar{\mathbf{z}}^{\mathbf{q}}, \tag{2.4}$$

where $\mathbf{p}, \mathbf{q} \in \mathbb{Z}^n$ and $|\mathbf{p}| + |\mathbf{q}| \geqslant 2$.

The value of $|\mathbf{p}| + |\mathbf{q}|$ is called the *order* of the corresponding expansion term.

DEFINITION 1. The Hamiltonian function $h(\mathbf{z}, \bar{\mathbf{z}})$ is called a *complex normal form* for semi-simple case if

(1) its quadratic part h_2 has the form $h_2 = \sum_{j=1}^{n} \lambda_j z_j \bar{z}_j$,
(2) the expansion (2.4) contains only the terms

$$h_{\mathbf{pq}} \mathbf{z}^{\mathbf{p}} \bar{\mathbf{z}}^{\mathbf{q}}, \tag{2.5}$$

which satisfy resonant condition

$$\langle \mathbf{p} - \mathbf{q}, \boldsymbol{\lambda} \rangle = 0.$$

The terms (2.5) that have $\mathbf{p} = \mathbf{q}$ are called *secular*, all others are called *strong resonant.* The Hamiltonian NF always contains secular terms, but strongly resonant terms are present if and only if the equation $\langle \mathbf{p}, \boldsymbol{\lambda} \rangle = 0$ has nontrivial solutions.

2.2. Properties of the normal form. Here we give some useful properties of the NF. For their proof see [**30**].

PROPERTY 1. If NF h is written in the form of power series $h = h_2 + f$, where h_2 is the quadratic part, then h_2 and f is commutable, i. e. $h_2 * f = 0$.

PROPERTY 2. If the quadratic part h_2 of the Hamiltonian $h = h_2 + f$ is normalized and the power series f consisting of the monomials of degree greater than two commutes with it $h_2 * f = 0$, then the series f contains only resonant terms.

Therefore, the condition $h_2 * f \equiv 0$ is a necessary and sufficient condition for the normal form.

We consider below an important from practical point of view case, when all eigenvalues λ_j are either real $\lambda_j = \gamma_j \in \mathbb{R}$ or pure imaginary $\lambda_j = \imath\omega_j$, $\omega_j \in \mathbb{R}$.

3. Normalization by the method of Lie series

3.1. Outlook of the normalization procedure. Here we describe normalization procedure.

- Firstly, the real Hamiltonian $H(\mathbf{x}, \mathbf{y})$ should be written in the complex form $H(\mathbf{z}, \bar{\mathbf{z}})$.
- One of the method of normalization is applied to $H(\mathbf{z}, \bar{\mathbf{z}})$ up to the definite order and we get it NF $h(\mathbf{Z}, \bar{\mathbf{Z}})$, which contains only resonant terms.
- Finally, the obtained complex NF $h(\mathbf{Z}, \bar{\mathbf{Z}})$ can be transformed into the real NF $h(\mathbf{X}, \mathbf{Y})$.

Here we consider a Hamiltonian system, which stationary point (SP) coincides with the origin. Applying scaling of the phase variables $\mathbf{x} \to \varepsilon\mathbf{x}$, $\mathbf{y} \to \varepsilon\mathbf{y}$ and independent variable $t \to \varepsilon^2 t$ to its Hamiltonian H near the SP one can write it in the form of power series in ε

$$H(\mathbf{x}, \mathbf{y}) = H^0 + F = H^0 + \sum_{j=1}^{\infty} \varepsilon^j H_j(\mathbf{x}, \mathbf{y}),$$

where H^0 is quadratic (unperturbed) form and H_j is a homogeneous form of order $j + 2$:

$$H_j = \sum_{|\mathbf{p}|+|\mathbf{q}|=j+2} H_{\mathbf{pq}} \mathbf{x}^{\mathbf{p}} \mathbf{y}^{\mathbf{q}}.$$

We are looking for the NF of the original Hamiltonian H as a power series

$$h(\mathbf{z}, \bar{\mathbf{z}}) = h^0 + f = h^0 + \sum_{j=1}^{\infty} \varepsilon^j h_j(\mathbf{z}, \bar{\mathbf{z}}), \text{ where } h^0 = \sum_{j=1} \lambda_j z_j \bar{z}_j$$

and homogeneous forms h_j, $j > 0$, contain only resonant terms $h_{\mathbf{pq}} \mathbf{z}^{\mathbf{p}} \bar{\mathbf{z}}^{\mathbf{q}}$, $|\mathbf{p}| + |\mathbf{q}| = j + 2$, such that $\langle \boldsymbol{\lambda}, \mathbf{p} - \mathbf{q} \rangle = 0$.

Transformation from the initial Hamiltonian H to its NF h is provided by Lie generator (generating function) G having form of a power series of a small parameter ε: $G = \sum_{j=1} \varepsilon^j G_j$:

$$h = H + \sum_{j=1}^{\infty} \frac{1}{j!} H * G^j.$$

Lie generator G produces a near identical transformation, so we have $h^0 = H^0$ and then

$$f = h^0 * G + M, \quad M = F + \sum_{j=1} \frac{1}{j!} H * G^j. \tag{3.1}$$

Using Property 1 and collecting terms in equation (3.1) with equal factors ε^j, $j = 0, 1, 2, \ldots$ one can rewrite it as the recurrent system of *homological equations*

$$h^0 * f_j = 0, \quad f_j = h^0 * G_j + M_j, \quad j = 1, 2, \ldots, \tag{3.2}$$

where the term M_j depends on H_k, F_k, G_k, $k < j$, obtained on the previous steps.

For small values of index j terms M_j are the following

$$\begin{aligned} M_1 =& F_1, \quad M_2 = F_2 + F_1 * G_1 + \frac{1}{2} H^0 * G_1^2, \\ M_3 =& F_3 + F_1 * G_2 + F_2 * G_1 + \frac{1}{2} H^0 * (G_1 * G_2 + G_2 * G_1) + \\ &+ \frac{1}{2} F_1 * G_1^2 + \frac{1}{6} H^0 * G_1^3. \end{aligned} \tag{3.3}$$

While index j increases the number N_M of terms in M_j grows exponentially, more precisely as $N_M = 2^j - 1$.

3.2. Solution of homological equations. There are two methods of solving the homological equations (3.2)

- *Algebraic method*, which was independently developed by G. Hori [**15**] and A. Deprit [**9**] and improved in subsequent works [**10, 16, 17, 19**]. Homological equations are solved as a system of linear algebraic equations of the coefficients of homogeneous forms f_j and G_j. This method has no restrictions on the structure of the quadratic part h^0 and can be applied in the case of multiple or zero eigenvalues as well. Nevertheless, because of the large number of monomials, the order of the corresponding systems grows rapidly.
- *Method of invariant normalization*, proposed by V.F. Zhuravlev [**29, 30**]. This method can be considered as subsequent averaging of functions M_j along the unperturbed solutions $\mathbf{z}(t, \mathbf{Z}, \bar{\mathbf{Z}})$ obtained from the unperturbed system with Hamiltonian H^0. It can be applied for the case of nonzero eigenvalues. In the case of multiples eigenvalues, this method also works, but requires a different scaling of the phase coordinates, which leads to

the fact that the terms $H_j(\mathbf{x},\mathbf{y})$ are no longer homogeneous forms (see [**8**]).

Here we give a brief description of the method of invariant normalization. For details see [**29**] or [**30**, Ch. 7].

Along the solutions $\mathbf{z}(t,\mathbf{Z},\bar{\mathbf{Z}})$, $\bar{\mathbf{z}}(t,\mathbf{Z},\bar{\mathbf{Z}})$ to the unperturbed canonical system

$$\dot{\mathbf{z}} = \frac{\partial H^0}{\partial \bar{\mathbf{z}}}, \quad \dot{\bar{\mathbf{z}}} = \frac{\partial H^0}{\partial \mathbf{z}} \tag{3.4}$$

we have the following identities:

- $h^0 * f_j = df_j/dt = 0$ according to Property 1;
- $h^0 * G_j = dG_j/dt$.

So, homological equations can be rewritten in the form

$$\frac{df_j}{dt} = 0, \quad M_j = f_j - \frac{dG_j}{dt}, \quad j = 1, 2, \ldots \tag{3.5}$$

Substituting the solutions $\mathbf{z}(t,\mathbf{Z},\bar{\mathbf{Z}})$, $\bar{\mathbf{z}}(t,\mathbf{Z},\bar{\mathbf{Z}})$ to the unperturbed system (3.4) into the function M_j one gets function:

$$m_j(t,\mathbf{Z},\bar{\mathbf{Z}}) = M_j(t,\mathbf{Z},\bar{\mathbf{Z}}).$$

According to equations (3.5) one obtains the following quadrature

$$\int_0^t m_j(t,\mathbf{Z},\bar{\mathbf{Z}})dt = tf_j(\mathbf{Z},\bar{\mathbf{Z}}) + G_j(\mathbf{Z},\bar{\mathbf{Z}}) + g(t). \tag{3.6}$$

Hence, on each step of the normalization procedure the next term of the NF f_j equals the coefficient at t, and the Lie generator term G_j equals the time-independent term in (3.6).

When all the eigenvalues $\boldsymbol{\lambda}$ are simple and purely imagine or real it is not necessary to integrate the left hand side of (3.6). All the functions M_j are homogeneous polynomials in $\mathbf{z},\bar{\mathbf{z}}$, so after the substitution we get $m_j = \sum_k C_k e^{\beta_k t} + C_0$, where β_k is a non zero linear combination of the eigenvalues λ_j, $j = 1, \ldots, n$ and C_0, C_k are polynomials in new variables $\mathbf{Z},\bar{\mathbf{Z}}$. It follows from (3.6) that

$$f_j = C_0, \quad G_j = \sum_k \frac{C_k}{\beta_k}.$$

3.3. Simplification of homological equation. It is possible to reduce approximately in 4 times the number of terms in functions M_j, $j = 2, 3, \ldots$. From the first equation of (3.1) for each $j = 2, 3, \ldots$ one can get that $h^0 * G_j = f_j - M_j$. The terms f_j and M_j are computed on the previous steps and can be used in subsequent ones:

$$h^0 * G_1 = f_1 - F_1 \Rightarrow \frac{1}{2} H^0 * G_1^2 = \frac{1}{2}(f_1 - F_1) * G_1.$$

Substituting this into M_2 in (3.3) we get

$$M_2 = F_2 + \frac{1}{2}(F_1 + f_1) * G_1.$$

So three Poisson brackets computations reduce to only one.

Let us introduce the following notations:

$$f_j^+ \equiv F_j + f_j, f_j^- \equiv F_j - f_j,$$
$$H * G_{j_1\cdots j_k}^k = H * G_{j_1\cdots j_{k-1}}^{k-1} * G_{j_k}.$$

Here we give simplified expressions of the functions M_j for $j = 3, 4, 5, 6$. All indices $k_1, k_2, k_3 \in \mathbb{N}$.

$$M_3 = F_3 + \frac{1}{2} \sum_{k_1+k_2=3} f_{k_1}^+ * G_{k_2} + \frac{1}{12} f_1^- * G_1^2,$$

$$M_4 = F_4 + \frac{1}{2} \sum_{k_1+k_2=4} f_{k_1}^+ * G_{k_2} + \frac{1}{12} \sum_{k_1+k_2+k_3=4} f_{k_1}^- * G_{k_2k_3}^2,$$

$$M_5 = F_5 + \frac{1}{2} \sum_{k_1+k_2=5} f_{k_1}^+ * G_{k_2} + \frac{1}{12} \sum_{k_1+k_2+k_3=5} f_{k_1}^- * G_{k_2k_3}^2 - \frac{1}{720} f_1^- * G_1^4,$$

$$M_6 = F_6 + \frac{1}{2} \sum_{k_1+k_2=6} f_{k_1}^+ * G_{k_2} + \frac{1}{12} \sum_{k_1+k_2+k_3=6} f_{k_1}^- * G_{k_2k_3}^2 -$$
$$- \frac{1}{720} \sum_{k_1+\cdots+k_4=6} f_{k_1}^- * G_{k_2k_3k_4}^4.$$

According to the above formulas the author made an assumption about the structure of the simplified homological equation. This assumption has not yet been strictly proven, but it has been verified using an algorithm implemented in the computer algebra system Maple for equations up to order 17.

Each subsequent term of the sequence $\{M_j\}$ is obtained from the previous one according to the following scheme:

(1) the term F_j from the expansion of the function H is given;
(2) the sum $\frac{1}{2} \sum_{k_1+k_2=j} f_{k_1}^+ * G_{k_2}$ is added;
(3) the previously written sums of multiples of Poisson brackets, composed of terms f_k^- and G_l, $k, l < j$, multiplied by the corresponding coefficient α_k are added. The multiplicity of Poisson brackets is always odd and not exceeding j, and the sum of indices of these functions involved in the calculation equals j.

More precise description of the simplified homological equations can be given in terms of the integer partitions.

DEFINITION 2. A partition $\nu(n)$ of a natural number n is any finite non-decreasing sequence of natural numbers ν_1, ν_2, ..., nu_k such that $\sum_{i=1}^k \nu_i = n$. Each partition can be written in the form $\nu = [1^{n_1} 2^{n_2} 3^{n_3} \dots]$, where n_i is the number of repetitions of term i in the partition; i.e. $\sum i n_i = n$. Let denote by ν_k such partition of n, which consists of exactly k terms.

STATEMENT 1. *For $j > 2$ function M_j is constructed in a such way:*

- *We take the term F_j and add the sum $\frac{1}{2} \sum_{k=1}^{j-1} f_k^+ * G_{j-k}$.*
- *For each k not greater than $[j/2]$ we compute the set $\mu_{2k+1}(j)$ of all permutations of any partition $\nu_{2k+1}(j)$, i.e. the set $\mu_{2k+1}(j)$ contains the tuple of $2k+1$ indices which sum is equal to j. For each such tuple $(i_1, \dots, i_{2k+1})$ of indices one has to compute all the Poisson brackets of form $f_{i_1}^- * G_{i_2 \cdots i_{2k+1}}^{2k}$.*
- *The sum all the computed above Poisson brackets is multiplied by the coefficient α_{2k}. These coefficients (first of them are given in Table 1) are*

well known Bernoulli numbers B_{2k} divided by factorial $(2k)!$:

$$\alpha_{2k} = \frac{B_{2k}}{(2k)!}.$$

They can be computed with the help of generating function

$$\mathfrak{g}(\varepsilon) = \frac{\varepsilon}{2} + \frac{\varepsilon}{e^{\varepsilon} - 1} - 1.$$

- *The final formula for M_j can be written in the following form*

$$M_j = F_j + \frac{1}{2}\sum_{k=1}^{j-1} f_k^+ * G_{j-k} + \sum_{k=1}^{[j/2]} \alpha_{2k} \sum_{(i_1,\ldots,i_{2k+1})\in\mu_{2k+1}^j} f_{i_1}^- * G_{i_2\cdots i_{2k+1}}^{2k}.$$

TABLE 1. Coefficients α_k

k	2	4	6	8	10	12
α_k	$\frac{1}{12}$	$-\frac{1}{720}$	$\frac{1}{30240}$	$-\frac{1}{1209600}$	$\frac{1}{47900160}$	$-\frac{691}{1307674368000}$

The main computational effort in obtaining the normal form is connected with multiple computation of Poisson brackets of the functions f_j^+, f_j^- and G_j. In order to optimize the computation when implementing the method, multilevel tables are constructed. They are sequentially filled with the results of computed multiple Poisson brackets of previously defined functions. This makes it possible to significantly reduce the computation time of the generator G_j from the homological equation.

REMARK 1. Since the homologous equation is the basis of the Lie series normalization method, its simplified form obtained in Statement 1 can also be used in the Depri-Hori method.

REMARK 2. It is evident that high order normalization of the Hamiltonian H is only possible with computer algebra systems. For example, such software [**30**, Ch. 7] was developed in CAS `Wolfram Mathematica` [**28**]. For computations given in Section 4 the author implemented the described above algorithm in CAS `Maplesoft Maple` (see, for example, [**27**]). Nevertheless, the invariant normalization method described in this section can be implemented without any problems in other open source CAS. For example, in `SageMath` [**26**], which essentially uses the `SymPy` symbolic computation package [**20**], or `Maxima` [**18**].

4. Application to the Hill problem

4.1. Hill problem Hamiltonian. The Hill problem [**14**], which equations describe satellite motion in the vicinity of the smaller of two massive bodies, has the following Hamiltonian [**1**, S. 3]

$$H(\mathbf{x}, \mathbf{y}) = \frac{1}{2}\left(y_1^2 + y_2^2\right) + x_2 y_1 - x_1 y_2 - x_1^2 + \frac{1}{2}x_2^2 - \frac{1}{r}, \tag{4.1}$$

where $r = \sqrt{x_1^2 + x_2^2}$, $\mathbf{x} = (x_1, x_2)$ is the coordinate vector, and $\mathbf{y} = (y_1, y_2)$ is the canonically conjugate momentum vector.

There is a canonical transformation, called the Levi-Civita regularization [**25**, Ch. 4], which reduces the Hamiltonian (4.1) with a singularity at the origin into polynomial form

$$\widetilde{H}\left(\mathbf{q},\mathbf{p},\tilde{h}\right) = \frac{\tilde{h}}{4}\left(q_1^2+q_2^2\right)+\frac{1}{2}\left(p_1^2+p_2^2\right)+\frac{1}{2}\left(q_1^2+q_2^2\right)\left(q_2p_1-q_1p_2\right)- \\ -\frac{1}{4}\left(q_1^2+q_2^2\right)\left(q_1^4-4q_1^2q_2^2+q_2^4\right)-\frac{1}{4}\equiv 0. \tag{4.2}$$

Here the parameter $\tilde{h}=-H(\mathbf{x},\mathbf{y})$. Since the Hamiltonian $H(\mathbf{x},\mathbf{y})$ does not depend explicitly on time, the value $-\tilde{h}$ is the value of the generalized energy integral. It is usually written in the form of the Jacobi constant

$$C = 3x_1^2-\frac{2}{r}-\dot{x}_1^2-\dot{x}_2^2 = 2\tilde{h}. \tag{4.3}$$

In the paper [**24**] the following transformation is suggested

$$\mathbf{q}=2^{3/4}|\tilde{h}|^{1/4}\mathbf{Q},\quad \mathbf{p}=2^{1/4}|\tilde{h}|^{3/4}\mathbf{P},\quad \nu=\sqrt{2/|\tilde{h}|}\,\tau,$$

defined for all values of $\tilde{h}\neq 0$. It reduces the regularized Hamiltonian (4.2) into the form

$$\mathcal{H}(\mathbf{Q},\mathbf{P}) = \frac{\delta}{2}\left(Q_1^2+Q_2^2\right)+\frac{1}{2}\left(P_1^2+P_2^2\right)+2\left(Q_1^2+Q_2^2\right)\left(Q_2P_1-Q_1P_2\right)- \\ -4\left(Q_1^2+Q_2^2\right)\left(Q_1^4-4Q_1^2Q_2^2+Q_2^4\right) \tag{4.4}$$

where $\delta=\pm 1$. Here δ is chosen depending on the sign of the parameter $\tilde{h}$, namely $\delta=\operatorname{sign}\tilde{h}$.

The structure of the function (4.4) is such that the origin of coordinates corresponds to the equilibrium point of the system of canonical equations defined by the Hamiltonian $\mathcal{H}(\mathbf{Q},\mathbf{P})$. The quadratic part of $\mathcal{H}_2$ has already been reduced to its normal form.

This allows us to apply the normalization algorithm described in Section 3 without additional transformations, previously writing the function $\mathcal{H}$ in the complex form.

From the point of view of applications of the Hill problem to problems of satellite dynamics, the case when $\delta=1$ is more interesting, since it corresponds to values of the Jacobi constant (4.3) $C>0$. Exactly, such values of C correspond to the motions of the natural satellites of the planets of the Solar System (see [**13**]).

Let us give a description of the Hill problem NF in regular variables Q_j, P_j, $j=1,2$, for the case $\delta=1$, i.e., when the quadratic form $\mathcal{H}_2$ has two pairs of semi-simple eigenvalues $\pm\imath$.

4.2. Normal form of the Hill problem near the origin. Here we introduce the complex variables $z_j,\bar{z}_j$, $j=1,2$, using the univalent canonical transformation

$$Q_j=\frac{1}{\sqrt{2\imath}}\left(z_j-\bar{z}_j\right),\quad P_j=\sqrt{\frac{\imath}{2}}\left(z_j+\bar{z}_j\right),\quad j=1,2.$$

To write the Hamiltonian function $\mathcal{H}$ in more compact form, let us introduce the following notations:

$$\rho_i=z_i\bar{z}_i,\ i=1,2,\quad h_2=\imath(\rho_1+\rho_2),\quad g_2=z_1\bar{z}_2-z_2\bar{z}_1. \tag{4.5}$$

Then the Hamiltonian function (4.4) is written in the form $\mathcal{H}(\mathbf{z},\bar{\mathbf{z}}) = \mathcal{H}_2 + \mathcal{H}_4 + \mathcal{H}_6$, where

$$
\begin{aligned}
\mathcal{H}_2 =& h_2,\\
\mathcal{H}_4 =& 1\left((z_1-\bar{z}_1)^2 + (z_2-\bar{z}_2)^2\right) g_2,\\
\mathcal{H}_6 =& -\frac{1}{2}\left((z_1-\bar{z}_1)^2 + (z_2-\bar{z}_2)^2\right) \times\\
& \times \left((z_1-\bar{z}_1)^4 - 4(z_1-\bar{z}_1)^2(z_2-\bar{z}_2)^2 + (z_2-\bar{z}_2)^4\right).
\end{aligned}
$$

The procedure of invariant normalization of the Hamiltonian $\mathcal{H}$ was implemented in the CAS `Maple` and is performed up to the 22th order inclusive.

The first terms of the normal form h up to the order 14 are the following:

(4.6)

$$
\begin{aligned}
h_4 =& -2h_2g_2,\\
h_6 =& -2h_2\left(5h_2^2 + 4g_2^2 + 30\rho_1\rho_2\right),\\
h_8 =& -5h_2g_2\left(15h_2^2 + 11g_2^2 + 96\rho_1\rho_2\right),\\
h_{10} =& -h_2\left(417h_2^4 + 678g_2^2h_2^2 + 445g_2^4 + 4\rho_1\rho_2\left(3336h_2^2 + 4668g_2^2 + 7860\rho_1\rho_2\right)\right),\\
h_{12} =& -\frac{h_2g_2}{2}\left(15955h_2^4 + 13832g_2^2h_2^2 + 7841g_2^4 + \right.\\
& \left. +4\rho_1\rho_2\left(31873h_2^2 + 24921g_2^2 + 76140\rho_1\rho_2\right)\right),\\
h_{14} =& -\frac{h_2}{24}\left(848407h_2^6 + 874685g_2^6 + 2666775h_2^4g_2^2 + 1725613h_2^2g_2^4 + \right.\\
& + 16\rho_1\rho_2\left(546163h_2^4 + 826144g_2^4 + 1373271h_2^2g_2^2 + \right.\\
& \left.\left. + 4\rho_1\rho_2\left(476834h_2^2 + 876771g_2^2 + 619290\rho_1\rho_2\right)\right)\right),\\
h_{16} =& -\frac{h_2g_2}{360}\left(387639075h_2^6 + 126523557g_2^6 + 529277675h_2^4g_2^2 + 264089597h_2^2g_2^4 + \right.\\
& + 8\rho_1\rho_2\left(510067595h_2^4 + 277494642g_2^4 + 553863605h_2^2g_2^2 + \right.\\
& \left.\left. + 600\rho_1\rho_2\left(2976121h_2^2 + 2509718g_2^2 + 3825180\rho_1\rho_2\right)\right)\right),
\end{aligned}
$$

where in the expressions for ρ_1, ρ_2, h_2 and g_2 all initial coordinates $z_j, \bar{z}_j$ should be replaced by the new ones $Z_j, \bar{Z}_j$, $j = 1, 2$. The other found terms of the NF h are rather cumbersome and are not given here.

The computed initial terms of the NF h expansion allow us to make some assumptions about its structure.

HYPOTHESIS 1. *Normal form h of the regularized Hamiltonian (4.4) of the Hill problem in the vicinity of the origin has the following properties.*

(1) *By virtue of the notation (4.5) the NF h is a function only of the values ρ_j, $j = 1, 2$, and g_2, i.e. $h = h(\rho_1, \rho_2, g_2)$.*
(2) *The terms of the expansion with numbers $4k$, $k \geqslant 1$ consist only of strongly resonant monomials, the terms with numbers $4k - 2$ contain both secular and strongly resonant monomials.*
(3) *All terms of the expansion starting from h_4 are factorized into multipliers, one of which is equal to h_2.*
(4) *The terms of the expansion with numbers $4k$, $k \geqslant 2$ are factorized into three multipliers, one of which is g_2 and the other is h_2.*

(5) *Each term of h_k, and hence the whole NF h, is invariant with respect to the permutation of $\rho_1 \leftrightarrow \rho_2$.*

The computed NF of the Hill problem near the origin allows one to investigate the dynamics of the original system in this region of the phase space. In particular, one can use the NF to search for the so-called *set of analyticity* [**4**, Part II], [**6**], on which the normalizing transform converges, and hence, solutions of the normalized Hamiltonian system with Hamiltonian h will correspond to solutions of the original system with Hamiltonian (4.1) or (4.4).

REMARK 3. It is also possible to compute NF of the Hill problem for the case $\delta = -1$, i. e. when the quadratic form $\mathcal{H}_2$ has two pairs of semi-simple eigenvalues ± 1. In this situation the complex variables $z_j, \bar{z}_j$, $j = 1, 2$, are introduced by canonical transformation

$$Q_j = \frac{1}{\sqrt{2}}\left(z_j - \bar{z}_j\right), \quad P_j = \sqrt{\frac{1}{2}}\left(z_j + \bar{z}_j\right), \quad j = 1, 2.$$

The computed NF coincides with the NF for the case $\delta = 1$, given by (4.6), where $h_2 = \rho_1 + \rho_2$.

In this case it is not necessary to compute the set of analyticity, because according to results of V.S. Samovol [**23**] there exists a smooth transformation that reduces the obtained Hamiltonian system in the NF into equivalent linear system.

4.3. Computation of the set of analyticity $\mathcal{A}$. Usually the normalizing transformation diverges. However, the normal form of a Hamiltonian system allows us to find sets in the phase space of the system on which convergence takes place. The analysis of the convergence domains makes it possible to find families of periodic and quasi-periodic solutions of the original system. Here we restrict ourselves to computation of the set $\mathcal{A}$.

Briefly recall the definition of the set $\mathcal{A}$ (for details, see [**4**, Part II], [**6**]). Let a Hamiltonian system with n degrees of freedom be reduced to the normal form h in the neighborhood of the equilibrium position coinciding with the origin of the coordinates. Then the set $\mathcal{A}$ is an affine manifold in the phase space with normal coordinates $z_j, \bar{z}_j$, $j = 1, \ldots, n$, given by the system of equations

$$\frac{\partial h}{\partial z_j} = \lambda_j a \bar{z}_j, \quad \frac{\partial h}{\partial \bar{z}_j} = \lambda_j a z_j, \quad j = 1, \ldots, n. \tag{4.7}$$

Here a is a free parameter.

Using the properties of the NF h of the regularized Hill problem formulated in Hypothesis 1, let us compute some subvariety of the set $\mathcal{A}$.

For the case of the Hill problem, i.e., for two degrees of freedom, taking into account the item 3 of Hypothesis 1 and the relations (4.5), the system (4.7) is rewritten as two subsystems. The first one has the form

$$\frac{\partial h}{\bar{z}_1 \partial z_1} = \frac{\partial h}{\partial \rho_1} + \frac{\partial h}{\partial g_2}\frac{\bar{z}_2}{\bar{z}_1} = \imath a, \tag{4.8a}$$

$$\frac{\partial h}{z_1 \partial \bar{z}_1} = \frac{\partial h}{\partial \rho_1} - \frac{\partial h}{\partial g_2}\frac{z_2}{z_1} = \imath a, \tag{4.8b}$$

and the second one correspondingly takes the form:

$$\frac{\partial h}{\bar z_2 \partial z_2} = \frac{\partial h}{\partial \rho_2} + \frac{\partial h}{\partial g_2}\frac{\bar z_1}{\bar z_2} = \imath a, \tag{4.9a}$$

$$\frac{\partial h}{z_2 \partial \bar z_2} = \frac{\partial h}{\partial \rho_2} - \frac{\partial h}{\partial g_2}\frac{z_1}{z_2} = \imath a. \tag{4.9b}$$

From the equations of the system (4.8) or the system (4.9) we get the following relation

$$\frac{\partial h}{\partial g_2}\left(z_1 \bar z_2 + z_2 \bar z_1\right) = 0,$$

from which it follows that either $\partial h/\partial g_2 = 0$ or

$$z_1 \bar z_2 + z_2 \bar z_1 = 0. \tag{4.10}$$

The difference of equations (4.8a) and (4.9b) (or (4.8b) and (4.9a)) can be represented as

$$\frac{\partial h}{\partial \rho_1} - \frac{\partial h}{\partial \rho_2} + \frac{\partial h}{\partial g_2}\cdot\frac{\rho_2 - \rho_1}{\bar z_1 z_2} = 0. \tag{4.11}$$

By virtue of the property 5 of Hypothesis 1 difference of partial derivatives $\partial h/\partial \rho_1 - \partial h/\partial \rho_2$ can be factorized into multipliers, one of which is the difference $\rho_1 - \rho_2$ or in variables $z_j, \bar z_j$ we obtain equation $z_1 \bar z_1 - z_2 \bar z_2 = 0$. The another multiplier of (4.11) does not give any local roots.

The detail analysis of the case $\partial h/\partial g_2 = 0$ (not presented in this paper) does not provide any roots of the systems (4.8) and (4.9). Therefore, we restrict ourselves to analyzing the solutions of the system obtained from the conditions (4.10) and $\rho_1 = \rho_2$:

$$\begin{cases} z_1 \bar z_1 - z_2 \bar z_2 = 0, \\ z_1 \bar z_2 + z_2 \bar z_1 = 0. \end{cases}$$

It is easy to get the equations $z_1\left(\bar z_1^2 + \bar z_2^2\right) = z_2\left(\bar z_1^2 + \bar z_2^2\right) = 0$. The root $z_1 = z_2 = 0$ gives a trivial real solution. The second multiplier gives a pair of solutions

$$z_2 = +\imath z_1, \quad \bar z_2 = -\imath \bar z_1, \tag{4.12}$$

$$z_2 = -\imath z_1, \quad \bar z_2 = +\imath \bar z_1. \tag{4.13}$$

Substituting the found solutions (4.12) and (4.13) respectively into the left-hand side of equation (4.8a), and limiting the terms to the 6th order of ρ_1, we get the relation

$$\imath \mp 8\rho_1 + 36\imath\rho_1^2 \pm 320\rho_1^3 - 2420\imath\rho_1^4 \pm 12576\rho_1^5 + \ldots = \imath a.$$

Obviously, there is no real a satisfying the last equality. This means that there are no periodic or quasi-periodic solutions on these components of the analyticity set $\mathcal{A}$.

Here we have an example of complex sets of analyticity, which intersect at the real point. This situation corresponds to Figure 5 in Section 3.3 of [**4**, Part II].

References

[1] A. B. Batkhin and N. V. Batkhina, *The Hill Problem*, Volgogradskoe nauchnoe izdatel'stvo, Volgograd, 2009, (in Russian).

[2] George D. Birkhoff, *Dynamical systems*, American Mathematical Society Colloquium Publications, Vol. IX, American Mathematical Society, Providence, R.I., 1966. With an addendum by Jurgen Moser. MR0209095

[3] A. D. Brjuno, *Analytic form of differential equations. I, II* (Russian), Trudy Moskov. Mat. Obšč. **25** (1971), 119–262; ibid. 26 (1972), 199–239. MR0377192

[4] A. D. Bruno, *Local Methods in Nonlinear Differential Equations*, Springer–Verlag, Berlin – Heidelberg – New York – London – Paris – Tokyo, 1989.

[5] A. D. Bryuno, *Ogranichennaya zadacha trekh tel* (Russian), "Nauka", Moscow, 1990. Ploskie periodicheskie orbity. [Plane periodic orbits]; With an English summary. MR1064935

[6] Galina Filipuk, Alberto Lastra, and Sławomir Michalik (eds.), *Formal and analytic solutions of diff. equations*, Springer Proceedings in Mathematics & Statistics, vol. 256, Springer, Cham, 2018. FASdiff, Alcalá de Henares, Spain, September 2017; Selected, revised contributions, DOI 10.1007/978-3-319-99148-1. MR3872184

[7] A. D. Bruno and A. B. Batkhin, *Survey of eight modern methods of Hamiltonian mechanics*, Axioms **10** (2021), no. 4. DOI: 10.3390/axioms10040293

[8] A. D. Bryuno and A. G. Petrov, *On the computation of the Hamiltonian normal form* (Russian), Dokl. Akad. Nauk **410** (2006), no. 4, 474–478. MR2446843

[9] André Deprit, *Canonical transformations depending on a small parameter*, Celestial Mech. **1** (1969/70), 12–30, DOI 10.1007/BF01230629. MR253611

[10] G. E. O. Giacaglia, *Perturbation methods in non-linear systems*, Applied Mathematical Sciences, Vol. 8, Springer-Verlag, New York-Heidelberg, 1972. MR0478875

[11] F. G. Gustavson, *On constructing formal integrals of a Hamiltonian system near ail equilibrium point*, Astronomical Journal **71** (1966), no. 8, 670–686.

[12] Àlex Haro, *An algorithm to generate canonical transformations: application to normal forms*, Phys. D **167** (2002), no. 3-4, 197–217, DOI 10.1016/S0167-2789(02)00459-1. MR1927473

[13] M. Hénon, *Numerical exploration of the restricted problem. VI. Hill's case: non-periodic orbits*, Astron. & Astr. (1970), no. 9, 24–36.

[14] G. W. Hill, *Researches in the Lunar Theory*, Amer. J. Math. **1** (1878), no. 3, 245–260, DOI 10.2307/2369313. MR1505166

[15] G. Hori, *Theory of general perturbation with unspecified canonical variable*, Publications of the Astronomical Society of Japan **18** (1966), no. 4, 287–296.

[16] Ahmed Aly Kamel, *Expansion formulae in canonical transformations depending on a small parameter*, Celestial Mech. **1** (1969/70), 190–199, DOI 10.1007/BF01228838. MR273870

[17] A. P. Markeev, *Libration Points in Celestial Mechanics and Cosmodynamics*, Nauka, Moscow, 1978, (in Russian).

[18] Maxima, *Maxima, a computer algebra system. version 5.45.1*, 2021.

[19] William A. Mersman, *A new algorithm for the Lie transformation*, Celestial Mech. **3** (1970/71), no. 1, 81–89, DOI 10.1007/BF01230434. MR289107

[20] A. Meurer, Ch. P. Smith, M. Paprocki et al. *SymPy: symbolic computing in Python*, PeerJ Computer Science **3** (2017), e103. DOI: 10.7717/peerj-cs.103x

[21] A. G. Petrov, *Asymptotic methods for solving the Hamilton equations by the parametrization of canonical transformations* (Russian, with Russian summary), Differ. Uravn. **40** (2004), no. 5, 626–638, 718, DOI 10.1023/B:DIEQ.0000043525.28771.28; English transl., Differ. Equ. **40** (2004), no. 5, 672–685. MR2162475

[22] A. G. Petrov, *On the invariant normalization of nonautonomous Hamiltonian systems* (Russian, with Russian summary), Prikl. Mat. Mekh. **68** (2004), no. 3, 402–413, DOI 10.1016/S0021-8928(04)00049-8; English transl., J. Appl. Math. Mech. **68** (2004), no. 3, 357–367. MR2092155

[23] V. S. Samovol, *Linearization of a system of differential equations in the neighborhood of a singular point* (Russian), Dokl. Akad. Nauk SSSR **206** (1972), 545–548. MR0315202

[24] C. Simó and T. J. Stuchi, *Central stable/unstable manifolds and the destruction of KAM tori in the planar Hill problem*, Phys. D **140** (2000), no. 1-2, 1–32, DOI 10.1016/S0167-2789(99)00211-0. MR1752804

[25] V. Szebehely, *Theory of Orbits. The Restricted Problem of Three Bodies*, Academic Press, New York and London, 1967.

[26] The Sage Developers, *Sagemath, the Sage Mathematics Software System (Version 9.1.1)*, 2020, https://www.sagemath.org.

[27] I. Thompson, *Understanding Maple*, Cambridge University Press, 2016.

[28] Stephen Wolfram, *The Mathematica book*, 3rd ed., Wolfram Media, Inc., Champaign, IL; Cambridge University Press, Cambridge, 1996. MR1404696

[29] V. F. Zhuravlëv, *Invariant normalization of nonautonomous Hamiltonian systems* (Russian, with Russian summary), Prikl. Mat. Mekh. **66** (2002), no. 3, 356–365, DOI 10.1016/S0021-8928(02)00044-8; English transl., J. Appl. Math. Mech. **66** (2002), no. 3, 347–355. MR1937451

[30] V. F. Zhuravlev, A. G. Petrov, and M. M. Shunderyuk, *Selected Problems of Hamiltonian Mechanics*, LENAND, Moscow, 2015, (in Russian).

Keldysh Institute of Applied Mathematics of RAS, Miusskaya sq. 4, Moscow, 125047, Russia

Email address: batkhin@gmail.com

Contemporary Mathematics
Volume **782**, 2023
https://doi.org/10.1090/conm/782/15719

A note on value distribution of solutions of certain second order ODEs

Ewa Ciechanowicz

ABSTRACT. While value distribution of solutions of Painlevé equations $P_1 - P_6$ has been thoroughly examined, so-called Painlevé sigma equations S_1, S_2 and S_4 are relatively less recognized. Various properties of S_1, S_2 and S_4 can be derived both from the form of the equations themselves and from their connection through Hamiltonian systems with P_1, P_2 and P_4, respectively. This includes distribution of zeros, asymptotic form of the second main theorem and behavior with respect to small target functions. Apart from usual Nevanlinna functionals, also Petrenko's function of deviation plays a role in the estimates. Additionally, results concerning distribution of values are presented for the modified third Painlevé equation in the final part of the paper.

1. Second order equations with meromorphic solutions

Although various second order equations have one or more meromorphic solutions, this note concentrates on a group of equations with all local solutions extending to a function meromorphic in the whole complex plane. It naturally leads to Painlevé equations and associated with them sigma equations [**4, 14, 15**].

The first Painlevé equation P_1

$$f'' = 6f^2 + z \tag{1.1}$$

can be represented as a Hamiltonian system with the Hamiltonian

$$H_1(p,q,z) = \frac{1}{2}p^2 - 2q^3 - zq \qquad \text{and} \qquad \begin{cases} q' = p \\ p' = 6q^2 + z \end{cases}, \tag{1.2}$$

where q fulfills P_1. If we put $\sigma_1(z) := H_1(p,q,z)$, then

$$\begin{cases} q = -\sigma_1' \\ p = -\sigma_1'' \end{cases},$$

and the function σ_1 fulfills the second order equation S_1

$$(\sigma'')^2 + 4(\sigma')^3 + 2z\sigma' - 2\sigma = 0. \tag{1.3}$$

Conversely, if σ solves S_1, then $q = -\sigma'$, $p = -\sigma''$ solve (1.2) [**14**].

2020 *Mathematics Subject Classification.* Primary 30D35; Secondary 34M05, 34M55.
Key words and phrases. Painlevé equation, meromorphic function, value distribution.

Let us now consider the Hamiltonian

$$H_2(p,q,z) := \frac{1}{2}p^2 - (q^2 + \frac{1}{2}z)p - (\alpha + \frac{1}{2})q,$$

where

$$\begin{cases} q' = p - q^2 - \frac{1}{2}z \\ p' = 2qp + \alpha + \frac{1}{2} \end{cases}$$

and α is a complex parameter. It connects the second Painlevé equation $P_2(\alpha)$

$$f'' = 2f^3 + zf + \alpha, \tag{1.4}$$

with another second order equation from Painlevé list

$$ff'' = \frac{1}{2}(f')^2 + 2f^3 - zf^2 - \frac{1}{2}(\alpha + \frac{1}{2})^2,$$

referred to as P_{34}. It means that if p is a solution of $P_2(\alpha)$, then q is a solution of $P_{34}(\alpha)$. If we put $\sigma_2(z) := H_2(p,q,z)$, then σ_2 is a solution of the equation $S_2(\alpha)$

$$(\sigma'')^2 + 4(\sigma')^3 + 2\sigma'(z\sigma' - \sigma) = \frac{1}{4}(\alpha + \frac{1}{2})^2 \tag{1.5}$$

called the second Painlevé sigma equation. Moreover, if for a fixed α the function σ solves S_2, then

$$\begin{cases} q = \frac{4\sigma'' + 2\alpha + 1}{8\sigma'} \\ p = -2\sigma' \end{cases} \tag{1.6}$$

are solutions of $P_2(\alpha)$ and $P_{34}(\alpha)$, respectively [**15**].

The fourth Painlevé equation $P_4(\alpha, \beta)$ is given by

$$f'' = \frac{f'^2}{2f} + \frac{3f^3}{2} + 4zf^2 + 2(z^2 - \alpha)f + \frac{\beta}{f}, \tag{1.7}$$

where α, β are arbitrary complex parameters. If we represent P_4 as a Hamiltonian system with the Hamiltonian

$$H_4(p,q,z) := 2qp^2 - (q^2 + 2zq + 2\theta_0)p + \theta_\infty q,$$

we get

$$\begin{cases} q' = 4qp - q^2 - 2zq - 2\theta_0 \\ p' = -2p^2 + 2qp + 2zp - \theta_\infty \end{cases}.$$

Thus, eliminating p in the system we obtain that q satisfies $P_4(\alpha, \beta)$ with

$$\alpha = 1 - \theta_0 + 2\theta_\infty, \qquad \beta = -2\theta_0^2,$$

and by eliminating q we obtain, that $-2p$ satisfies $P_4(\alpha, \beta)$ with

$$\alpha = 2\theta_0 - \theta_\infty - 1, \qquad \beta = -2\theta_\infty^2.$$

The function $\sigma_4(z) := H_4(p,q,z)$ satisfies the equation $S_4(\theta_0, \theta_\infty)$

$$(\sigma'')^2 - 4(z\sigma' - \sigma)^2 + 4\sigma'(\sigma' + 2\theta_0)(\sigma' + 2\theta_\infty) = 0, \tag{1.8}$$

and conversely, if σ is a solution of S_4, then

$$q = \frac{\sigma'' - 2z\sigma' + 2\sigma}{2(\sigma' + 2\theta_\infty)}, \qquad p = \frac{\sigma'' + 2z\sigma' - 2\sigma}{4(\sigma' + 2\theta_0)}$$

are solutions of the Hamiltonian system for P_4 [**15**].

In this note we consider some of value distribution properties of transcendental meromorphic functions solving S_1, S_2 or S_4. In particular, we mention lack of exceptional values in the sense of Nevanlinna for solutions of S_2, asymptotic form of the second main theorem and behavior towards small target functions for solutions of all the equations. In the last part of the paper we also discuss the existence of exceptional and asymptotic values for solutions of the equation $\tilde{P}_3$. by means of applying the notion of Petrenko's deviation.

2. Auxiliary results

We apply the standard notations of value distribution theory [**8**, **13**]. Thus

$$N(r,f) = \int_0^r (n(t,f) - n(0,f))\frac{dt}{t} + n(0,f)\log r$$

denotes the integrated counting function. Here $n(r,f)$ counts the number of poles, including multiplicity, in the disk of radius r. Similarly,

$$\overline{N}(r,f) = \int_0^r (\overline{n}(t,f) - \overline{n}(0,f))\frac{dt}{t} + \overline{n}(0,f)\log r$$

counts poles of f ignoring multiplicity. Moreover,

$$N(r,a,f) := N(r, \frac{1}{f-a}), \quad \overline{N}(r,a,f) := \overline{N}(r, \frac{1}{f-a}) \ (a \in \mathbb{C})$$

count a-points with and without multiplicity respectively. We also put

$$N_1(r,a,f) := N(r,a,f) - \overline{N}(r,a,f).$$

Next,

$$m(r,f) = m(r,\infty,f) := \frac{1}{2\pi}\int_0^{2\pi} \log^+ |f(re^{i\phi})|d\phi, \quad \log^+ x = \max(0,\log x),$$

is the mean proximity function and

$$m(r,a,f) := m(r, \frac{1}{f-a})$$

denotes the mean proximity to $a \in \mathbb{C}$. The function $T(r,f) := m(r,f) + N(r,f)$ is called *the characteristic function of* f We define (Nevanlinna) deficiency of f at a value $a \in \overline{\mathbb{C}}$, by the formula

$$\delta(a,f) = \liminf_{r\to\infty} \frac{m(r,a,f)}{T(r,f)} = 1 - \limsup_{r\to\infty} \frac{N(r,a,f)}{T(r,f)}.$$

Let us also recall that the order and lower order of growth of a meromorphic function $f(z)$ are defined, respectively, as

$$\varrho(f) := \limsup_{r\to\infty} \frac{\log T(r,f)}{\log r}, \qquad \mu(f) := \liminf_{r\to\infty} \frac{\log T(r,f)}{\log r}.$$

We say that $\phi : (0,+\infty) \to \mathbb{R}$ is $S(r,f)$ if

$$\phi(r) = o(T(r,f)) \quad (r \to \infty, \, r \notin E),$$

where E is a set of finite linear measure.

We start with the following, slightly generalized, version of the lemma on the logarithmic derivative.

LEMMA 2.1 ([**7**, Prop. B.5, p. 276]). *For an arbitrary nonconstant meromorphic function $f(z)$, and for an arbitrary $k \in \mathbb{N}$,*

$$m(r, \frac{f^{(k)}}{f}) = S(r, f).$$

In particular, if $\varrho(f) < \infty$,

$$m(r, \frac{f^{(k)}}{f}) = O(\log r).$$

The next result is helpful with establishing the asymptotic form of the second main theorem.

LEMMA 2.2 ([**7**, Lem. 11.1, p. 69]). *Let $f(z)$ be an arbitrary transcendental meromorphic function satisfying*

$$m(r, f') = O(\log r) \qquad \text{and} \qquad m(r, \frac{f}{f'}) = O(\log r), \qquad r \to \infty.$$

Then

$$m(r, \frac{1}{f}) + N(r, \frac{1}{f'}) = N(r, f') + O(\log r), \qquad r \to \infty.$$

Let us bring up the well-known lemma of Clunie as formulated in [**7**, App. B]. The lemma plays crucial role in discussing non-deficiency of most values for meromorphic solutions of differential equations.

LEMMA 2.3. *Let f be a transcendental meromorphic function of finite order such that*

$$f^{p+1} = Q(z, f), \; p \in \mathbb{N},$$

where $Q(z, u)$ is a polynomial in u and its derivatives with meromorphic coefficients b_μ $(\mu \in M)$. If the total degree of $Q(z, u)$ as a polynomial in u and its derivatives does not exceed p, then

$$m(r, f) = O(\sum_{\mu \in M} m(r, b_\mu)) + O(\log r) \qquad (r \to \infty).$$

The quantity

$$\beta(a, f) = \liminf_{r \to \infty} \frac{\mathcal{L}(r, a, f)}{T(r, f)}$$

is called *deviation of a meromorphic function with respect to $a \in \overline{\mathbb{C}}$*, where

$$\mathcal{L}(r, a, f) := \begin{cases} \max\limits_{|z|=r} \log^+ |f(z)| & \text{for } a = \infty, \\ \max\limits_{|z|=r} \log^+ \left| \frac{1}{f(z)-a} \right| & \text{for } a \neq \infty. \end{cases}$$

If $\beta(a, f) > 0$ we say, that a is *deficient in the sense of Petrenko*. For all $a \in \overline{\mathbb{C}}$ the inequality

$$\delta(a, f) \leq \beta(a, f)$$

follows easily from the respective definitions. Thus for each value deficient in the sense of Nevanlinna also $\beta(a, f) > 0$. For meromorphic functions of finite lower order we have upper bounds of deviations similar to the deficiency relations following from the theorems of Nevanlinna. Indeed, it was proved by Petrenko in [**16**] that for a function f of finite lower order μ,

$$\beta(a,f) \le B(\mu) := \begin{cases} \frac{\pi\mu}{\sin \pi\mu} & \text{if } \mu \le 0.5\,, \\ \pi\mu & \text{if } \mu > 0.5 \end{cases}$$

and the set of deficient values in this sense is at most countable. The estimates are sharp. In general, the sets of deficient values in the sense of Nevanlinna and in the sense of Petrenko may differ even in case of functions of finite order and of regular growth.

The following results concerning deviation are modified versions of the (generalized) lemma on the logarithmic derivative and Clunie lemma [**2, 11**].

LEMMA 2.4. *Let f be a meromorphic function. Then, possibly except for r in a set of finite linear measure, for $k = 1, 2, \ldots$ we have*

$$\mathcal{L}\left(r, \infty, \frac{f^{(k)}}{f}\right) = O(\log(rT(r,f)))\,, \qquad (r \to \infty),$$

where $f^{(k)}$ means the k-th derivative of f.

LEMMA 2.5. *Let f be a transcendental meromorphic solution of*

$$f^n P(z,f) = Q(z,f), \tag{2.1}$$

where n is a positive integer, $P(z,f)$, $Q(z,f)$ are polynomials in f and its derivatives with meromorphic coefficients a_ν, b_ν, respectively, which are small with respect to f in the sense that

$$\mathcal{L}(r,\infty,a_\nu) = S(r,f), \qquad \mathcal{L}(r,\infty,b_\nu) = S(r,f).$$

If the total degree d of $Q(z,f)$ as a polynomial in f and its derivatives is $d \le n$, then

$$\mathcal{L}(r,\infty,P(z,f)) = S(r,f).$$

3. Deficiencies of transcendental solutions of S_2

It is an easy conclusion following from meromorphicity of solutions of P_2 and P_{34} that all local solutions of S_2 can be extended to meromorphic functions in the complex plane. As far as Nevanlinna deficiencies go, for a solution f of $P_2(\alpha)$ with $\alpha \neq 0$, we have $\delta(a,f) = 0$ for every value a. In case of $P_2(0)$, the same holds both for all non-zero values and for the value zero, which has been recently shown by Steinmetz [**20**].

Similarly, for a transcendental solution of $P_{34}(\alpha)$ we have $\delta(a,f) = 0$ for all values if $\alpha \neq -1/2$ and for all nonzero values if $\alpha = -1/2$ [**2**]. In [**21**] Steinmetz proved nondeficiency of zero also in the special case. It leads to the following result.

THEOREM 3.1. *For transcendental meromorphic solutions of $S_2(\alpha)$ the equality $\delta(a,\sigma) = 0$ holds for all $a \in \overline{\mathbb{C}}$.*

PROOF. The proof of the statement for all values a in case $\alpha \neq -1/2$ and for all nonzero values in case $\alpha = -1/2$ can be found in [**1**]. We conduct the proof for $\alpha = -1/2$ and $a = 0$. Let us then consider the equation

$$(\sigma'')^2 + 4(\sigma')^3 + 2\sigma'(z\sigma' - \sigma) = 0.$$

It follows from equalities (1.6) that, in particular,

$$\sigma' = -\frac{1}{2}p,$$

where p is a solution of P_{34}. This way we obtain

$$m\left(r,\frac{1}{\sigma}\right) \leq m\left(r,\frac{\sigma'}{\sigma}\right) + m\left(r,\frac{1}{\sigma'}\right) \leq m\left(r,\frac{\sigma'}{\sigma}\right) + m\left(r,\frac{1}{p}\right) + O(1).$$

Recall that solutions of S_2 are of finite order. This way it follows from Lemma 2.1 and Theorem 4 in [**21**] that there exists a sequence (r_n) tending to infinity such that

$$m(r,\frac{1}{\sigma}) \leq m(r,\frac{1}{p}) + O(\log r) = o(T(r,p)) + O(\log r) = o(T(r,\sigma)) + O(\log r)$$

holds for $r = r_n$. It follows that $\delta(0,\sigma) = 0$. □

4. Second main theorem for σ-equations

Let us now recall the second main theorem of Nevanlinna.

THEOREM 4.1 ([**13**]). *Let $\{a_k\}_{\nu=1}^{k}$ be a finite set of pairwise distinct complex numbers and let f be a meromorphic function. The inequality*

$$\begin{aligned} m(r,f) + \sum_{\nu=1}^{k} m(r,a_\nu,f) \leq & 2T(r,f) - (N(r,\frac{1}{f'}) + 2N(r,f) - N(r,f')) \\ & + O(\log T(r,f) + \log r) \end{aligned}$$

holds true for all $r \to \infty$, possibly except for r in a set E of finite linear measure.

It follows from properties of solutions of P_1 that solutions of S_1 are transcendental meromorphic functions of finite order and the equality $\delta(a,\sigma) = 0$ holds for all $a \in \overline{\mathbb{C}}$ [**1**]. Similar behavior characterizes solutions of S_2 and S_4, with the exception that for S_4 zero can actually occur as an exceptional value of some solutions.

In case of both Painlevé main equations (see: [**7**, Ch. 3, §11]) and Painlevé sigma equations the inequality in the second main theorem translates into an asymptotic equality.

THEOREM 4.2. *Let σ be a solution of S_1 or S_2. Then the second main theorem of Nevanlinna for σ has the form of asymptotic equality*

$$N(r,\frac{1}{\sigma'}) + N_1(r,\sigma) = 2T(r,\sigma) + O(\log r).$$

PROOF. First, let σ denote a solution of S_1. Since $m(r,\sigma) = O(\log r)$ [**1**], it follows from Lemma 2.1 that also $m(r,\sigma') = O(\log r)$. Moreover, S_1 can be presented in the form

$$\frac{\sigma}{\sigma'} = \frac{1}{2}\frac{\sigma''}{\sigma'} \cdot \sigma'' + 2(\sigma')^2 + z.$$

Applying simple properties of the mean proximity function and Lemma 2.1 we obtain for $r \to \infty$,

$$\begin{aligned} m(r,\frac{\sigma}{\sigma'}) &\leq m(r,\frac{\sigma''}{\sigma'}) + m(r,\sigma'') + 2m(r,\sigma') + O(\log r) \\ &\leq 3m(r,\sigma) + 2m(r,\frac{\sigma''}{\sigma'}) + 3m(r,\frac{\sigma'}{\sigma}) + O(\log r) = O(\log r). \end{aligned}$$

It follows that the conditions of Lemma 2.2 hold. Taking into consideration that all the poles of solutions of S_1 are simple, applying the first main theorem we get

$$\begin{aligned} N(r,\frac{1}{\sigma'})+N_1(r,\sigma) &= m(r,\sigma')+N(r,\sigma')+O(1)=N(r,\sigma')+O(\log r) \\ &= 2N(r,\sigma)+O(\log r)=2T(r,\sigma)+O(\log r). \end{aligned}$$

Let now σ be a transcendental solution of $S_2(\alpha)$. If $\alpha=-\frac{1}{2}$ the equation S_2 can be represented in the form

$$\frac{\sigma}{\sigma'}=\frac{1}{2}\left(\frac{\sigma''}{\sigma'}\right)^2+2\sigma'+z.$$

Thus we get for $r\to\infty$,

$$m(r,\frac{\sigma}{\sigma'})\le 2m(r,\frac{\sigma''}{\sigma'})+m(r,\sigma')+O(\log r)\le 2m(r,\frac{\sigma''}{\sigma'})+m(r,\frac{\sigma'}{\sigma})+m(r,\sigma)+O(\log r).$$

As σ is of finite order and we know that $m(r,\sigma)=O(\log r)$ [**1**], it follows from Lemma 2.1 that

$$m(r,\frac{\sigma}{\sigma'})=O(\log r)\qquad (r\to\infty).$$

If $\alpha\ne-\frac{1}{2}$, we differentiate which transforms S_2 into

$$\frac{\sigma}{\sigma'}=\frac{\sigma'''}{\sigma'}+\sigma\sigma'+z+2.$$

Applying Lemma 2.1 and the fact that both

$$m(r,\sigma)=O(\log r)\qquad\text{and}\qquad m(r,\sigma')=O(\log r)$$

we get

$$m(r,\frac{\sigma}{\sigma'})=O(\log r)\qquad (r\to\infty)$$

also in this case. Let us recall that the poles of σ are simple, so $N_1(r,\sigma)=0$. As the conditions of Lemma 2.2 are fulfilled, we have

$$\begin{aligned} N(r,\frac{1}{\sigma'})+N_1(r,\sigma) &= N(r,\sigma')+O(\log r) \\ &= 2N(r,\sigma)+O(\log r) \\ &= 2T(r,\sigma)+O(\log r). \end{aligned}$$

□

Solutions of S_4 also admit an asymptotic equality in the second main theorem.

THEOREM 4.3. *Let σ be a transcendental solution of S_4. Then the second main theorem of Nevanlinna for σ has the form of asymptotic equality*

$$m(r,\frac{1}{\sigma})+N(r,\frac{1}{\sigma'})+N_1(r,\sigma)=2T(r,\sigma)+O(\log r).$$

PROOF. Consider first the case $\theta_0\theta_\infty=0$. Assume that $\theta_\infty=0$. Then S_4 has the form

$$(\sigma'')^2-4(z\sigma'-\sigma)^2+4(\sigma')^2(\sigma'+2\theta_0)=0,$$

which can be written as

$$\frac{\sigma}{\sigma'}=\frac{1}{4}(\frac{\sigma''}{\sigma})^2\frac{\sigma'}{\sigma}+(\frac{\sigma'}{\sigma})^2\sigma+2\theta_0\frac{\sigma'}{\sigma}+2z.$$

Since $m(r,\sigma) = O(\log r)$ [**1**], applying Lemma 2.1 and simple properties of the mean proximity function, we get

$$m(r, \frac{\sigma}{\sigma'}) = O(\log r), \qquad r \to \infty.$$

As the conditions of Lemma 2.2 are fulfilled, we obtain

$$m(r, \frac{1}{\sigma}) + N(r, \frac{1}{\sigma'}) = N(r,\sigma') + O(\log r), \qquad r \to \infty.$$

The poles of σ are simple, so $N_1(r,\sigma) = 0$. It follows that

$$\begin{aligned} m(r, \frac{1}{\sigma}) + N(r, \frac{1}{\sigma'}) + N_1(r,\sigma) &= N(r,\sigma') + O(\log r) \\ &= 2N(r,\sigma) + O(\log r) \\ &= 2T(r,\sigma) + O(\log r). \end{aligned}$$

Consider now the case $\theta_0\theta_\infty \neq 0$. Let $\sigma(z)$ be a transcendental solution of (1.7). Denote

$$G(z) := (\sigma'')^2 - 4(z\sigma' - \sigma)^2 + 4\sigma'(\sigma' + 2\theta_0)(\sigma' + 2\theta_\infty).$$

Differentiating twice we obtain

$$G'(z) = \sigma''' + 4[2(\theta_0 + \theta_\infty) - z^2]\sigma' + 4z\sigma + 6(\sigma')^2 + 8\theta_0\theta_\infty,$$

and

$$G''(z) = \sigma^{(4)} - 4z\sigma' + 4[2(\theta_0 + \theta_\infty) - z^2]\sigma'' + 4\sigma + 12\sigma'\sigma'' = 0.$$

Computing $zG'(z) - G''(z)$ leads to the equation

$$z\sigma^{(4)} - \sigma''' - 8(\theta_0 + \theta_\infty - z^2)\sigma' + 4[2(\theta_0 + \theta_\infty) - z^2]z\sigma'' + 12z\sigma'\sigma'' + 6(\sigma')^2 - 8\theta_0\theta_\infty = 0.$$

This equation can be easily transformed into

$$\begin{aligned} \frac{\sigma}{\sigma'} = \frac{\sigma}{\theta_0\theta_\infty}\Big(\frac{z}{8}\frac{\sigma^{(4)}}{\sigma'} - \frac{1}{8}\frac{\sigma'''}{\sigma'} + (\theta_0 + \theta_\infty - 4z^2)z\frac{\sigma''}{\sigma'} + \frac{3}{2}z\sigma'' + z^2 - \theta_0 - \theta_\infty\Big) \\ + \frac{3}{4\theta_0\theta_\infty}\sigma'\sigma''. \end{aligned}$$

Applying Lemma 2.1, the fact that $m(r,\sigma) = O(\log r)$ and simple properties of the mean proximity function we obtain

$$m(r, \frac{\sigma}{\sigma'}) = O(\log r).$$

This way the conditions of Lemma 2.2 are fulfilled. As the poles of σ are simple, we obtain

$$\begin{aligned} m(r, \frac{1}{\sigma}) + N(r, \frac{1}{\sigma'}) + N_1(r,\sigma) &= N(r,\sigma') + O(\log r) \\ &= 2N(r,\sigma) + O(\log r) \\ &= 2T(r,\sigma) + O(\log r). \end{aligned}$$

□

5. Small targets of σ-equations

For meromorphic functions a, f, we say that a *is small with respect to* f if $T(r,a) = S(r,f)$. Meromorphic functions small with respect to a meromorphic function f are sometimes called *small targets of* f [**7**, Ch.3, p.72]. Behavior of solutions of P_1, P_2 and P_4 towards small targets was described in [**18, 19**]. Below we present (partial) results for respective sigma equations

THEOREM 5.1. *Let $\sigma(z)$ be a transcendental solution of S_1, S_2 or S_4 and let $a(z)$ be a small target of $\sigma(z)$. If $a(z)$ does not solve S_1, S_2 or S_4, respectively, then*

$$m(r, \frac{1}{\sigma - a}) = S(r,\sigma) \qquad \textit{and} \qquad \delta(a,\sigma) = 0.$$

PROOF. Let us first assume that $\sigma(z)$ is a solution of S_1 and that $a(z)$ is a small target of $\sigma(z)$, which does not solve S_1. In the equation S_1 we substitute

$$u(z) := \frac{1}{\sigma(z) - a(z)}.$$

Notice that $a(z)$ is a small target of $u(z)$. As

$$\sigma(z) = \frac{1}{u(z)} + a(z),$$

we obtain the following differential equation of u with coefficients being polynomials in a and its derivatives

$$\begin{aligned} A(z)u^6 &= 2u^5 + [12(a')^2 + 2z]u^4u' + 2a''u^4u'' - 4a''u^3(u')^2 \\ &\quad - 12u^2(u')^2 - u^2(u'')^2 + 4u(u')^2u'' - 4(u')^4 + 4(u')^3, \end{aligned}$$

where

$$A(z) := (a'')^2 + 4(a')^3 + 2za' - 2a.$$

By assumption, $a(z)$ is not a solution of S_1, so $A(z) \not\equiv 0$. Moreover, from the fact that $a(z)$ is a small target of σ we conclude that the equality

$$T(r, A(z)^{-1}) = S(r,\sigma) = S(r,u)$$

holds, which means that $m(r, A(z)^{-1}) = S(r,u)$. This way we obtain the equation

$$\begin{aligned} u^6 = A(z)^{-1}[2u^5 &+ (12(a')^2 + 2z)u^4u' + 2a''u^4u'' - 4a''u^3(u')^2 \\ &-12u^2(u')^2 - u^2(u'')^2 + 4u(u')^2u'' - 4(u')^4 + 4(u')^3]. \end{aligned}$$

It means that the conditions of Lemma 2.3 are met with $p = 5$. It follows that

$$m(r,u) = S(r,u),$$

which is equivalent to

$$m(r, \frac{1}{\sigma - a}) = S(r,\sigma),$$

so $\delta(a,\sigma) = 0$.

Let us now assume that $\sigma(z)$ is a solution of S_2 and that $a(z)$ is a small target of $\sigma(z)$, which does not solve S_2. Similarly as before, in the equation S_2 we substitute

$$u(z) := \frac{1}{\sigma(z) - a(z)}.$$

This way we obtain the following differential equation of u with coefficients being polynomials in a and its derivatives

$$u^6 = B(z)^{-1}[2a'u^5 + (12(a')^2 + 4za' - 2a)u^4u' + 2a''u^4u'' - 4a''u^3(u')^2 - 2u^3u' - 2(z+6)u^2(u')^2 - u^2(u'')^2 + 4u(u')^2u'' - 4(u')^4 + 4(u')^3],$$

where $B(z) := (a'')^2 + 4(a')^3 + 2z(a')^2 - 2aa' - \frac{1}{4}(\alpha + \frac{1}{2})^2$. By assumption, $a(z)$ is not a solution of S_2, so $B(z) \not\equiv 0$ and, similarly as in the previous case we may arrive at the conclusion that the coefficients of the equation fulfill the conditions of Lemma 2.3. It follows that

$$m(r, u) = S(r, u),$$

which is equivalent to

$$m(r, \frac{1}{\sigma - a}) = S(r, \sigma),$$

so $\delta(a, \sigma) = 0$.

Let us finally assume that $\sigma(z)$ is a solution of S_4 and that $a(z)$ is a small target of $\sigma(z)$, which does not solve S_4. Again, in the equation S_4 we substitute

$$u(z) := \frac{1}{\sigma(z) - a(z)}.$$

This way we obtain the following differential equation of u with coefficients being polynomials in a and its derivatives

$$u^6 = C(z)^{-1}[8(a - za')u^5 + 4[4\theta_0\theta_\infty + 3(a')^2 + 2za + (4(\theta_0 + \theta_\infty) - 2z^2)a']u^4u' + 2a''u^4u'' - 2a''u^3(u')^2 + 4u^4 + 4[z^2 - 3a' - 2(\theta_0 + \theta_\infty)]u^2(u')^2 + 8zu^3u' - u^2(u'')^2 + 4u(u')^2u'' - 4(u')^4 + 4(u')^3],$$

where

$$C(z) := (a'')^2 + 4[2(\theta_0 + \theta_\infty) - z^2](a')^2 + 8zaa' - 4a^2 + 4(a')^3 + 16\theta_0\theta_\infty a'.$$

Since, by assumption, $a(z)$ is not a solution of S_4, so $C(z) \not\equiv 0$ and, by a similar reasoning as before, we conclude that

$$m(r, \frac{1}{\sigma - a}) = S(r, \sigma)$$

and $\delta(a, \sigma) = 0$. □

We can also formulate another result concerning rational small targets of the sigma equations in the following way.

THEOREM 5.2. *Let $\sigma(z)$ be a transcendental solution of S_1, S_2 or S_4 and let $a(z)$ be a rational function. If $a(z)$ does not solve S_1, S_2 or S_4, respectively, then $\beta(a, \sigma) = 0$.*

PROOF. First, let us first assume that $\sigma(z)$ is a solution of S_1 and that $a(z)$ is a rational function. Then $a(z)$ does not solve S_1 since S_1 does not have rational solutions. As in the proof of Theorem 5.1 we substitute $u(z) := (\sigma(z) - a(z))^{-1}$ and put $A(z) := (a'')^2 + 4(a')^3 + 2za' - 2a$. We obtain the equation

$$A(z)u^6 = 2u^5 + [12(a')^2 + 2z]u^4u' + 2a''u^4u'' - 4a''u^3(u')^2 - 12u^2(u')^2 - u^2(u'')^2 + 4u(u')^2u'' - 4(u')^4 + 4(u')^3,$$

which has rational coefficients. It means that the coefficients fulfill the conditions of Lemma 2.5, which leads us to the conclusion that

$$\mathcal{L}(r,u) = S(r,u),$$

which is equivalent to

$$\mathcal{L}(r, \frac{1}{\sigma - a}) = S(r,\sigma),$$

so $\beta(a,\sigma) = 0$.

The proof of the remaining two cases can be conducted along the same lines as in the proof of Theorem 5.1, with application of Lemma 2.5 instead of Lemma 2.3. □

Let us recall now the notions of an asymptotic value and an asymptotic spot. A pair $\{a, \Gamma\}$, where $a \in \overline{\mathbb{C}}$ and Γ is a continuous curve

$$\Gamma : z = z(t),\ 0 \le t < \infty, \qquad z(t) \to \infty \text{ for } t \to \infty,$$

such that

$$\lim_{z\to\infty, z\in\Gamma} f(z) = \lim_{t\to\infty} f(z(t)) = a,$$

is called *an asymptotic spot of f*. Also a is called *an asymptotic value of f*. By the Denjoy-Carleman-Ahlfors theorem [**6**], an entire function of finite lower order μ cannot have more than $\max\{[2\mu], 1\}$ different asymptotic spots. Similarly, $a(z)$ is an asymptotic function of $f(z)$ if zero is an asymptotic value of $f(z) - a(z)$. Estimates of the possible number of asymptotic functions similar to Denjoy-Carleman Ahlfors theorem have been formulated in attempt to prove so-called Denjoy conjecture (see: [**5**]). The number of asymptotic values of an entire function of infinite lower order or a meromorphic function even of finite order may be infinite.

If the function moves along the curve towards the value a with a fixed higher speed, for instance comparable with characteristic $T(r, f)$, we say that a is *a strong asymptotic value.*

Definition 5.3 ([**12**]). We say that $a \in \overline{\mathbb{C}}$ is *an α_0-strong asymptotic value* of a meromorphic function f, if there exists a continuous curve $\Gamma : z = z(t),\ 0 \le t < \infty,\ z(t) \to \infty$ as $t \to \infty$, such, that

$$\liminf_{t\to\infty} \frac{\log|f(z(t)) - a|^{-1}}{T(|z(t)|, f)} = \alpha(a) \ge \alpha_0 > 0, \qquad \text{if} \quad a \ne \infty;$$

$$\liminf_{t\to\infty} \frac{\log|f(z(t))|}{T(|z(t)|, f)} \ge \alpha_0 > 0, \qquad \text{if} \quad a = \infty.$$

It is easy to notice that, if a is an α_0-strong asymptotic value of f, then the magnitude of Petrenko's deviation

$$\beta(a, f) \ge \alpha_0. \tag{5.1}$$

It means that a is also a deficient value in the sense of Petrenko. Marchenko proved that the number k of distinct α_0-strong asymptotic spots of a meromorphic function of finite lower order λ is finite and the inequality

$$k \le \left[\frac{2B(\lambda)}{\alpha_0}\right]$$

holds (see: [**12**]). The estimates of the number of strong asymptotic values of solutions of P_1, P_2, P_{34}, P_4 and S_1, S_2, S_4. were shown in [**1**]. It should be noticed that for certain solutions of P_4 and S_4 zero may in fact be strongly asymptotic. Similarly to the notion of a strong asymptotic value we may formulate a definition of a strong asymptotic function. The estimates of the number of strong asymptotic functions of meromorphic functions can be found for example in [**3**].

The inequality (5.1) means that Theorem 5.2 leads to the following conclusion.

COROLLARY 5.4. *Let $\sigma(z)$ be a transcendental solution of S_2 or S_4 and let $a(z)$ be a rational function. Then, if $a(z)$ does not solve S_2 or S_4, respectively, then $a(z)$ is not an α_0-strong asymptotic function of σ for any value $\alpha_0 > 0$.*

In case of the first Painlevé σ- equation we may formulate a slightly sharper conclusion.

COROLLARY 5.5. *Solutions of Painlevé σ-equation S_1 do not have rational strong asymptotic functions.*

6. The third Painlevé equation

A modifed third Painlevé equation $\tilde{P}_3$

$$f'' = \frac{(f')^2}{f} + \alpha f^2 + \gamma f^3 + \beta e^z + \frac{\delta e^{2z}}{f} \tag{6.1}$$

has meromorphic solutions only [**9**]. Unlike previously mentioned equations, $\tilde{P}_3$ has transcendental coefficients if $(\beta, \delta) \neq (0,0)$. Therefore instead of focusing on transcendental solutions in general we shall apply the notion of admissible solutions. We say that f *is an admissible solution of a differential equation* if all the coefficients of the equation are small with respect to f. In case of admissible solutions of $\tilde{P}_3$, in fact we get $S(r,f) = O(r)$ without an exceptional set [**7, 10**]. Nevanlinna deficiencies of solutions of $\tilde{P}_3$ were obtained by Shimomura in [**17**]. The following result holds for Petrenko's deviations.

THEOREM 6.1. *Let f be an admissible solution of $\tilde{P}_3(\alpha,\beta,\gamma,\delta)$ with $(\alpha,\beta) \neq (0,0)$ and $(\gamma,\delta) \neq (0,0)$. For all values a in $\overline{\mathbb{C}}$ we have*

$$\mathcal{L}(r,a,f) = S(r,f) \qquad (r \to \infty)$$

and $\beta(a,f) = 0$.

PROOF. Let f be an admissible solution of P_3. As

$$T(r,\beta e^z) \leq \frac{r}{\pi} + \log^+|\beta| \qquad \text{and} \qquad T(r,\delta e^{2z}) \leq \frac{2r}{\pi} + \log^+|\delta|,$$

it means that, in general we have

$$O(r) = S(r,f).$$

Since

$$\mathcal{L}(r,\infty,\beta e^z) = r + \log^+|\beta| \qquad \text{and} \qquad \mathcal{L}(r,\infty,\delta e^{2z}) = 2r + \log^+|\delta|,$$

we can see that the coefficients of the equation (6.1) are small with respect to f also in the sense mentioned in Lemma 2.5.

Let $(\alpha,\gamma) \neq (0,0)$. and $(\beta,\delta) \neq (0,0)$. Assume that $\gamma \neq 0$. The equation (6.1) can be presented in the form

$$(-\gamma)f^4 = \alpha f^3 + (f')^2 - ff'' + \beta e^z f + \delta e^{2z}.$$

This way the conditions of Lemma 2.5 are fulfilled with $n = 3$. It follows that

$$\mathcal{L}(r, \infty, f) = S(r, f) \qquad (r \to \infty).$$

Let now $\gamma = 0$. Then the equation (6.1) can be written as

$$(-\alpha) f^3 = (f')^2 - f f'' + \beta e^z f + \delta e^{2z},$$

and the conditions of Lemma 2.5 are fulfilled with $n = 2$. Again, it follows that

$$\mathcal{L}(r, \infty, f) = S(r, f) \qquad (r \to \infty).$$

Now for a fixed value $a \in \mathbb{C}$ we substitute $f = \frac{1}{g} + a$. The equation (6.1) assumes the form of an equation with respect to g

$$\begin{aligned}(\alpha a^3 + \gamma a^4 + \beta a e^z + \delta e^{2z}) g^4 = 2a(g')^2 g - a g'' g^2 + (g')^2 - g'' g - \gamma \\ - (\alpha + 4\gamma a) g - (3\alpha a + 6\gamma a^2) g^2 \\ - (3\alpha a^2 + 4\gamma a^3 + \beta e^z) g^3.\end{aligned} \tag{6.2}$$

By the first theorem of Nevanlinna,

$$T(r, g) = T(r, f) + O(1),$$

so we can see that if f is an admissible solution of (6.1) then g is an admissible solution of (6.2), also in the sense of Lemma 2.5. Applying again Lemma 2.5 we obtain

$$\mathcal{L}(r, a, f) = \mathcal{L}(r, \infty, \frac{1}{f - a}) = \mathcal{L}(r, \infty, g) = S(r, g) = S(r, f) \qquad (r \to \infty).$$

□

COROLLARY 6.2. *If f is an admissible solution of $\tilde{P}_3(\alpha, \beta, \gamma, \delta)$ with $(\alpha, \gamma) \neq (0, 0)$ and $(\beta, \delta) \neq (0, 0)$ then f has no strong asymptotic values in $\overline{\mathbb{C}}$.*

Let us now consider the situation when one pair (α, γ) or (β, δ) of coefficients of $\tilde{P}_3$ equals $(0, 0)$. It is easy to check, by a suitable substitution, that the pairs of coefficients (α, γ) and (β, δ) in (6.1) have a symmetric role (see, for example [**9**]). Let us then concentrate on the case $(\alpha, \gamma) = (0, 0)$. If f is an admissible solution of $\tilde{P}_3(\alpha, \beta, \gamma, \delta)$ with $(\alpha, \gamma) = (0, 0)$ and $(\beta, \delta) \neq (0, 0)$, that is the equation

$$f f'' = (f')^2 + \beta e^z f + \delta e^{2z}, \tag{6.3}$$

then for every value $a \in \mathbb{C}$ we have $\mathcal{L}(r, a, f) = S(r, f)$ and $\beta(a, f) = 0$.

Indeed, let f be an admissible solution of (6.3) and for a fixed value $a \in \mathbb{C}$ put $f = \frac{1}{g} + a$. The equation (6.3) transforms into an equation with respect to g:

$$(\delta e^{2z} + \beta a e^z) g^4 = 2a(g')^2 g - a g'' g^2 + (g')^2 - g'' g - \beta e^z g^3.$$

Assume first that $\delta = 0$. Since β and δ are not both zeros, Lemma 2.5 applies with $n = 3$ if $a \neq 0$ and with $n = 2$ if $a = 0$. If, on the other hand, $\delta \neq 0$ then the conditions of Lemma 2.5 hold with $n = 3$. In all cases we obtain

$$\mathcal{L}(r, a, f) = \mathcal{L}(r, \infty, g) = S(r, g) = S(r, f).$$

It follows that for every value $a \in \mathbb{C}$ there is $\beta(a, f) = 0$ and, as a result f has no finite strong asymptotic values.

Finally, notice that $\tilde{P}_3(0,0,0,0)$, that is the equation

$$ff'' = (f')^2$$

has a solution $f(z) = e^z$. As $\beta(\infty, e^z) = \beta(0, e^z) = \pi$, this solution has two strong asymptotic values ($\alpha_0 = \pi$), each of them connected with one strong asymptotic spot.

References

[1] Ewa Ciechanowicz, *Painlevé σ-equations s_1, s_2, s_4 and their value distribution*, Filomat **34** (2020), no. 6, 1959–1973. MR4207728

[2] Ewa Ciechanowicz and Galina Filipuk, *Meromorphic solutions of $P_{4,34}$ and their value distribution*, Ann. Acad. Sci. Fenn. Math. **41** (2016), no. 2, 617–638, DOI 10.5186/aasfm.2016.4146. MR3525388

[3] Ewa Ciechanowicz and Ivan Ivanovich Marchenko, *On deviations, small functions and strong asymptotic functions of meromorphic functions*, Kodai Math. J. **40** (2017), no. 2, 379–404, DOI 10.2996/kmj/1499846603. MR3680567

[4] P. A. Clarkson, *Painlevé transcendents*, NIST Digital Library of Mathematical Functions, `http://dlmf.nist.gov`.

[5] P. C. Fenton, *Entire functions having asymptotic functions*, Bull. Austral. Math. Soc. **27** (1983), no. 3, 321–328, DOI 10.1017/S0004972700025831. MR715310

[6] A. A. Gol′dberg and I. V. Ostrovskiĭ, *Raspredelenie znacheniĭ meromorfnykh funktsiĭ* (Russian), Izdat. "Nauka", Moscow, 1970. MR0280720

[7] Valerii I. Gromak, Ilpo Laine, and Shun Shimomura, *Painlevé differential equations in the complex plane*, De Gruyter Studies in Mathematics, vol. 28, Walter de Gruyter & Co., Berlin, 2002, DOI 10.1515/9783110198096. MR1960811

[8] W. K. Hayman, *Meromorphic functions*, Oxford Mathematical Monographs, Clarendon Press, Oxford, 1964. MR0164038

[9] Aimo Hinkkanen and Ilpo Laine, *Solutions of a modified third Painlevé equation are meromorphic*, J. Anal. Math. **85** (2001), 323–337, DOI 10.1007/BF02788086. MR1869614

[10] Ilpo Laine, *Nevanlinna theory and complex differential equations*, De Gruyter Studies in Mathematics, vol. 15, Walter de Gruyter & Co., Berlin, 1993, DOI 10.1515/9783110863147. MR1207139

[11] I. I. Marchenko, *On an analogue of the second fundamental theorem for the uniform metric* (Russian, with English, Russian and Ukrainian summaries), Mat. Fiz. Anal. Geom. **5** (1998), no. 3-4, 212–227. MR1668973

[12] I. I. Marchenko, *On strong asymptotic spots of meromorphic functions of finite lower order* (Russian, with English, Russian and Ukrainian summaries), Mat. Fiz. Anal. Geom. **11** (2004), no. 4, 484–491. MR2114007

[13] R. Nevanlinna, *Eindeutige analytische Funktionen* (German), Die Grundlehren der mathematischen Wissenschaften, Band 46, Springer-Verlag, Berlin-New York, 1974. Zweite Auflage; Reprint. MR0344426

[14] Kazuo Okamoto, *Polynomial Hamiltonians associated with Painlevé equations. I*, Proc. Japan Acad. Ser. A Math. Sci. **56** (1980), no. 6, 264–268. MR581468

[15] Kazuo Okamoto, *Studies on the Painlevé equations. III. Second and fourth Painlevé equations, $P_{\rm II}$ and $P_{\rm IV}$*, Math. Ann. **275** (1986), no. 2, 221–255, DOI 10.1007/BF01458459. MR854008

[16] V. P. Petrenko, *Growth of meromorphic functions of finite lower order* (Russian), Izv. Akad. Nauk SSSR Ser. Mat. **33** (1969), 414–454. MR0251226

[17] Shun Shimomura, *Value distribution of Painlevé transcendents of the third kind*, Complex Variables Theory Appl. **40** (1999), no. 1, 51–62, DOI 10.1080/17476939908815208. MR1742870

[18] Shun Shimomura, *Value distribution of Painlevé transcendents of the first and second kind*, J. Anal. Math. **82** (2000), 333–346, DOI 10.1007/BF02791233. MR1799669

[19] Shun Shimomura, *On deficiencies of small functions for Painlevé transcendents of the fourth kind*, Ann. Acad. Sci. Fenn. Math. **27** (2002), no. 1, 109–120. MR1884353

[20] Norbert Steinmetz, *Sub-normal solutions to Painlevé's second differential equation*, Bull. Lond. Math. Soc. **45** (2013), no. 2, 225–235, DOI 10.1112/blms/bds073. MR3064409

[21] Norbert Steinmetz, *Remarks on Painlevé's differential equation* P_{34}, J. Anal. Math. **141** (2020), no. 1, 383–395, DOI 10.1007/s11854-020-0118-3. MR4174048

Institute of Mathematics, University of Szczecin, 70-451 Szczecin, ul. Wielkopolska 15
Email address: ewa.ciechanowicz@usz.edu.pl

Contemporary Mathematics
Volume **782**, 2023
https://doi.org/10.1090/conm/782/15720

Relations between different Hamiltonian forms of the third Painlevé equation

Galina Filipuk, Adam Ligęza, and Alexander Stokes

ABSTRACT. In this paper we study different Hamiltonian systems with polynomial and rational Hamiltonians associated with the generic third Painlevé equation and present explicit birational transformations relating them.

1. Introduction

A standard form of the third Painlevé equation is

$$\text{(1.1)} \qquad \mathrm{P}_{\mathrm{III}}: \quad \frac{d^2x}{dt^2} = \frac{1}{x}\left(\frac{dx}{dt}\right)^2 - \frac{1}{t}\frac{dx}{dt} + \frac{1}{t}(\alpha x^2 + \beta) + \gamma x^3 + \frac{\delta}{x},$$

where x is a function of t and α, β, γ, δ with $\gamma\delta \neq 0$ are arbitrary complex parameters. There exist in the literature multiple non-autonomous Hamiltonian systems equivalent to $\mathrm{P}_{\mathrm{III}}$ obtained from various viewpoints, but the relations between these have not been fully established. In this paper we continue the program announced in [**DFLS**], which aims to compare the different Hamiltonian systems associated with each Painlevé equation and find explicit birational (or algebraic) transformations between them. We draw on a variety of methods, including the comparison of linear systems from the isomonodromy approach as well as the method presented in [**DFLS**], which uses tools from the geometric theory of Painlevé equations developed by Okamoto and Sakai [**Ok1, Sak**]. Obtaining such relations between different Hamiltonian systems related to Painlevé equations is sometimes quite subtle, for example for the fourth Painlevé equation studied in [**DFLS**] the relations are very much non-trivial and the geometric method is more effective than a direct approach. In this paper we consider the case of the third Painlevé equation, more specifically the generic case where $\gamma\delta \neq 0$ as above, for which the relations between various Hamiltonian systems turn out to be more straightforward and less interesting on the level of the geometric theory, so we omit details and present only the birational transformations themselves.

The Hamiltonians and the associated systems of differential equations equivalent to $\mathrm{P}_{\mathrm{III}}$ that we plan to compare in this paper are the Okamoto (polynomial) Hamiltonian [**Ok2, Ok3**], the Kajiwara-Noumi-Yamada (KNY) Hamiltonian [**KNY**], the Filipuk-Żołądek rational Hamiltonian [**FZ**] and two Hamiltonians obtained from the isomonodromic deformation of linear systems: the Jimbo-Miwa

2020 *Mathematics Subject Classification.* Primary 34M55.

Hamiltonian [**JM**] and the Its-Prokhorov Hamiltonian [**IP**]. We discuss only the Its-Prokhorov Hamiltonian since up to a function of t and scaling of parameters they are the same.

2. Main results

In the papers [**Ok2**] and [**Ok3**], the third Painlevé equation appears as the result of the monodromy preserving deformation of a scalar second order linear differential equation (with coefficients that depend on t) with certain singularities in the complex plane. The third Painlevé equation is obtained in the form of a Hamiltonian system with polynomial Hamiltonian [**Ok1, Ok2**], given by

$$\text{(2.1)}\quad \mathrm{H}^{\mathrm{Ok}}(f,g,t)=\frac{1}{t}(2f^2g^2+(\theta_0+\theta_\infty)\eta_\infty tf-((1+2\theta_0)f-2\eta_0 t+2\eta_\infty tf^2)g),$$

where $f=f(t)$ and $g=g(t)$, t is an independent variable, and θ_0, θ_∞, η_0, η_∞ are complex constants. The corresponding Hamiltonian system is

$$\text{(2.2)}\quad \begin{cases} \dfrac{df}{dt}=\dfrac{\partial\,\mathrm{H}^{\mathrm{Ok}}}{\partial g}=\dfrac{4f^2g-2\eta_\infty tf^2-(1+2\theta_0)f+2\eta_0 t}{t},\\[2mm] \dfrac{dg}{dt}=-\dfrac{\partial\,\mathrm{H}^{\mathrm{Ok}}}{\partial f}=-\dfrac{4fg^2-(1+2\theta_0+4\eta_\infty tf)g+\eta_\infty(\theta_0+\theta_\infty)t}{t}.\end{cases}$$

Eliminating the function g from the system above, we obtain $\mathrm{P}_{\mathrm{III}}$ for the function f with parameters

$$\text{(2.3)}\quad \alpha=-4\eta_\infty\theta_\infty,\quad \beta=4\eta_0(1+\theta_0),\quad \gamma=4\eta_\infty^2,\quad \delta=-4\eta_0^2.$$

The KNY Hamiltonian [**KNY**, (8.237)] is given by

$$\text{(2.4)}\quad \mathrm{H}^{\mathrm{KNY}}(q,p,s)=\frac{1}{s}(p(p-1)q^2+(a_1+a_2)qp+sp-a_2q),$$

where $q=q(s)$ and $p=p(s)$, s is an independent variable and a_1 and a_2 are complex constants. The corresponding Hamiltonian system is given by

$$\text{(2.5)}\quad \begin{cases} \dfrac{dq}{ds}=\dfrac{\partial\,\mathrm{H}^{\mathrm{KNY}}}{\partial p}=\dfrac{q^2p+(p-1)q^2+(a_1+a_2)q+s}{s},\\[2mm] \dfrac{dp}{ds}=-\dfrac{\partial\,\mathrm{H}^{\mathrm{KNY}}}{\partial q}=-\dfrac{2(p-1)qp+(a_1+a_2)p-a_2)s}{s}.\end{cases}$$

The KNY Hamiltonian system is very convenient to use as a reference example for the geometric approach, since [**KNY**] provides a comprehensive collection of relevant data. However the relation to the third Painlevé equation is not as straightforward here as in the Okamoto case. According to [**KNY**], one needs to eliminate first the variable p, then introduce the function $x(t)$ by $q(s)=tx(t)$ with $s=t^2$ and then one obtains the third Painlevé equation for $x(t)$ with parameters

$$\text{(2.6)}\quad \alpha=-4(a_1-a_2),\quad \beta=4(1-a_1-a_2),\quad \gamma=4,\quad \delta=-4.$$

The Filipuk-Żołądek (rational) Hamiltonian [**FZ**] is given by

$$\text{(2.7)}\quad \mathrm{H}^{\mathrm{FZ}}(x,y,t)=\frac{x^2y^2}{2t}-\alpha x+\frac{\beta}{x}-\frac{\gamma tx^2}{2}+\frac{\delta t}{2x^2},$$

where $x = x(t)$ and $y = y(t)$, t is an independent variable and α, β, γ, δ are complex constants. The corresponding Hamiltonian system is given by

$$(2.8)\qquad \begin{cases} \dfrac{dx}{dt} = \dfrac{\partial \mathrm{H}^{\mathrm{FZ}}}{\partial y} = \dfrac{x^2 y}{t}, \\ \dfrac{dy}{dt} = -\dfrac{\partial \mathrm{H}^{\mathrm{FZ}}}{\partial x} = \alpha - \dfrac{xy^2}{t} + \dfrac{\beta}{x^2} + \gamma t x + \dfrac{\delta t}{x^3}. \end{cases}$$

Eliminating the function y between these equations we immediately get the third Painlevé equation for the function x, with parameters $\alpha, \beta, \gamma, \delta$.

In [**IP**], the authors obtained Hamiltonians related to the Painlevé equations from the study of monodromy preserving deformations of 2×2 linear systems, with the $\mathrm{P_{III}}$ case being

$$(2.9)\quad \mathrm{H}^{\mathrm{IP}}(f, g, t) = \frac{1}{t}(2f^2g^2 + (2t - 2tf^2 + (4\theta_\infty - 1)f)g - 2tf(\theta_0 + \theta_\infty) + (\theta_\infty^2 - \theta_0^2)).$$

The corresponding Hamiltonian system gives the third Painlevé equation for the function f, with parameters $\alpha = 8\theta_0$, $\beta = 4 - 8\theta_\infty$, $\gamma = -\delta = 4$. For the fourth Painlevé equation, the corresponding Its-Prokhorov Hamiltonian was different from that of Okamoto. However, in this case we see that if we take Okamoto's Hamiltonian with η_1 and η_∞ equal to one, and set $\theta_\infty^{\mathrm{IP}} = -\theta_0^{\mathrm{Ok}}/2$, $\theta_0^{\mathrm{IP}} = -\theta_\infty^{\mathrm{Ok}}/2$, then up to a function of t both Hamiltonians (and, hence, corresponding systems) coincide.

Hence, we are left with two nontrivial cases of relations between systems to establish: first between the rational Hamiltonian H^{FZ} and the Okamoto Hamiltonian H^{Ok}, and second between the rational Hamiltonian H^{FZ} and the KNY Hamiltonian $\mathrm{H}^{\mathrm{KNY}}$. In what follows we will present these, completing the picture of how different Hamiltonian forms of $\mathrm{P_{III}}$ are related, as outlined in Figure 1.

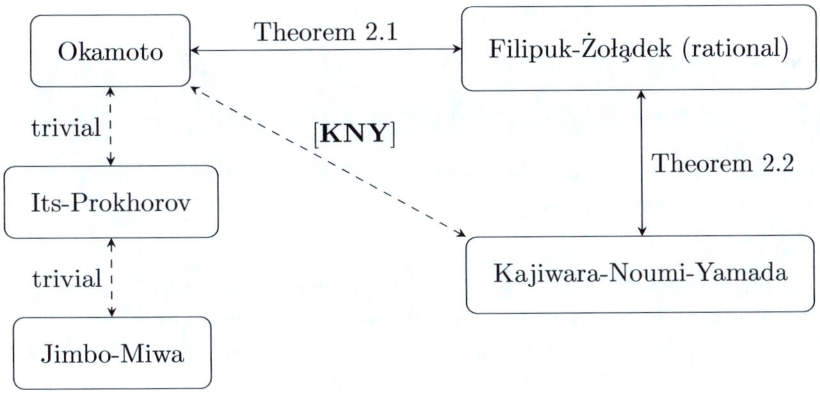

FIGURE 1. Relationships between different Hamiltonian forms of $\mathrm{P_{III}}$

THEOREM 2.1. *Let $x = x(t)$ and $y = y(t)$ be solutions of the Filipuk-Żołądek system* (2.8) *and $f = f(t)$ and $g = g(t)$ be solutions of the Okamoto system* (2.2). *Under the identification of parameters as in* (2.3), *the following birational transformation is valid:*

$$(2.10)\qquad x = f, \quad y = \frac{4f^2g - 2\eta_\infty tf^2 - (2\theta_0 + 1)f + 2\eta_0 t}{f^2},$$

with inverse given explicitly by

$$f = x, \quad g = \frac{x^2y + 2\eta_\infty tx^2 + (2\theta_0 + 1)x - 2\eta_0 t}{4x^2}. \tag{2.11}$$

We have the following equality of the 2-forms defining the non-autonomous Hamiltonian structures of the two systems, under the transformation above:

$$dy \wedge dx - d\,\mathrm{H}^{\mathrm{FZ}} \wedge dt = 4\left(dg \wedge df - d\,\mathrm{H}^{\mathrm{Ok}} \wedge dt\right). \tag{2.12}$$

The Hamiltonian functions themselves are related modulo functions of t by

$$\mathrm{H}^{\mathrm{FZ}}(x, y, t) = 4\,\mathrm{H}^{\mathrm{Ok}}(f, g, t) + 2\eta_\infty f + \frac{2\eta_0}{f}. \tag{2.13}$$

THEOREM 2.2. *Let $q = q(s)$ and $p = p(s)$, be solutions of the KNY system (2.5) and $x = x(t)$ and $x = y(t)$ be solutions of system (2.8) with $\gamma = -\delta = 4$. Then with $s = t^2$ the following birational transformation holds:*

$$q = tx, \quad p = \frac{2x^2y + 4tx^2 + (\beta - 2)x - 4t}{8tx^2}, \tag{2.14}$$

with parameters related according to

$$\alpha = -4(a_1 - a_2), \quad \beta = 4(1 - a_1 - a_2), \tag{2.15}$$

or, equivalently, $a_1 = (4 - \alpha - \beta)/8$, $a_2 = (4 + \alpha - \beta)/8$. The inverse transformation is given by

$$x = \frac{q}{\sqrt{s}}, \quad y = \frac{\sqrt{s}(8pq^2 - 4q^2 - 2(1 - 2a_1 - 2a_2)q + 4s)}{2q^2}. \tag{2.16}$$

We again have the 2-forms defining the non-autonomous Hamiltonian structures of the systems equal under the identification above:

$$dy \wedge dx - d\,\mathrm{H}^{\mathrm{FZ}} \wedge dt = 4\left(dp \wedge dq - d\,\mathrm{H}^{\mathrm{KNY}} \wedge ds\right). \tag{2.17}$$

The Hamiltonian functions themselves are related modulo functions of t by

$$\mathrm{H}^{\mathrm{KNY}}(q, p, s) = \frac{1}{8t}\,\mathrm{H}^{\mathrm{FZ}}(x, y, t) + \frac{xy}{8t^2} - \frac{1}{2tx}. \tag{2.18}$$

REMARK 2.3. Note that the exterior derivative in the 2-forms referred to in both of these Theorems is on the 3-dimensional extended phase space so t and s are treated as variables.

Both of the above transformations may be verified by direct calculation. The transformations and relations between parameters may be obtained using the method presented in [**DFLS**], but the geometry of these systems and the calculations establishing this are rather straightforward in this case so we omit them.

We make remarks at this point concerning the relation of our results to the degenerate cases of the third Painlevé equation, in which the geometry of the associated surfaces change. The generic case where $\gamma, \delta \neq 0$ has surface type $D_6^{(1)}$ according to Sakai's classification [**Sak**], whereas if $\gamma = 0$, $\delta \neq 0$ then the surface type is $D_7^{(1)}$. The KNY Hamiltonian in the $D_7^{(1)}$ case is

$$K_1 = \frac{1}{t}(q^2p^2 + q + pt + a_1qp), \tag{2.19}$$

which leads to the third Painlevé equation with parameters

$$\alpha = -8, \quad \beta = -4(a_1 - 1), \quad \gamma = 0, \quad \delta = -4. \tag{2.20}$$

If both $\gamma = 0$ and $\delta = 0$, then the surface type is $D_8^{(1)}$ and the KNY Hamiltonian further degenerates to

$$K_2 = \frac{1}{t}(q^2p^2 + qp + q + t/q), \tag{2.21}$$

which gives the third Painlevé equation with parameters

$$\alpha = -\beta = -8, \quad \gamma = \delta = 0. \tag{2.22}$$

The change in geometry in these cases means that the birational transformations given in Theorem 2.2 may cease to work. This is indeed the case, but we can make adjustments to obtain transformations that are valid in these degenerate cases.

PROPOSITION 2.4. *The $\gamma = 0$, $\delta = -4$ case of the Filipuk-Żołądek Hamiltonian system* (2.8) *is equivalent to the KNY form of the $D_7^{(1)}$ third Painlevé equation with Hamiltonian* (2.19), *under the birational transformation*

$$q = tx, \quad p = \frac{x^2y - 2a_1x + x - 2t}{4tx^2}, \tag{2.23}$$

with parameters related according to (2.20).

PROPOSITION 2.5. *The $\gamma = 0$, $\delta = 0$ case of the Filipuk-Żołądek Hamiltonian system* (2.8) *is equivalent to the KNY form of the $D_8^{(1)}$ third Painlevé equation with Hamiltonian* (2.21), *under the birational transformation*

$$q = tx, \quad p = \frac{xy - 1}{4tx}, \tag{2.24}$$

with parameters related according to (2.22).

References

[DFLS] A. Dzhamay, G. Filipuk, A. Ligęza, and A. Stokes, *Different Hamiltonians for the Painlevé P_{IV} equation and their identification using a geometric approach*, submitted, available at arXiv:2109.06428.

[FZ] Henryk Żołądek and Galina Filipuk, *Painlevé equations, elliptic integrals and elementary functions*, J. Differential Equations **258** (2015), no. 4, 1303–1355, DOI 10.1016/j.jde.2014.10.018. MR3294349

[IP] A. R. Its and A. Prokhorov, *On some Hamiltonian properties of the isomonodromic tau functions*, Rev. Math. Phys. **30** (2018), no. 7, 1840008, 38, DOI 10.1142/S0129055X18400081. MR3833049

[JM] Michio Jimbo and Tetsuji Miwa, *Monodromy preserving deformation of linear ordinary differential equations with rational coefficients. II*, Phys. D **2** (1981), no. 3, 407–448, DOI 10.1016/0167-2789(81)90021-X. MR625446

[KNY] Kenji Kajiwara, Masatoshi Noumi, and Yasuhiko Yamada, *Geometric aspects of Painlevé equations*, J. Phys. A **50** (2017), no. 7, 073001, 164, DOI 10.1088/1751-8121/50/7/073001. MR3609039

[Ok1] Kazuo Okamoto, *Sur les feuilletages associés aux équations du second ordre à points critiques fixes de P. Painlevé* (French), Japan. J. Math. (N.S.) **5** (1979), no. 1, 1–79, DOI 10.4099/math1924.5.1. MR614694

[Ok2] Kazuo Okamoto, *Polynomial Hamiltonians associated with Painlevé equations. I*, Proc. Japan Acad. Ser. A Math. Sci. **56** (1980), no. 6, 264–268. MR581468

[Ok3] Kazuo Okamoto, *Polynomial Hamiltonians associated with Painlevé equations. II. Differential equations satisfied by polynomial Hamiltonians*, Proc. Japan Acad. Ser. A Math. Sci. **56** (1980), no. 8, 367–371. MR596006

[Sak] Hidetaka Sakai, *Rational surfaces associated with affine root systems and geometry of the Painlevé equations*, Comm. Math. Phys. **220** (2001), no. 1, 165–229, DOI 10.1007/s002200100446. MR1882403

Institute of Mathematics, University of Warsaw, ul. Banacha 2, 02-097, Warsaw, Poland

Email address: filipuk@mimuw.edu.pl

Institute of Mathematics, University of Warsaw, ul. Banacha 2, 02-097, Warsaw, Poland

Email address: a.ligeza@mimuw.edu.pl

Graduate School of Mathematical Sciences, The University of Tokyo, 3-8-1 Komaba, Meguro-ku, Tokyo 153-8914, Japan

Email address: alexander.stokes.14@ucl.ac.uk

Contemporary Mathematics
Volume **782**, 2023
https://doi.org/10.1090/conm/782/15721

Degeneration structure of the Voros coefficients of the generalized hypergeometric differential equations with a large parameter

Takashi Aoki and Shofu Uchida

1. Introduction

The notion of Voros coefficients plays an important role in studying the parametric Stokes phenomena, calculation of the monodromy and the relations between Borel resummed WKB solutions and classical special functions [**4, 5, 8, 10, 14, 16, 18–20**]. In the previous works of the first author and his collaborators [**6, 7**], the Voros coefficients for Gauss hypergeometric differential equation at the singular point were given and used in order to study the relation between ${}_2F_1$ and the WKB solutions. Moreover, the Voros coefficients for ${}_NF_{N-1}$ at the origin and infinity are given in [**9**].

In this article, the Voros coefficients of the generalized hypergeometric differential equations for ${}_pF_q$ $(p \neq q+1)$ are defined and their explicit forms are given at the origin and the infinity. Furthermore, we have the degeneration structure of the Voros coefficients of the generalized hypergeometric differential equations with a large parameter. The degeneration of the Voros coefficients and that of the generalized hypergeometric differential equations are compatible.

2. WKB solutions

Let p and q be integers greater than or equal to 0 and $N := \max\{p, q+1\}$. We consider the following equation which contains the large parameter:

$$ {}_pP_q\psi = 0, \tag{2.1} $$

where we set

$$ {}_pP_q = \begin{cases} \eta^{-(q+1)}\left(\prod_{j=1}^{q}(\vartheta_x + b_j)\right)\partial_x - \eta^{-p}\left(\prod_{i=1}^{p}(\vartheta_x + a_i)\right) & (p, q \neq 0), \\ \eta^{-(q+1)}\left(\prod_{j=1}^{q}(\vartheta_x + b_j)\right)\partial_x - 1 & (p = 0), \\ \eta^{-1}\partial_x - \eta^{-p}\left(\prod_{i=1}^{p}(\vartheta_x + a_i)\right) & (q = 0), \end{cases} \tag{2.2} $$

$a_i = a_{i,0} + a_{i,1}\eta$ $(i = 1, 2, \ldots, p)$ and $b_j = b_{j,0} + b_{j,1}\eta$ $(j = 1, 2, \ldots, q)$. We call (2.1) the generalized hypergeometric differential equation with the large parameter. The differential operator (2.2) is a WKB type operator defined on $\mathbb{C}$ in the sence of [**1**, Definition 2.1]. For an m-tuple $\mathbf{c} = (c_1, c_2, \ldots, c_m)$ of parameters, we denote by

Supported by JSPS KAKENHI Grant No. 18K03385.

$\mathfrak{s}_\ell(\boldsymbol{c})$ the elementary symmetric polynomial of degree ℓ of $c_1, c_2, \ldots, c_m$. If $\ell > m$ or $\ell \le -1$, we set $\mathfrak{s}_\ell(\boldsymbol{c}) = 0$. Let ξ denote the dual variable of x and $\zeta = \xi/\eta$. The total symbol $\sigma({}_pP_q)(x, \zeta, \eta)$ (cf. [**9**]) is written in the form:

$$\begin{aligned}(2.3)\quad & \sigma({}_pP_q)(x,\zeta,\eta) \\ &= \sum_{k=1}^{N}\left(\sum_{j=k}^{N}\left(\eta^{-(q+1)}\begin{Bmatrix} j-1 \\ k-1\end{Bmatrix}\mathfrak{s}_{q+1-j}(\boldsymbol{b}_q) - \eta^{-p}\begin{Bmatrix} j \\ k\end{Bmatrix}\mathfrak{s}_{p-j}(\boldsymbol{a}_p)x\right)\right)x^{k-1}\eta^k\zeta^k \\ &\quad - \eta^{-p}\mathfrak{s}_p(\boldsymbol{a}_p),\end{aligned}$$

where $\boldsymbol{a}_p = (a_1, a_2, \ldots, a_p)$ and $\boldsymbol{b}_q = (b_1, b_2, \ldots, b_q)$. Then the principal symbol $\sigma_0({}_pP_q)(x, \zeta)$ is witten in the form:

$$(2.4)\qquad \sigma_0({}_pP_q)(x,\zeta) = \sum_{k=1}^{N}\Big(\mathfrak{s}_{q+1-k}(\boldsymbol{b}_{q,1}) - \mathfrak{s}_{p-k}(\boldsymbol{a}_{p,1})x\Big)x^{k-1}\zeta^k - \mathfrak{s}_p(\boldsymbol{a}_{p,1}),$$

where $\boldsymbol{a}_{p,1} = (a_{1,1}, a_{2,1}, \ldots, a_{p,1})$ and $\boldsymbol{b}_{q,1} = (b_{1,1}, b_{2,1}, \ldots, b_{q,1})$. Let $\mathcal{P}^{(p,q)}$ denote the set

$$(2.5)\qquad \{(x_*, \zeta_*) \mid \sigma_0({}_pP_q)(x_*, \zeta_*) = \partial_\zeta\sigma_0({}_pP_q)(x_*, \zeta_*) = 0\}.$$

This set can be regarded as a subset of $\mathbb{P}^2_{\mathbb{C}}$. In the same way as in [**9**, Lemma 2.4], if $p \neq q+1$, we can prove that $\#\mathcal{P}^{(p,q)} = p + q$ for generic $\boldsymbol{a}_{p,1}$ and $\boldsymbol{b}_{q,1}$. Let $\mathcal{P}^{(p,q)}_{\mathrm{tp}}$ be the projection of (2.5) to the x-space. Assume that the singular points of the equation (2.1) do not belong to $\mathcal{P}^{(p,q)}_{\mathrm{tp}}$. We consider the Puiseux expansion at $\varrho \in \{0, \infty\}$ of each characteristic root. For the case of $\varrho = 0$, we substitute $\zeta = \sum_{k=m}^{\infty} c_{k/\nu}x^{k/\nu}$ into (2.4). Looking at the degree of the leading term of (2.4) with respect to x, we can determine m and ν. Then the leading coefficient $c_{m/\nu}$ can be also determined. The coefficients of higher-order terms $c_{k/\nu}$ $(k > m)$ are determined recursively. Similarly, we can find the Puiseux expansions of the characterictic roots at $x = \infty$.

Definition 2.1. We take local numbering of the characteristic roots of ${}_pP_q$ as follows: For the singular point ϱ, the roots are denoted by $\zeta^{(\varrho)}_m$ $(m = 1, 2, \ldots, N)$ so that

(i) If $p < q + 1$,

$$\begin{aligned}\zeta^{(0)}_\ell &= -\frac{b_{\ell,1}}{x} + O(1) \qquad (\ell = 1, 2, \ldots, q), \\ \zeta^{(0)}_{q+1} &= \frac{a_{1,1}a_{2,1}\cdots a_{p,1}}{b_{1,1}\cdots b_{q,1}} + O(x), \\ \zeta^{(\infty)}_m &= -\frac{a_{m,1}}{x} + O\left(\frac{1}{x^2}\right) \qquad (p \neq 0,\ m = 1, 2, \ldots, p), \\ \zeta^{(\infty)}_m &= \frac{e^{\frac{2\pi\sqrt{-1}}{q-p+1}(m-p)}}{x^{1-\frac{1}{q-p+1}}} + O\left(\frac{1}{x}\right) \qquad (m = p+1, \ldots, q+1)\end{aligned}$$

hold.

(ii) If $q+1<p$,

$$\zeta_\ell^{(0)} = -\frac{b_{\ell,1}}{x} + O(1) \qquad (q \neq 0,\ \ell = 1,2,\dots,q),$$

$$\zeta_m^{(0)} = \frac{e^{\frac{2\pi\sqrt{-1}}{p-q-1}(m-q)}}{x^{1+\frac{1}{p-q-1}}} + O\left(\frac{1}{x}\right) \qquad (m = q+1,\dots,p-1),$$

$$\zeta_p^{(0)} = \frac{a_{1,1}a_{2,1}\cdots a_{p,1}}{b_{1,1}\cdots b_{q,1}} + O(x),$$

$$\zeta_m^{(\infty)} = -\frac{a_{m,1}}{x} + O\left(\frac{1}{x^2}\right) \qquad (m = 1,2,\dots,p)$$

hold.

Taking suitable branch cuts connecting simple turning points, we can regard this numbering is defined globally. The characterictic vareity $\mathrm{Ch}({}_pP_q)$ can be regarded as a compact Riemann surface $\Sigma^{(p,q)}$. There is a natural projection $\pi : \Sigma^{(p,q)} \to \mathbb{P}^1_{\mathbb{C}}$, which is an N-covering map. By using the Riemann–Hurwitz formula, the genus of $\Sigma^{(p,q)}$ equals 0.

A WKB solution ψ of (2.1) is a formal solution of the form

$$\psi = \exp\left(\int S dx\right), \tag{2.6}$$

$$S = \eta S_{-1} + S_0 + \eta^{-1} S_1 + \cdots = \sum_{\ell=-1}^{\infty} \eta^{-\ell} S_\ell \tag{2.7}$$

(see [**1**, Definition 3.2]). The leading term of the equation $\mathrm{Ri}({}_pP_q)(S) = 0$ with respect to η^{-1} determines S_{-1}, where

$$\mathrm{Ri}({}_pP_q)(S) = \exp\left(-\int S dx\right) {}_pP_q \exp\left(\int S dx\right). \tag{2.8}$$

Therefore S_{-1} should satisfy $\sigma_0({}_pP_q)(x, S_{-1}) = 0$. Hence we can take one of the characteristic roots as S_{-1}. The higher-order terms S_j ($j = 0,\ 1,\ 2,\ \dots$) can be determined recursively and uniquely outside the turning points if S_{-1} is chosen. Thus obtained N formal solutions of $\mathrm{Ri}({}_pP_q)(S) = 0$ are denoted by $S^{(\varrho,m)}$ ($m = 1,\ 2,\ \dots,\ N$) so that this numbering is consistent with Definition 2.1.

3. Factorization

We assume that there is a simple turning point x_* of ${}_pP_q$ with the characteristic value ζ_*. Then $\sigma_0({}_pP_q)(x,\zeta)$ is uniquely decomposed holomorphically in the form

$$\sigma_0({}_pP_q)(x,\zeta) = l(x,\zeta) r(x,\zeta), \tag{3.1}$$

where $r(x,\zeta)$ is a Weierstrass polynomial of degree 2 in ζ with the center at (x_*,ζ_*) and $l(x_*,\zeta_*) \neq 0$. By the definition, $r(x,\zeta)$ has the form

$$r(x,\zeta) = (\zeta - \zeta_*)^2 + f_1(x)(\zeta - \zeta_*) + f_2(x), \tag{3.2}$$

where $f_j(x)$ vanishes at $x = x_*$ for $j = 1,2$. We may assume $\zeta_j^{(\varrho)}$ and $\zeta_k^{(\varrho)}$ are the roots of $r(x,\zeta) = 0$. We set

$${}_p\tilde{P}_q = \frac{1}{x^{N-1}(\mathfrak{s}_{q+1-N}(\boldsymbol{b}_q) - \mathfrak{s}_{p-N}(\boldsymbol{a}_p)x)} {}_pP_q \tag{3.3}$$

Then $\sigma({}_p\tilde{P}_q)(x,\zeta)$ becomes a monic polynomial with respect to ζ. It follows from [**1**, Theorem 5.1] that there uniquely exist differential operators L and R of WKB type near x_* which satisfy

$$
{}_p\tilde{P}_q = LR \tag{3.4}
$$

and

(i) The principal symbol $\sigma_0(R)(x,\zeta)$ of R coincides with $r(x,\zeta)$.
(ii) For each $j>0$, the coefficient $\sigma_j(R)(x,\zeta)$ of η^{-j} of the symbol of R is of degree at most one in ζ.
(iii) The principal symbol $\sigma_0(L)(x,\zeta)$ of L does not vanish at (x_*,ζ_*).

Thus constructed operator R has the form

$$
R = (\eta^{-1}\partial_x)^2 + A(x,\eta)\eta^{-1}\partial_x + B(x,\eta) \tag{3.5}
$$

with

$$
A(x,\eta) = A_0(x) + A_1(x)\eta^{-1} + A_2(x)\eta^{-2} + \cdots, \tag{3.6}
$$

$$
B(x,\eta) = B_0(x) + B_1(x)\eta^{-1} + B_2(x)\eta^{-2} + \cdots \tag{3.7}
$$

and L is an $(N-2)$-th order operator

$$
L = \sum_{k=0}^{N-2} L_k(x,\eta)(\eta^{-1}\partial_x)^k \tag{3.8}
$$

with

$$
L_k(x,\eta) = L_{k,0}(x) + L_{k,1}(x)\eta^{-1} + L_{k,2}(x)\eta^{-2} + \cdots. \tag{3.9}
$$

Relation (3.4) determines A, B and L_k $(0 \le k \le N-2)$, if we choose the leading terms of them, which satisfy a system of algebraic equations. We consider a WKB solution

$$
\phi = \exp\left(\int T dx\right), \tag{3.10}
$$

$$
T = \eta T_{-1} + T_0 + \eta^{-1}T_1 + \cdots = \sum_{\ell=-1}^{\infty} \eta^{-\ell}T_\ell \tag{3.11}
$$

of $R\phi = 0$. Here T is determined by the Riccati equation associated with this equation:

$$
\mathrm{Ri}(R)(T) = \eta^{-2}(T^2 + T') + A\eta^{-1}T + B = 0. \tag{3.12}
$$

In the same way for $p = q+1$, we have the following relation for $N \ge 3$:

$$
\mathrm{Ri}({}_p\tilde{P}_q)(S) = \sum_{k=0}^{N-2} \frac{\eta^{-k}}{k!} \mathrm{Ri}\left(:\partial_\zeta^k \sigma(L)(x,\zeta):\right)(S)\partial_x^k \,\mathrm{Ri}(R)(S). \tag{3.13}
$$

Thus $S = T$ if $S_{-1} = T_{-1}$.

4. Definition of the Voros coefficients

We assume

$$(4.1)\quad S_{-1}^{(\varrho,j)} = \zeta_j^{(\varrho)} = \frac{-A_0 + \sqrt{A_0^2 - 4B_0}}{2}, \qquad S_{-1}^{(\varrho,k)} = \zeta_k^{(\varrho)} = \frac{-A_0 - \sqrt{A_0^2 - 4B_0}}{2}$$

for suitable choice of the branch of the square root. We set

$$(4.2)\qquad S_{\mathrm{even}}^{(\varrho,j,k)} = \frac{1}{2}\left(S^{(\varrho,j)} + S^{(\varrho,k)}\right), \qquad S_{\mathrm{odd}}^{(\varrho,j,k)} = \frac{1}{2}\left(S^{(\varrho,j)} - S^{(\varrho,k)}\right).$$

Moreover, we denote $S_{\mathrm{odd},\ell}^{(\varrho,j,k)}$ by the coefficient of $\eta^{-\ell}$ of $S_{\mathrm{odd}}^{(\varrho,j,k)}$ and set $S_{\mathrm{odd},\leq 0}^{(\varrho,j,k)} = \eta S_{\mathrm{odd},-1}^{(\varrho,j,k)} + S_{\mathrm{odd},0}^{(\varrho,j,k)}$.

DEFINITION 4.1. The WKB solutions $\psi_\pm^{(\tau)}$ of (2.1) normalized at the turning point τ of type (j,k) are defined by

$$(4.3)\qquad \psi_\pm^{(\tau)} = \frac{1}{\sqrt{S_{\mathrm{odd}}^{(\varrho,j,k)}}} \exp\left(-\frac{1}{2}\eta \int_\tau^x A dx\right) \exp\left(\pm \int_\tau^x S_{\mathrm{odd}}^{(\varrho,j,k)} dx\right).$$

The WKB solutions $\psi_\pm^{(\varrho)}$ of (2.1) normalized at the singular point ϱ are defined by

$$(4.4)\qquad \begin{aligned} \psi_\pm^{(\varrho)} &= \frac{1}{\sqrt{S_{\mathrm{odd}}^{(\varrho,j,k)}}} \exp\left(-\frac{1}{2}\eta \int_\tau^x A dx\right) \\ &\quad \exp\left(\pm \int_\varrho^x \left(S_{\mathrm{odd}}^{(\varrho,j,k)} - S_{\mathrm{odd},\leq 0}^{(\varrho,j,k)}\right) dx \pm \int_\tau^x S_{\mathrm{odd},\leq 0}^{(\varrho,j,k)} dx\right). \end{aligned}$$

DEFINITION 4.2. The Voros coefficient $V_\varrho^{(j,k)}$ at $x = \varrho$ of type (j,k) is defined by

$$(4.5)\qquad V_\varrho^{(j,k)} = \int_\varrho^\tau \left(S_{\mathrm{odd}}^{(\varrho,j,k)} - S_{\mathrm{odd},\leq 0}^{(\varrho,j,k)}\right) dx.$$

The Voros coefficient relates two kinds of the normalization of WKB solutions defined in the preceding subsection. We have the following formal relations:

$$(4.6)\qquad \psi_\pm^{(\varrho)} = \exp\left(V_\varrho^{(j,k)}\right) \psi_\pm^{(\tau)}.$$

The Voros coefficient is rewritten as

$$(4.7)\qquad V_\varrho^{(j,k)} = \frac{1}{2} \int_{\gamma_\varrho^{(j,k)}} (S - \eta S_{-1} - S_0)\, dx.$$

Here we consider that S is defined on the Riemann surface $\Sigma^{(p,q)}$ of S_{-1} and $\gamma_\varrho^{(j,k)}$ is a path on $\Sigma^{(p,q)}$ starting from the singular point ϱ on the j-th sheet, going to and detouring τ counterclockwise on the base space and back to the singular point ϱ on the k-th sheet. The Voros coefficients $V_\varrho^{(j,k)}$ depend on the equivalence class of the path $\gamma_\varrho^{(j,k)}$.

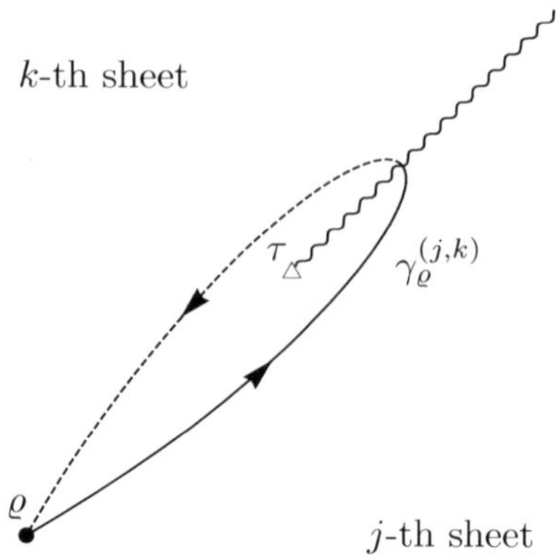

FIGURE 1. The path $\gamma_\varrho^{(j,k)}$.

5. Explicit forms of the Voros coefficients

Let κ_1 and κ_0 be linear homogeneous functions of $a_{1,1}, \ldots, a_{p,1}, b_{1,1}, \ldots, b_{q,1}$ and $a_{1,0}, \ldots, a_{p,0}, b_{1,0}, \ldots, b_{q,0}$, respectively. We set

$$\boldsymbol{B}_\ell(\kappa) = \frac{B_\ell(\kappa_0)}{\kappa_1^{\ell-1}} \qquad (\kappa = \kappa_0 + \kappa_1 \eta), \tag{5.1}$$

where $B_\ell(\kappa_0)$ denotes the ℓ-th Bernoulli polynomial defined by

$$\frac{x e^{x\kappa_0}}{e^x - 1} = \sum_{\ell=0}^{\infty} \frac{B_\ell(\kappa_0)}{\ell!} x^\ell. \tag{5.2}$$

The explicit forms of the Voros coefficients $V_\varrho^{(j,k)}$ ($\varrho = 0, \infty$) can be obtained by the same method as in [**9**, Theorem 4.9]. More specifically, we can have the differential-difference equations for the Voros coefficients by using the ladder operators. We can solve them by using formal differential operators of infinite order.

THEOREM 5.1. *Let ϱ be 0 or ∞. If there exists a simple turning point of type (j,k) $(j<k)$ of* (2.1), *then the Voros coefficients $V_\varrho^{(j,k)}$ are written in the form:*

$$\frac{1}{2}\sum_{\ell=2}^{\infty} \frac{(-1)^{\ell+1}\eta^{1-\ell}}{\ell(\ell-1)} V_{\varrho,\ell}^{(j,k)}, \tag{5.3}$$

where

$$V_{0,\ell}^{(j,k)} = \begin{cases} \sum_{\substack{1\le m\le q\\ m\ne j}} \boldsymbol{B}_\ell(b_j-b_m) + \sum_{1\le m\le q} \boldsymbol{B}_\ell(b_m) \\ \quad + \boldsymbol{B}_\ell(b_j-1) - \sum_{1\le i\le p} (\boldsymbol{B}_\ell(b_j-a_i)+\boldsymbol{B}_\ell(a_i)) & (k=q+1), \\ \sum_{\substack{1\le m\le q\\ m\ne j}} \boldsymbol{B}_\ell(b_j-b_m) - \sum_{\substack{1\le m\le q\\ m\ne k}} \boldsymbol{B}_\ell(b_k-b_m) - \boldsymbol{B}_\ell(b_k-1) \\ \quad + \boldsymbol{B}_\ell(b_j-1) - \sum_{1\le i\le p} (\boldsymbol{B}_\ell(b_j-a_i)-\boldsymbol{B}_\ell(b_k-a_i)) & (k\le q), \end{cases}$$

$$V_{\infty,\ell}^{(j,k)} = \begin{cases} \sum_{\substack{1\le i\le p\\ i\ne j}} \boldsymbol{B}_\ell(a_i-a_j) - \sum_{\substack{1\le i\le p\\ i\ne k}} \boldsymbol{B}_\ell(a_i-a_k) \\ \quad + \boldsymbol{B}_\ell(a_j) - \boldsymbol{B}_\ell(a_k) + \sum_{1\le m\le q} (\boldsymbol{B}_\ell(b_m-a_k)-\boldsymbol{B}_\ell(b_m-a_j)) & (k\le p), \\ \sum_{\substack{1\le i\le p\\ i\ne j}} \boldsymbol{B}_\ell(a_i-a_j) + \boldsymbol{B}_\ell(a_j) - \sum_{1\le m\le q} \boldsymbol{B}_\ell(b_m-a_j) & (k+1\le p\le q+1) \end{cases}$$

for $p < q+1$ *and*

$$V_{0,\ell}^{(j,k)} = \begin{cases} \sum_{\substack{1\le m\le q\\ m\ne j}} \boldsymbol{B}_\ell(b_j-b_m) + \sum_{1\le m\le q} \boldsymbol{B}_\ell(b_m) \\ \quad + \boldsymbol{B}_\ell(b_j-1) - \sum_{1\le i\le p} (\boldsymbol{B}_\ell(b_j-a_i)+\boldsymbol{B}_\ell(a_i)) & (j\le q, k=p), \\ \sum_{1\le m\le q} \boldsymbol{B}_\ell(b_m) - \sum_{1\le i\le p} \boldsymbol{B}_\ell(a_i) & (q+1\le j, k=p), \\ \sum_{\substack{1\le m\le q\\ m\ne j}} \boldsymbol{B}_\ell(b_j-b_m) - \sum_{\substack{1\le m\le q\\ m\ne k}} \boldsymbol{B}_\ell(b_k-b_m) - \boldsymbol{B}_\ell(b_k-1) \\ \quad + \boldsymbol{B}_\ell(b_j-1) - \sum_{1\le i\le p} (\boldsymbol{B}_\ell(b_j-a_i)-\boldsymbol{B}_\ell(b_k-a_i)) & (j\le q, k\le q), \\ 0 \quad (q+1\le j, q+1\le k\le p-1), \\ \sum_{\substack{1\le m\le q\\ m\ne j}} \boldsymbol{B}_\ell(b_j-b_m) + \boldsymbol{B}_\ell(b_j-1) - \sum_{1\le i\le p} \boldsymbol{B}_\ell(b_j-a_i) & (j\le q, q+1\le k\le p-1), \\ -\sum_{\substack{1\le m\le q\\ m\ne k}} \boldsymbol{B}_\ell(b_k-b_m) - \boldsymbol{B}_\ell(b_k-1) + \sum_{1\le i\le p} \boldsymbol{B}_\ell(b_k-a_i) & (q+1\le j, k\le q), \end{cases}$$

$$V_{\infty,\ell}^{(j,k)} = \sum_{\substack{1\le i\le p\\ i\ne j}} \boldsymbol{B}_\ell(a_i-a_j) - \sum_{\substack{1\le i\le p\\ i\ne k}} \boldsymbol{B}_\ell(a_i-a_k) + \boldsymbol{B}_\ell(a_j) - \boldsymbol{B}_\ell(a_k) + \sum_{1\le m\le q} (\boldsymbol{B}_\ell(b_m-a_k)-\boldsymbol{B}_\ell(b_m-a_j))$$

for $q+1 < p$.

We consider the Borel summability of $V_\varrho^{(j,k)}$. We set

$$\tilde{V}(\kappa_0,\kappa_1) = \frac{1}{2}\sum_{\ell=2}^{\infty} \frac{(-1)^{\ell+1}\eta^{1-\ell}}{\ell(\ell-1)} \boldsymbol{B}_\ell(\kappa) \qquad (\kappa=\kappa_0+\kappa_1\eta). \tag{5.4}$$

As we saw in the preceding theorem, $V_\varrho^{(j,k)}$ are expressed as a finite sum of the formal series of the form $\tilde{V}(\kappa_0,\kappa_1)$ for suitable choice of (κ_0,κ_1). Hence it is sufficient to consider the Borel summability of $\tilde{V}(\kappa_0,\kappa_1)$. Its summability is well known: the formal series $\tilde{V}(\kappa_0,\kappa_1)$ of η^{-1} is Borel summable if $\mathrm{Re}(\kappa_1)\ne 0$ and the Borel sum of $\tilde{V}(\kappa_0, \kappa_1)$ equals

$$(5.5)\qquad \begin{cases} \dfrac{1}{2}\log\left(\dfrac{\sqrt{2\pi}(\kappa_1\eta)^{\kappa_0+\kappa_1\eta-\frac{1}{2}}}{e^{\kappa_1\eta}\Gamma(\kappa_0+\kappa_1\eta)}\right) & (\mathrm{Re}(\kappa_1)>0), \\ \dfrac{1}{2}\log\left(\dfrac{(-\kappa_1\eta)^{\kappa_0+\kappa_1\eta-\frac{1}{2}}\Gamma(1-(\kappa_0+\kappa_1\eta))}{\sqrt{2\pi}e^{\kappa_1\eta}}\right) & (\mathrm{Re}(\kappa_1)<0) \end{cases}$$

(see [**7**, Lemma 4.6]). This expression is obtained by the Binet formula ([**11**]). Thus, we have

COROLLARY 5.2. *Let ϱ be 0 or ∞. Also let*

$$(5.6)\qquad D(\kappa_1) = \{(a_{1,1},\ldots,a_{p,1},b_{1,1},\ldots,b_{q,1}) \mid \mathrm{Re}(\kappa_1)\neq 0\},$$

where κ_1 is a linear homogeneous function of $a_{1,1}, \ldots, a_{p,1}, b_{1,1}, \ldots, b_{q,1}$. The Voros coefficient $V_\varrho^{(j,k)}$ of type (j,k) $(j<k)$ is Borel summable if $(a_{1,1},\ldots,a_{p,1}, b_{1,1},\ldots,b_{q,1})$ belongs to $D_\varrho^{(j,k)}$. Here we set

$$D_0^{(j,k)} = \begin{cases} \displaystyle\bigcap_{\substack{1\le m\le q\\ m\neq j}} D(b_{j,1}-b_{m,1}) \cap \bigcap_{1\le m\le q} D(b_{m,1}) \\ \displaystyle\qquad \cap \bigcap_{1\le i\le p}(D(b_{j,1}-a_{i,1})\cap D(a_{i,1})) \qquad (k=q+1), \\ \displaystyle\bigcap_{\substack{1\le m\le q\\ m\neq j}} D(b_{j,1}-b_{m,1}) \cap \bigcap_{\substack{1\le m\le q\\ m\neq k}} D(b_{k,1}-b_{m,1}) \cap D(b_{k,1}) \\ \displaystyle\qquad \cap D(b_{j,1}) \cap \bigcap_{1\le i\le p}(D(b_{j,1}-a_{i,1})\cap D(b_{k,1}-a_{i,1})) \qquad (k\le q), \end{cases}$$

$$D_\infty^{(j,k)} = \begin{cases} \displaystyle\bigcap_{\substack{1\le i\le p\\ i\neq j}} D(a_{i,1}-a_{j,1}) \cap \bigcap_{\substack{1\le i\le p\\ i\neq k}} D(a_{i,1}-a_k) \\ \displaystyle\qquad \cap D(a_{j,1}) \cap D(a_k) \cap \bigcap_{1\le m\le q}(D(b_{m,1}-a_k)\cap D(b_{m,1}-a_{j,1})) \quad (k\le p), \\ \displaystyle\bigcap_{\substack{1\le i\le p\\ i\neq j}} D(a_{i,1}-a_{j,1}) \cap D(a_{j,1}) \cap \bigcap_{1\le m\le q} D(b_{m,1}-a_{j,1}) \qquad (k+1\le p\le q+1) \end{cases}$$

for $p < q+1$ *and*

$$D_0^{(j,k)} = \begin{cases} \bigcap_{\substack{1\le m\le q\\ m\neq j}} D(b_{j,1}-b_{m,1}) \cap \bigcap_{1\le m\le q} D(b_{m,1}) \\ \qquad \cap \bigcap_{1\le i\le p} (D(b_{j,1}-a_{i,1}) \cap D(a_{i,1})) \qquad (j\le q, k=p), \\ \bigcap_{1\le m\le q} D(b_{m,1}) \cap \bigcap_{1\le i\le p} D(a_{i,1}) \qquad (q+1\le j, k=p), \\ \bigcap_{\substack{1\le m\le q\\ m\neq j}} D(b_{j,1}-b_{m,1}) \cap \bigcap_{\substack{1\le m\le q\\ m\neq k}} D(b_{k,1}-b_{m,1}) \cap D(b_{k,1}) \\ \qquad \cap D(b_{j,1}) \cap \bigcap_{1\le i\le p} (D(b_{j,1}-a_{i,1}) \cap D(b_{k,1}-a_{i,1}))\ (j\le q, k\le q), \\ \mathbb{C}^{p+q} \qquad (q+1\le j, q+1\le k\le p-1), \\ \bigcap_{\substack{1\le m\le q\\ m\neq j}} D(b_{j,1}-b_{m,1}) \cap D(b_{j,1}) \cap \bigcap_{1\le i\le p} D(b_{j,1}-a_{i,1}) \quad (j\le q, q+1\le k\le p-1), \\ \bigcap_{\substack{1\le m\le q\\ m\neq k}} D(b_{k,1}-b_{m,1}) \cap D(b_{k,1}) \cap \bigcap_{1\le i\le p} D(b_{k,1}-a_{i,1}) \quad (q+1\le j, k\le q), \end{cases}$$

$$D_\infty^{(j,k)} = \bigcap_{\substack{1\le i\le p\\ i\neq j}} D(a_{i,1}-a_{j,1}) \cap \bigcap_{\substack{1\le i\le p\\ i\neq k}} D(a_{i,1}-a_{k,1}) \\ \cap D(a_{j,1}) \cap D(a_{k,1}) \cap \bigcap_{1\le m\le q} (D(b_{m,1}-a_{k,1}) \cap D(b_{m,1}-a_{j,1}))$$

for $q+1 < p$.

We consider the degeneration structure of (2.1). Let RAd denotes the reduced adjoint operator ([**14, 17**]). Then we can show that the following relations hold by a direct calculation:

(i) If $p \ge 1$ and $q \ge 0$,

$$\lim_{a_{p,1}\to\infty} \frac{1}{a_p\eta^{-1}} {}_pP_q\left(\begin{matrix}\boldsymbol{a}_p\\ \boldsymbol{b}_q\end{matrix}; \frac{x}{a_p\eta^{-1}}, \partial_x\right) = {}_{p-1}P_q\left(\begin{matrix}\boldsymbol{a}_{p-1}\\ \boldsymbol{b}_q\end{matrix}; x, \partial_x\right). \tag{5.7}$$

(ii) If $p \ge 0$ and $q \ge 1$,

$$\lim_{b_{q,1}\to\infty} {}_pP_q\left(\begin{matrix}\boldsymbol{a}_p\\ \boldsymbol{b}_q\end{matrix}; b_q\eta^{-1}x, \partial_x\right) = {}_pP_{q-1}\left(\begin{matrix}\boldsymbol{a}_p\\ \boldsymbol{b}_{q-1}\end{matrix}; x, \partial_x\right). \tag{5.8}$$

(iii) If $p \ge 0$ and $q \ge 1$,

$$\mathrm{RAd}(x^{a_p})_pP_q\left(\begin{matrix}\boldsymbol{a}_p\\ \boldsymbol{b}_q\end{matrix}; x, \partial_x\right) = (-1)^q {}_{q+1}P_{p-1}\left(\begin{matrix}\boldsymbol{\alpha}_{q+1}\\ \boldsymbol{\beta}_{p-1}\end{matrix}; (-1)^{p+q-1}t, \partial_t\right), \tag{5.9}$$

where $t = \frac{1}{x}$, $\alpha_i = a_p - b_i + 1$ $(1 \le i \le q)$, $\alpha_{q+1} = a_{p+1}$ and $\beta_j = a_{p+1} - a_j + 1$ $(1 \le j \le p-1)$.

(iv) If $N \ge 2$,

(5.10)

$${}_NP_{N-1}\left(\begin{matrix}\boldsymbol{a}_N\\ \boldsymbol{b}_{N-1}\end{matrix}; x, \partial_x\right) = \eta^{-1}\,\mathrm{RAd}(\partial_x^{-\mu})\,\mathrm{RAd}(x^{\lambda})_{N-1}P_{N-2}\left(\begin{matrix}\boldsymbol{\alpha}_{N-1}\\ \boldsymbol{\beta}_{N-2}\end{matrix}; x, \partial_x\right),$$

where $a_i = \alpha_i - \lambda - \mu$ $(1 \le i \le N-1)$, $a_N = 1 - \mu$, $b_j = \beta_j - \lambda - \mu$ $(1 \le j \le N-2)$, $b_{N-1} = 1 - \lambda - \mu$, $\alpha_i = \alpha_{i,0} + \alpha_{i,1}\eta$ $(1 \le i \le N-1)$, $\beta_j = \beta_{j,0} + \beta_{j,1}\eta$ $(1 \le j \le N-2)$, $\lambda = \lambda_0 + \lambda_1\eta$, and $\mu = \mu_0 + \mu_1\eta$.

We write $V_\varrho^{(j,k)}(\boldsymbol{a}_p, \boldsymbol{b}_q)$ for the Voros coefficient at $x = \varrho$ of type (j,k) for (2.1). Using Theorem 5.1 and [**9**, Theorem 3.9], we obtain

COROLLARY 5.3. *Let ϱ be 0 or ∞. Also let $V_\varrho^{(j,k)}(\boldsymbol{a}_p, \boldsymbol{b}_q)$ be the Voros coefficient for (2.1) at ϱ. Then we have*

$$\lim_{a_{p,1}\to\infty} V_\varrho^{(j,k)}(\boldsymbol{a}_p, \boldsymbol{b}_q) = V_\varrho^{(j,k)}(\boldsymbol{a}_{p-1}, \boldsymbol{b}_q) \qquad (p \le q+1),$$
$$\lim_{b_{q,1}\to\infty} V_\varrho^{(j,k)}(\boldsymbol{a}_p, \boldsymbol{b}_q) = V_\varrho^{(j,k)}(\boldsymbol{a}_p, \boldsymbol{b}_{q-1}) \qquad (q+1 \le p).$$

COROLLARY 5.4. *Let $p < q+1$. We set $\boldsymbol{\alpha}_{q+1} = (a_p - b_1 + 1, \dots, a_p - b_q + 1, a_p)$, $\boldsymbol{\beta}_{p-1} = (a_p - a_1 + 1, \dots, a_p - a_{p-1} + 1)$. Then we have*

$$V_0^{(j,k)}(\boldsymbol{a}_p, \boldsymbol{b}_q) = V_\infty^{(j,k)}(\boldsymbol{\alpha}_{q+1}, \boldsymbol{\beta}_{p-1}),$$
$$V_\infty^{(j,k)}(\boldsymbol{a}_p, \boldsymbol{b}_q) = V_0^{(\tilde\sigma(j),\tilde\sigma(k))}(\boldsymbol{\alpha}_{q+1}, \boldsymbol{\beta}_{p-1}).$$

Here $\tilde\sigma$ is given by the following product of the adjacent transpositions:

$$\tilde\sigma = (q\ q+1)(q-1\ q)\cdots(p\ p+1). \tag{5.11}$$

COROLLARY 5.5. *The following relations hold:*

$$V_0^{(j,k)}(\boldsymbol{a}_N, \boldsymbol{b}_{N-1})\Big|_{\mu=\lambda=0} = V_0^{(j,k)}(\boldsymbol{\alpha}_{N-1}, \boldsymbol{\beta}_{N-2}) \qquad (j,k \ne N-1, N),$$
$$\left(V_0^{(j,N-1)}(\boldsymbol{a}_N, \boldsymbol{b}_{N-1})\Big|_{\mu=0}\right)\Big|_{\lambda=0} = V_0^{(j,N-1)}(\boldsymbol{\alpha}_{N-1}, \boldsymbol{\beta}_{N-2}) \quad (j \ne N-1, N),$$
$$\left(V_0^{(j,N)}(\boldsymbol{a}_N, \boldsymbol{b}_{N-1})\Big|_{\lambda=0}\right)\Big|_{\mu=0} = V_0^{(j,N-1)}(\boldsymbol{\alpha}_{N-1}, \boldsymbol{\beta}_{N-2}) \quad (j \ne N-1, N),$$
$$V_\infty^{(j,k)}(\boldsymbol{a}_N, \boldsymbol{b}_{N-1})\Big|_{\mu=\lambda=0} = V_\infty^{(j,k)}(\boldsymbol{\alpha}_{N-1}, \boldsymbol{\beta}_{N-2}) \qquad (j,k \ne N).$$

Here we set $a_i = \alpha_i - \lambda - \mu$ $(1 \le i \le N-1)$, $a_N = 1 - \mu$, $b_j = \beta_j - \lambda - \mu$ $(1 \le j \le N-2)$, $b_{N-1} = 1-\lambda-\mu$, $\alpha_i = \alpha_{i,0}+\alpha_{i,1}\eta$ $(1 \le i \le N-1)$, $\beta_j = \beta_{j,0}+\beta_{j,1}\eta$ $(1 \le j \le N-2)$, $\lambda = \lambda_0 + \lambda_1\eta$ and $\mu = \mu_0 + \mu_1\eta$.

REMARK 5.6. Let $\operatorname{res}_\zeta(f,g)$ be the resultant of f and g with respect ζ. By using elementary operations of determinant (cf. [**9**]), we have

(5.12)

$$\lim_{a_{p,1}\to\infty}\left((a_{p,1})^{(N-1)^2}\operatorname{res}_\zeta\left(\sigma_0({}_pP_q)(x,\zeta), \partial_\zeta\sigma_0({}_pP_q)(x,\zeta)\right)\Big|_{x\to\frac{x}{a_{p,1}}}\right)$$
$$= \operatorname{res}_\zeta\left(\sigma_0({}_{p-1}P_q)(x,\zeta), \partial_\zeta\sigma_0({}_{p-1}P_q)(x,\zeta)\right) \qquad (p \ge 1, q \ge 0),$$

(5.13)

$$\lim_{b_{q,1}\to\infty}\left(\frac{1}{(b_{q,1})^{(N-1)^2}}\operatorname{res}_\zeta\left(\sigma_0({}_pP_q)(x,\zeta), \partial_\zeta\sigma_0({}_pP_q)(x,\zeta)\right)\Big|_{x\to b_{q,1}x}\right)$$
$$= \operatorname{res}_\zeta\left(\sigma_0({}_pP_{q-1})(x,\zeta), \partial_\zeta\sigma_0({}_pP_{q-1})(x,\zeta)\right) \qquad (p \ge 0, q \ge 1).$$

Furthermore, we obtain

(5.14)

$$\operatorname{res}_{\zeta}\left(\sigma_0({}_N P_{N-1})(x,\zeta), \partial_{\zeta}\sigma_0({}_N P_{N-1})(x,\zeta)\right)\Big|_{\mu=\lambda=0}$$
$$= (-1)^{N-1}x^{2N-1}\left(\mathfrak{s}_{N-1}(\boldsymbol{\alpha}_{N-1,1})\right)^2 \operatorname{res}_{\zeta}\left(\sigma_0({}_{N-1}P_{N-2})(x,\zeta), \partial_{\zeta}\sigma_0({}_{N-1}P_{N-2})(x,\zeta)\right)$$

for $N \geq 2$. Here the parameters of (5.14) should be set the same as in Corollary 5.5.

REMARK 5.7. If $p \neq q+1$ and $N = 3$, we have

(i-reg) If $p = 2, q = 2$ and $\varrho = 0$,

$$\begin{aligned} V_{0,\ell}^{(1,2)} &= -\boldsymbol{B}_\ell(b_1 - a_1) - \boldsymbol{B}_\ell(b_1 - a_2) + \boldsymbol{B}_\ell(b_2 - a_1) + \boldsymbol{B}_\ell(b_2 - a_2) \\ &\quad + \boldsymbol{B}_\ell(b_1 - 1) - \boldsymbol{B}_\ell(b_2 - 1) + \boldsymbol{B}_\ell(b_1 - b_2) - \boldsymbol{B}_\ell(b_2 - b_1), \\ V_{0,\ell}^{(1,3)} &= -\boldsymbol{B}_\ell(a_1) - \boldsymbol{B}_\ell(a_2) - \boldsymbol{B}_\ell(b_1 - a_1) - \boldsymbol{B}_\ell(b_1 - a_2) \\ &\quad + \boldsymbol{B}_\ell(b_1 - 1) + \boldsymbol{B}_\ell(b_1) + \boldsymbol{B}_\ell(b_2) + \boldsymbol{B}_\ell(b_1 - b_2), \\ V_{0,\ell}^{(2,3)} &= -\boldsymbol{B}_\ell(a_1) - \boldsymbol{B}_\ell(a_2) - \boldsymbol{B}_\ell(b_2 - a_1) - \boldsymbol{B}_\ell(b_2 - a_2) \\ &\quad + \boldsymbol{B}_\ell(b_2 - 1) + \boldsymbol{B}_\ell(b_2) + \boldsymbol{B}_\ell(b_1) + \boldsymbol{B}_\ell(b_2 - b_1). \end{aligned}$$

(ii-reg) If $p = 3, q = 1$ and $\varrho = \infty$,

$$\begin{aligned} V_{\infty,\ell}^{(1,2)} &= \boldsymbol{B}_\ell(a_1) - \boldsymbol{B}_\ell(a_2) - \boldsymbol{B}_\ell(b_1 - a_1) + \boldsymbol{B}_\ell(b_1 - a_2) \\ &\quad + \boldsymbol{B}_\ell(a_3 - a_1) - \boldsymbol{B}_\ell(a_3 - a_2) + \boldsymbol{B}_\ell(a_2 - a_1) - \boldsymbol{B}_\ell(a_1 - a_2), \\ V_{\infty,\ell}^{(1,3)} &= \boldsymbol{B}_\ell(a_1) - \boldsymbol{B}_\ell(a_3) - \boldsymbol{B}_\ell(b_1 - a_1) + \boldsymbol{B}_\ell(b_1 - a_3) \\ &\quad + \boldsymbol{B}_\ell(a_2 - a_1) - \boldsymbol{B}_\ell(a_2 - a_3) + \boldsymbol{B}_\ell(a_3 - a_1) - \boldsymbol{B}_\ell(a_1 - a_3), \\ V_{\infty,\ell}^{(2,3)} &= \boldsymbol{B}_\ell(a_3) - \boldsymbol{B}_\ell(a_1) - \boldsymbol{B}_\ell(b_1 - a_2) + \boldsymbol{B}_\ell(b_1 - a_3) \\ &\quad + \boldsymbol{B}_\ell(a_1 - a_2) - \boldsymbol{B}_\ell(a_2 - a_3) + \boldsymbol{B}_\ell(a_3 - a_2) - \boldsymbol{B}_\ell(a_1 - a_3). \end{aligned}$$

(iii-reg) If $p = 1, q = 2$ and $\varrho = 0$,

$$\begin{aligned} V_{0,\ell}^{(1,2)} &= \boldsymbol{B}_\ell(b_1 - 1) - \boldsymbol{B}_\ell(b_2 - 1) + \boldsymbol{B}_\ell(b_1 - b_2) - \boldsymbol{B}_\ell(b_2 - b_1) \\ &\quad - \boldsymbol{B}_\ell(b_1 - a_1) + \boldsymbol{B}_\ell(b_2 - a_1) - \boldsymbol{B}_\ell(b_2 - a_1) + \boldsymbol{B}_\ell(b_1 - a_2), \\ V_{0,\ell}^{(1,3)} &= -\boldsymbol{B}_\ell(a_1) - \boldsymbol{B}_\ell(b_1 - a_1) \\ &\quad + \boldsymbol{B}_\ell(b_1 - 1) + \boldsymbol{B}_\ell(b_1) + \boldsymbol{B}_\ell(b_2) + \boldsymbol{B}_\ell(b_1 - b_2), \\ V_{0,\ell}^{(2,3)} &= -\boldsymbol{B}_\ell(a_1) - \boldsymbol{B}_\ell(b_2 - a_1) \\ &\quad + \boldsymbol{B}_\ell(b_2 - 1) + \boldsymbol{B}_\ell(b_2) + \boldsymbol{B}_\ell(b_1) + \boldsymbol{B}_\ell(b_2 - b_1). \end{aligned}$$

(iv-reg) If $p = 3, q = 0$ and $\varrho = \infty$,

$$\begin{aligned} V_{\infty,\ell}^{(1,2)} &= \boldsymbol{B}_\ell(a_1) - \boldsymbol{B}_\ell(a_2) \\ &\qquad + \boldsymbol{B}_\ell(a_3 - a_1) - \boldsymbol{B}_\ell(a_3 - a_2) + \boldsymbol{B}_\ell(a_2 - a_1) - \boldsymbol{B}_\ell(a_1 - a_2), \\ V_{\infty,\ell}^{(1,3)} &= \boldsymbol{B}_\ell(a_1) - \boldsymbol{B}_\ell(a_3) \\ &\qquad + \boldsymbol{B}_\ell(a_2 - a_1) - \boldsymbol{B}_\ell(a_2 - a_3) + \boldsymbol{B}_\ell(a_3 - a_1) - \boldsymbol{B}_\ell(a_1 - a_3), \\ V_{\infty,\ell}^{(2,3)} &= \boldsymbol{B}_\ell(a_3) - \boldsymbol{B}_\ell(a_1) \\ &\qquad + \boldsymbol{B}_\ell(a_1 - a_2) - \boldsymbol{B}_\ell(a_2 - a_3) + \boldsymbol{B}_\ell(a_3 - a_2) - \boldsymbol{B}_\ell(a_1 - a_3). \end{aligned}$$

(v-reg) If $p = 0, q = 2$ and $\varrho = 0$,

$$\begin{aligned} V_{0,\ell}^{(1,2)} &= \boldsymbol{B}_\ell(b_1 - 1) - \boldsymbol{B}_\ell(b_2 - 1) + \boldsymbol{B}_\ell(b_1 - b_2) - \boldsymbol{B}_\ell(b_2 - b_1), \\ V_{0,\ell}^{(1,3)} &= \boldsymbol{B}_\ell(b_1 - 1) + \boldsymbol{B}_\ell(b_1) + \boldsymbol{B}_\ell(b_2) + \boldsymbol{B}_\ell(b_1 - b_2), \\ V_{0,\ell}^{(2,3)} &= \boldsymbol{B}_\ell(b_2 - 1) + \boldsymbol{B}_\ell(b_2) + \boldsymbol{B}_\ell(b_1) + \boldsymbol{B}_\ell(b_2 - b_1). \end{aligned}$$

(i-irr) If $p = 2, q = 2$ and $\varrho = \infty$,

$$\begin{aligned} V_{\infty,\ell}^{(1,2)} &= \boldsymbol{B}_\ell(a_1) - \boldsymbol{B}_\ell(a_2) + \boldsymbol{B}_\ell(a_2 - a_1) - \boldsymbol{B}_\ell(a_1 - a_2) \\ &\qquad - \boldsymbol{B}_\ell(b_1 - a_1) + \boldsymbol{B}_\ell(b_1 - a_2) - \boldsymbol{B}_\ell(b_2 - a_1) + \boldsymbol{B}_\ell(b_2 - a_2), \\ V_{\infty,\ell}^{(1,3)} &= \boldsymbol{B}_\ell(a_1) + \boldsymbol{B}_\ell(a_2 - a_1) - \boldsymbol{B}_\ell(b_1 - a_1) - \boldsymbol{B}_\ell(b_2 - a_1), \\ V_{\infty,\ell}^{(2,3)} &= \boldsymbol{B}_\ell(a_2) + \boldsymbol{B}_\ell(a_1 - a_2) - \boldsymbol{B}_\ell(b_1 - a_2) - \boldsymbol{B}_\ell(b_2 - a_2). \end{aligned}$$

(ii-irr) If $p = 3, q = 1$ and $\varrho = 0$,

$$\begin{aligned} V_{0,\ell}^{(1,2)} &= \boldsymbol{B}_\ell(b_1 - 1) - \boldsymbol{B}_\ell(b_1 - a_1) - \boldsymbol{B}_\ell(b_1 - a_2) - \boldsymbol{B}_\ell(b_1 - a_3), \\ V_{0,\ell}^{(1,3)} &= -\boldsymbol{B}_\ell(a_1) - \boldsymbol{B}_\ell(a_2) - \boldsymbol{B}_\ell(a_3) + \boldsymbol{B}_\ell(b_1) \\ &\qquad + \boldsymbol{B}_\ell(b_1 - 1) - \boldsymbol{B}_\ell(b_1 - a_1) - \boldsymbol{B}_\ell(b_1 - a_2) - \boldsymbol{B}_\ell(b_1 - a_3), \\ V_{0,\ell}^{(2,3)} &= -\boldsymbol{B}_\ell(a_1) - \boldsymbol{B}_\ell(a_2) - \boldsymbol{B}_\ell(a_3) + \boldsymbol{B}_\ell(b_1). \end{aligned}$$

(iii-irr) If $p = 1, q = 2$ and $\varrho = \infty$,

$$\begin{aligned} V_{\infty,\ell}^{(1,2)} &= \boldsymbol{B}_\ell(a_1) - \boldsymbol{B}_\ell(b_1 - a_1) - \boldsymbol{B}_\ell(b_2 - a_1), \\ V_{\infty,\ell}^{(1,3)} &= \boldsymbol{B}_\ell(a_1) - \boldsymbol{B}_\ell(b_1 - a_1) - \boldsymbol{B}_\ell(b_2 - a_1), \\ V_{\infty,\ell}^{(2,3)} &= 0. \end{aligned}$$

(iv-irr) If $p = 3, q = 0$ and $\varrho = 0$,

$$\begin{aligned} V_{0,\ell}^{(1,2)} &= 0, \\ V_{0,\ell}^{(1,3)} &= -\boldsymbol{B}_\ell(a_1) - \boldsymbol{B}_\ell(a_2) - \boldsymbol{B}_\ell(a_3), \\ V_{0,\ell}^{(2,3)} &= -\boldsymbol{B}_\ell(a_1) - \boldsymbol{B}_\ell(a_2) - \boldsymbol{B}_\ell(a_3). \end{aligned}$$

(v-irr) If $p = 0, q = 2$ and $\varrho = \infty$,

$$V_{\infty,\ell}^{(1,2)} = 0,$$
$$V_{\infty,\ell}^{(1,3)} = 0,$$
$$V_{\infty,\ell}^{(2,3)} = 0.$$

References

[1] Takashi Aoki, Takahiro Kawai, Tatsuya Koike, and Yoshitsugu Takei, *On the exact WKB analysis of operators admitting infinitely many phases*, Adv. Math. **181** (2004), no. 1, 165–189, DOI 10.1016/S0001-8708(03)00049-5. MR2020659

[2] Takashi Aoki, Takahiro Kawai, Tatsuya Koike, and Yoshitsugu Takei, *On the exact WKB analysis of microdifferential operators of WKB type* (English, with English and French summaries), Ann. Inst. Fourier (Grenoble) **54** (2004), no. 5, 1393–1421, xiv–xv, xx. MR2127852

[3] Takashi Aoki, Takahiro Kawai, and Yoshitsugu Takei, *New turning points in the exact WKB analysis for higher-order ordinary differential equations* (English, with English and French summaries), Analyse algébrique des perturbations singulières, I (Marseille-Luminy, 1991), Travaux en Cours, vol. 47, Hermann, Paris, 1994, pp. xiii, xv, 69–84. MR1296472

[4] Takashi Aoki, Toshinori Takahashi, and Mika Tanda, *Exact WKB analysis of confluent hypergeometric differential equations with a large parameter*, Exponential analysis of differential equations and related topics, RIMS Kôkyûroku Bessatsu, B52, Res. Inst. Math. Sci. (RIMS), Kyoto, 2014, pp. 165–174. MR3408181

[5] Takashi Aoki, Toshinori Takahashi, and Mika Tanda, *Relation between the hypergeometric function and WKB solutions*, Microlocal analysis and singular perturbation theory, RIMS Kôkyûroku Bessatsu, B61, Res. Inst. Math. Sci. (RIMS), Kyoto, 2017, pp. 1–7. MR3699865

[6] T. Aoki, T. Takahashi, and M. Tanda, *Voros coefficients of the Gauss hypergeometric differential equation with a large parameter*, Integral Transforms Spec. Funct. **32** (2021), no. 5-8, 336–345, DOI 10.1080/10652469.2020.1838504. MR4280687

[7] Takashi Aoki, Toshinori Takahashi, and Mika Tanda, *The hypergeometric function, the confluent hypergeometric function and WKB solutions*, J. Math. Soc. Japan **73** (2021), no. 4, 1019–1062, DOI 10.2969/jmsj/84528452. MR4329021

[8] Takashi Aoki and Mika Tanda, *Parametric Stokes phenomena of the Gauss hypergeometric differential equation with a large parameter*, J. Math. Soc. Japan **68** (2016), no. 3, 1099–1132, DOI 10.2969/jmsj/06831099. MR3523540

[9] Takashi Aoki and Shofu Uchida, *Voros coefficients at the origin and at the infinity of the generalized hypergeometric differential equations with a large parameter*, SIGMA Symmetry Integrability Geom. Methods Appl. **18** (2022), Paper No. 002, 23, DOI 10.3842/SIGMA.2022.002. MR4357620

[10] É. Delabaere, H. Dillinger, and F. Pham, *Résurgence de Voros et périodes des courbes hyperelliptiques* (French, with English and French summaries), Ann. Inst. Fourier (Grenoble) **43** (1993), no. 1, 163–199. MR1209700

[11] H. van Haeringen and L. P. Kok, *Table errata: Higher transcendental functions, Vol. I [McGraw-Hill, New York, 1953; MR* **15**, *419] by A. Erdélyi, W. Magnus, F. Oberhettinger and F. G. Tricomi*, Math. Comp. **41** (1983), no. 164, 778. MR717721

[12] Otto Forster, *Lectures on Riemann surfaces*, Graduate Texts in Mathematics, vol. 81, Springer-Verlag, New York-Berlin, 1981. Translated from the German by Bruce Gilligan. MR648106

[13] Naofumi Honda, Takahiro Kawai, and Yoshitsugu Takei, *Virtual turning points*, SpringerBriefs in Mathematical Physics, vol. 4, Springer, Tokyo, 2015, DOI 10.1007/978-4-431-55702-9. MR3380962

[14] Kohei Iwaki and Tatsuya Koike, *On the computation of Voros coefficients via middle convolutions*, Exponential analysis of differential equations and related topics, RIMS Kôkyûroku Bessatsu, B52, Res. Inst. Math. Sci. (RIMS), Kyoto, 2014, pp. 55–70. MR3408175

[15] Takahiro Kawai and Yoshitsugu Takei, *Algebraic analysis of singular perturbation theory*, Translations of Mathematical Monographs, vol. 227, American Mathematical Society, Providence, RI, 2005. Translated from the 1998 Japanese original by Goro Kato; Iwanami Series in Modern Mathematics, DOI 10.1090/mmono/227. MR2182990

[16] Tatsuya Koike and Yoshitsugu Takei, *On the Voros coefficient for the Whittaker equation with a large parameter—some progress around Sato's conjecture in exact WKB analysis*, Publ. Res. Inst. Math. Sci. **47** (2011), no. 1, 375–395, DOI 10.2977/PRIMS/39. MR2827731

[17] Toshio Oshima, *Fractional calculus of Weyl algebra and Fuchsian differential equations*, MSJ Memoirs, vol. 28, Mathematical Society of Japan, Tokyo, 2012. MR2986408

[18] Hujun Shen and Harris J. Silverstone, *Observations on the JWKB treatment of the quadratic barrier*, Algebraic analysis of differential equations from microlocal analysis to exponential asymptotics, Springer, Tokyo, 2008, pp. 237–250, DOI 10.1007/978-4-431-73240-2_20. MR2758911

[19] Yoshitsugu Takei, *Sato's conjecture for the Weber equation and transformation theory for Schrödinger equations with a merging pair of turning points*, Differential equations and exact WKB analysis, RIMS Kôkyûroku Bessatsu, B10, Res. Inst. Math. Sci. (RIMS), Kyoto, 2008, pp. 205–224. MR2498081

[20] A. Voros, *The return of the quartic oscillator: the complex WKB method* (English, with French summary), Ann. Inst. H. Poincaré Sect. A (N.S.) **39** (1983), no. 3, 211–338. MR729194

Department of Mathematics, Kindai University, Higashi–Osaka 577-8502, Japan
Email address: `aoki@math.kindai.ac.jp`

Graduate School of Science and Engineering, Kindai University, Higashi–Osaka 577-8502, Japan
Email address: `uchida@math.kindai.ac.jp`

Contemporary Mathematics
Volume **782**, 2023
https://doi.org/10.1090/conm/782/15722

Riemann-Liouville transform and linear differential equations on the Riemann sphere

Toshio Oshima

Abstract. We study the Riemann-Liouville transformation of solutions to linear differential equations on the Riemann sphere. The transformation corresponds to the middle convolution of the equations. Under the transformation, we examine the asymptotic behavior of the solutions at the singular points of the equations. When the singular points are regular, we studied it in [**O1**] and solved a connection problem for the general rigid Fuchsian equations. In this paper we mainly study the case when some singular points are irregular.

1. Introduction

Classical special functions such as Gauss hypergeometric functions, Bessel functions and Whittaker functions are solutions to linear differential equations on the Riemann sphere and the analysis of the equations gives fundamental properties of the special functions. Our purpose is to analyze solutions to linear differential equations on the Riemann sphere as in the case of the classical special functions.

N. Katz [**Kz**] defined and studied the middle convolution and the addition of local systems. They are invertible transformations keeping the index of rigidity of the system. The index gives the number of accessory parameters which cannot be determined by the local structure of the system. The corresponding transformations of the solution are given by Riemann-Liouville integral and multiplication by a function. The equation is called rigid if there is no accessory parameter.

Any rigid Fuchsian equation is reduced to and constructed from the trivial equation $u' = 0$ by successive applications of additions and middle convolutions, which is proved by [**Kz**] for Fuchsian systems of the first order and by [**O1**] for scalar higher order linear ordinary differential equations together with the definition of the middle convolution of the equations. In this case the addition corresponds to the multiplication by the function $(x-c)^\lambda$ and a study of the Riemann-Liouville integral in [**O1**] gives a global information of the solution such as a connection problem describing the relation between analytic continuations of local solutions.

We will generalize some results in [**O1**] when an equation has unramified irregular singularities. In particular, we study the Riemann-Liouville integral of the

2020 *Mathematics Subject Classification.* Primary 34M40; Secondary 34M30, 34M35.
Key words and phrases. Irregular singularity, middle convolution, connection problem.
The author was supported in part by Grant-in-Aid for Scientific Researches (A), No. 18K03341, Japan Society of Promotion.

solutions and clarify the change of the asymptotic behavior of the solutions under the integral transform at the singular points. Note that the rigid single equation without ramified irregular singularities is reduced to or constructed from the trivial equation by successive applications of additions and middle convolutions (cf. [**O6**]). These additions correspond to the multiplications by any functions ϕ such that $\frac{\phi'}{\phi}$ are rational functions.

Here we give some standard notation used in this paper. We mainly consider functions of one variable x and then $\mathbb{C}[x]$, $\mathbb{C}(x)$, $\mathbb{C}[\![x]\!]$ and $W[x]$ are the space of polynomials, the space of rational functions, the space of formal power series and the ring of linear differential operators with polynomial coefficients, respectively. Then $W[x]$ is generated by x and $\frac{d}{dx}$ with the relation $\frac{d}{dx}x = x\frac{d}{dx} + 1$. Moreover we put $W(x) := \mathbb{C}(x) \otimes W[x]$.

Let $\mathrm{C}^m(a,b)$ denote the space of functions f on the interval $(a,b) \subset \mathbb{R}$ such that $f^{(j)}$ exist and define continuous functions on (a,b) for $j = 0, \ldots, m$. The space $\mathrm{C}^m(a,b]$ is similarly defined. We also use the notation

$$
\begin{aligned}
&\partial = \tfrac{d}{dx}, \quad \vartheta = x\partial,\\
&\operatorname{Re} z = a, \ \operatorname{Im} z = b \quad (z = a + bi, \ a, \, b \in \mathbb{R}),\\
&\mathbb{Z}_{>0} = \{k \in \mathbb{Z} \mid k > 0\},\\
&(\lambda)_m = \lambda(\lambda+1)\cdots(\lambda+m-1) \quad (\lambda \in \mathbb{C}, \ m \in \mathbb{Z}_{>0}) \text{ and } (\lambda)_0 = 1,\\
&{}_mF_k(\alpha_1,\ldots,\alpha_m;\beta_1,\ldots,\beta_k;x) = \sum_{n=0}^{\infty} \frac{(\alpha_1)_n\cdots(\alpha_m)_n}{(\beta_1)_n\cdots(\beta_k)_n(1)_n} x^n \in \mathbb{C}[\![x]\!],\\
&F(\alpha,\beta;\gamma;x) = {}_2F_1(\alpha,\beta;\gamma;x).
\end{aligned}
$$

Under this notation the addition $\operatorname{Ad}(\phi)$ of the equation $Pu = 0$ corresponds to the map $u \mapsto \phi u$ and then $x \mapsto x$ and $\partial \mapsto \partial - \frac{\phi'}{\phi}$ in $W(x)$. And the middle convolution mc_μ with $\mu \in \mathbb{C}$ corresponds to the map $u \mapsto \partial^{-\mu}u$ and then $\vartheta \mapsto \vartheta - \mu$ and $\partial \mapsto \partial$ in $W[x]$. We can define $\operatorname{Ad}(\partial^{-\mu})$ in the ring $\mathbb{C}[\partial^{-1}] \otimes W[x]$, which is isomorphic to the ring $W(x)$ by the correspondence $(x,\partial) \mapsto (-\partial, x)$.

In §2 Riemann-Liouville transform of a function $u(x)$ is defined by a regularization of the Riemann-Liouville integral

$$
(I_a^\mu u)(x) = \frac{1}{\Gamma(\mu)} \int_a^x u(t)(x-t)^{\mu-1}dt
$$

and we examine its fundamental properties. In §3 we prove Theorem 3.1 and Theorem 3.7 which describe the asymptotic of Riemann-Liouville transform $I_a^\mu u$ of a function $u(x)$ in terms of the asymptotic of $u(x)$. Namely, when a and b are singularities of $u(x)$, we examine the asymptotic of $(I_a^\mu u)(x)$ when $x \to a$ or $x \to b$. We show that the equation $Pu = 0$ implies $\bigl(\mathrm{mc}_\mu(P)\bigr)(I_a^\mu u) = 0$ in §4. In §5 we review the linear differential equations on the Riemann sphere. Important concepts are characteristic exponents, a generalized Riemann scheme, index of rigidity and middle convolution of the equation. In §6 we give some examples to understand the results in this paper.

Using versal additions introduced in [**O1**, §2.3], we define and construct in [**O6**] a versal unfolding of any rigid differential equation on $P^1_{\mathbb{C}}$ whose irregular singularities are unramified. Hence the equation is naturally realized as a confluent limit of rigid Fuchsian equations and we can study the original equation by the

Fuchsian equations and therefore in this paper we also consider regular singular points in parallel. Moreover in [**O5**] we regard the singular points as new variables and then the versal differential equation can be extended to a versal KZ equation with several variables, which will be useful to study the confluence of singularities.

2. Definition of Riemann-Liouville transform

In this section we define Riemann-Liouville transform by a regularization of Riemann-Liouville integral and show fundamental properties of the transform. We will apply the transform to solutions to linear differential equations with irregular singularities. Other regularizations are given in [**O1**, p.5] and [**Ha**, §2] mainly for the purpose to study Fuchsian differential equations.

DEFINITION 2.1. For $\mu \in \mathbb{C}$ and $a \in \mathbb{C}$ the Riemann-Liouville transform (Euler transformation) I_a^μ of a holomorphic function $u(x)$ is defined by the complex integral

$$(2.1)\quad (I^\mu_{a+e^{i\theta_0}0}u)(x) := \frac{1}{\Gamma(\mu)}\int_a^x u(t)(x-t)^{\mu-1}dt = \frac{1}{\Gamma(\mu)}\int_{a+e^{i\theta_0}0}^x u(t)(x-t)^{\mu-1}dt$$

along a piecewise C^1-path

$$(2.2)\quad \begin{aligned}&L : [\alpha,\beta] \ni t \mapsto L(t) \in \mathbb{C},\ L(\alpha)=a,\ L(\beta)=x,\ \theta_0 = \operatorname{Arg} L'(\alpha),\\ &\alpha = t_0 < t_1 < \cdots < t_m = \beta \text{ and } L|_{[t_{j-1},t_j]} \in \mathrm{C}^1[t_{j-1},t_j] \quad (j=1,\dots,m),\\ &L(s) \neq L(t) \quad \text{for } \alpha \le s < t \le \beta.\end{aligned}$$

Sometimes the above $I^\mu_{a+e^{i\theta_0}0}$ will be denoted by I_a^μ and the resulting function $I_a^\mu u$ will be denoted by $I_a^\mu(u)$. We assume that $u(x)$ is holomorphic along the path $L(t)$ for $t \in (\alpha,\beta]$ and may have a singularity at the starting point $a = L(\alpha)$ of the path. When $u(x)$ is bounded along L and $\operatorname{Re}\mu > 0$, $(I_a^\mu u)(x)$ is naturally defined and holomorphic in a neighborhood of $L(t)$ for $t \in (\alpha,\beta]$.

Here $(I_a^\mu u)(L(t))$ is defined by the path $L|_{[\alpha,t]}$ and the condition $L(s) \neq L(t)$ for $\alpha \le s < t \le \beta$ is essential which assures that $(I_a^\mu u)(x)$ is holomorphic along the path L.

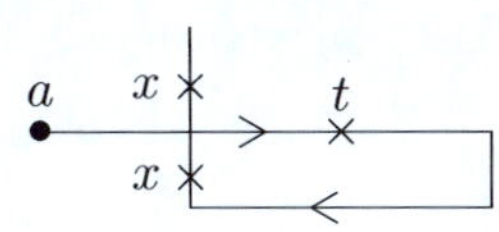

We also assume that L can be replaced by a piecewise linear path, namely,

$$(2.3)\quad L(t) = \tfrac{t_j-t}{t_j-t_{j-1}}L(t_{j-1}) + \tfrac{t-t_{j-1}}{t_j-t_{j-1}}L(t_j) \text{ for } t\in[t_{j-1},t_j] \text{ and } j=1,\dots,m.$$

We will study the case that $u(x)$ is a solution to a differential equation $Pu=0$ with $P \in W[x]$ and a is a singular point of the equation. In this case (2.3) can be always assumed.

When a, $x \in \mathbb{R}$ and $x < a$, we may consider

$$(\check{I}_a^\mu u)(x) = \frac{1}{\Gamma(\mu)}\int_x^a u(t)(t-x)^{\mu-1}dt,$$

which has a clear meaning.

If we use the path L satisfying $\operatorname{Im} L(t) \ge 0$ in the case a, $x \in \mathbb{R}$, we have

$$(2.4)\qquad \check{I}_a^\mu u = e^{-\mu\pi i} I_a^\mu u,$$

which follows from

$$\int_a^x u(t)(x-t)^{\mu-1}dt = \int_a^x u(t)\big(e^{\pi i}(t-x)\big)^{\mu-1}dt$$
$$= e^{\mu\pi i}\int_x^a u(t)(t-x)^{\mu-1}dt.$$

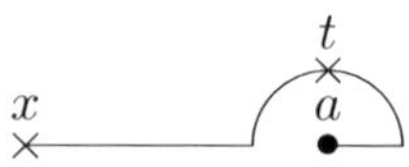

For $K>0$ we have

$$(I^\mu_{a+e^{i\theta_0}0}u)(Kx) = K^\mu I^\mu_{\frac{a}{K}+e^{i\theta_0}0}\big(u(Kx)\big). \tag{2.5}$$

EXAMPLE 2.2. First we consider a typical example:

$$u(x) = x^\lambda(c-x)^{\lambda'}e^{\frac{C'}{c-x}-\frac{C}{x}} \quad (c>0,\ \lambda,\ \lambda',\ C,\ C'\in\mathbb{C}), \tag{2.6}$$

$$v(x) := (I^\mu_0 u)(x) = \frac{1}{\Gamma(\mu)}\int_0^x t^\lambda(c-t)^{\lambda'}e^{\frac{C'}{1-t}-\frac{C}{t}}(x-t)^{\mu-1}dt. \tag{2.7}$$

If $\operatorname{Re}\mu>0$ and one of the conditions

$$\operatorname{Re}C>0 \tag{2.8}$$

or

$$C=0 \text{ and } \operatorname{Re}\lambda>-1 \tag{2.9}$$

is valid, then $(I^\mu_0 u)(x)$ is naturally defined for $x\in[0,c)$ and gives a holomorphic function on $\mathbb{C}\setminus\big((-\infty,0)\cup[c,\infty)\big)$. By the argument in this section we see that the condition $\operatorname{Re}\mu>0$ can be removed and the condition for λ in (2.9) is replaced by $\lambda\notin\mathbb{Z}_{<0}$. In fact $u(x)$ holomorphically depends on $(\lambda,\mu)\in(\mathbb{C}\setminus\mathbb{Z}_{<0})\times\mathbb{C}$.

If the path L satisfies $\operatorname{Re}\frac{C}{L'(\alpha)}>0$, we can define $v(x)$. In particular if $\operatorname{Re}C<0$, we may take $-L'(\alpha)>0$ and $v(x)$ is naturally a holomorphic function on $\mathbb{C}\setminus\big((-\infty,0]\cup[c,\infty)\big)$.

Since $\frac{u'(x)}{u(x)}=\frac{a_0(x)}{a_1(x)}$ with $a_0(x),\,a_1(x)\in\mathbb{C}[x]$, $u(x)$ satisfies $Pu=0$ with $P=a_1(x)\partial-a_0(x)$. Here

$$\frac{u'(x)}{u(x)} = \frac{\lambda}{x}+\frac{C}{x^2}-\frac{\lambda'}{c-x}-\frac{C'}{(c-x)^2}. \tag{2.10}$$

Then $v=I^\mu_0 u$ satisfies a differential equation $Qv=0$ with $Q=\operatorname{mc}_\mu(P)$, where $\operatorname{mc}_\mu(P)$ is given by (4.2) and (4.1). If $C=C'=0$, the differential equation $Qu=0$ is reduced to Gauss hypergeometric equation and

$$u(x) = \frac{\Gamma(\lambda+1)}{\Gamma(\lambda+\mu+1)}x^{\lambda+\mu}F(-\lambda',\lambda+1;\mu+1;x), \tag{2.11}$$

which is obtained by an easy calculation (cf. [**O1**, (1.47)] and (3.4)).

If $\lambda'=C'=0$, it is reduced to Kummer's equation by a simple transformation. If $CC'\neq 0$, it is a confluence of Jordan-Pochhammer's equation of order 4.

We will examine the asymptotics of $(I^\mu_0 u)(x)$ for $x\to+0$ and $x\to c-0$ and $x\to\infty$. In fact we show general theorems, which give these asymptotics.

Now we return to study $I^\mu_a u$ in the general case. First we assume that $u(x)$ is bounded on the path L and $\operatorname{Re}\mu>0$ and $\operatorname{Re}\mu'>0$. We will show

$$I^{\mu'}_a\circ I^\mu_a u = I^{\mu+\mu'}_a u. \tag{2.12}$$

Since $(I_a^{\mu'} \circ I_a^{\mu} u)(x)$ and $(I_a^{\mu+\mu'} u)(x)$ holomorphically depend on the variable x in the path L, it is sufficient to prove (2.12) for $x = L(t)$ for $t \in [t_0, t_1]$. Hence we may assume that $L(t) = \frac{t_1 - t}{t_1 - t_0} L(t_0) + \frac{t - t_0}{t_1 - t_0} L(t_1)$. By a change of the variable we may moreover assume that $a = 0$ and $x \in \mathbb{R}_{>0}$. Then

$$\begin{aligned}(I_0^{\mu'} \circ I_0^{\mu} u)(x) &= \frac{1}{\Gamma(\mu')} \int_0^x (x-t)^{\mu'-1} \Big(\frac{1}{\Gamma(\mu)} \int_0^t u(s)(t-s)^{\mu-1} ds \Big) dt \\ &= \frac{1}{\Gamma(\mu)\Gamma(\mu')} \int_0^x u(s) \int_s^x (x-t)^{\mu'-1}(t-s)^{\mu-1} dt\, ds,\end{aligned}$$

$$\begin{aligned}\int_s^x (x-t)^{\mu-1}(t-s)^{\mu'-1} dt &= \int_0^{x-s} (x-s-t)^{\mu'-1} t^{\mu-1} dt \\ &= (x-s)^{\mu+\mu'-1} \int_0^1 (1-t_1)^{\mu'-1} t_1^{\mu-1} dt_1 \quad (t = (x-s)t_1) \\ &= (x-s)^{\mu+\mu'-1} \frac{\Gamma(\mu)\Gamma(\mu')}{\Gamma(\mu+\mu')}\end{aligned}$$

and therefore we have (2.1).

By the same reason as above we may assume $a = 0$ and $x \in (0, c)$ with $c > 0$ and $L = [0, x]$ to prove several relations satisfied by $I_a^{\mu} u$.

Put $C_0^m[0, c) := \{u \in C^m[0, c) \mid (\partial^j u)(0) = 0 \quad (0 \le j \le m)\}$ for $m \in \mathbb{Z}_{\ge 0}$. If $u \in C_0^m[0, c)$ and $\operatorname{Re} \mu > 0$,

$$\int_a^x u(t)(x-t)^{\mu-1} dt = \int_a^x u(x-t) t^{\mu-1} dt$$

and $I_0^{\mu}(u) \in C_0^m[0, c)$. Here we note that $(I_0^{\mu} u)(x)$ holomorphically depends on μ.

Suppose $u \in C_0^1[0, c)$. Since

$$\begin{aligned}\tfrac{d}{dt}(u(t)(x-t)^{\mu}) &= u'(t)(x-t)^{\mu} - \mu u(t)(x-t)^{\mu-1} \\ &= u'(t)(x-t)^{\mu} - \tfrac{d}{dx}(u(t)(x-t)^{\mu}) \\ &= xu'(t)(x-t)^{\mu-1} - tu'(t)(x-t)^{\mu-1} - \mu u(t)(x-t)^{\mu-1}\end{aligned}$$

we have

$$I_0^{\mu}(u) = I_0^{\mu+1}(u') = \partial I_0^{\mu+1}(u)$$

for $\operatorname{Re} \mu > 0$ and

$$I_0^{\mu}(\vartheta u) = (\vartheta - \mu) I_0^{\mu}(u) \tag{2.13}$$

for $\operatorname{Re} \mu > 1$.

Hence for $u \in C_0^m[0, c)$ with $m \in \mathbb{Z}_{\ge 0}$ we have

$$I_0^{\mu}(u) = \partial^m I_0^{\mu+m}(u) = I_0^{\mu+m}(\partial^m u) \tag{2.14}$$

and define $I_0^{\mu}(u)$ for $\operatorname{Re} \mu > -m$ by this equality. In particular

$$I_0^0(u) = I_0^1(\partial u) = u$$

for $u \in C_0^1[0, c)$. If $0 < \operatorname{Re} \mu < m$ and $u \in C_0^m[0, c)$,

$$\begin{aligned}&I_0^{-\mu} I_0^{\mu}(u) = I_0^{m-\mu} \partial^m I_0^{\mu}(u) = \partial^m I_0^{m-\mu} I_0^{\mu}(u) = \partial^m I_0^m(u) = u, \\ &I_0^{\mu}\, I_0^{-\mu}(u) = I_0^{\mu} \circ I_0^{m-\mu}(\partial^m u) = I_0^m(\partial^m u) = u,\end{aligned}$$

namely, if $|\operatorname{Re}\mu| < m$ and $u \in C_0^m[0,c)$, we have

$$I_0^{-\mu} \circ I_0^{\mu}(u) = I_0^0(u) = u. \tag{2.15}$$

Let $k \in \mathbb{Z}_{\geq 0}$. If $\operatorname{Re}\lambda > m$ and $\phi(x) \in C^\infty[0,c)$, then $\phi(x)x^\lambda \log^k x \in C_0^m[0,x)$ and we may apply I_0^λ to u. In particular

$$I_0^\mu(x^\lambda) = \frac{\Gamma(\lambda+1)}{\Gamma(\lambda+\mu+1)} x^{\lambda+\mu} \tag{2.16}$$

if $\operatorname{Re}\lambda > -1$ and $\operatorname{Re}\mu > 0$. Applying $\frac{d^k}{d\lambda^k}$ to the both side of the equation, we have

$$I_0^\mu(x^\lambda \log^k x) = \sum_{j=0}^{k} \binom{k}{j} \left(\frac{d^{k-j}}{d\lambda^{k-j}} \frac{\Gamma(\lambda+1)}{\Gamma(\lambda+\mu+1)} \right) x^{\lambda+\mu} \log^j x. \tag{2.17}$$

For $\lambda \in \mathbb{C} \setminus \mathbb{Z}_{<0}$ The right hand side of the above is holomorphic for $(\lambda,\mu) \in (\mathbb{C}\setminus\mathbb{Z}_{<0}) \times \mathbb{C}$ and we define $I_0^\mu(x^\lambda \log^k x)$ by this equality. Owing to this definition, we have (2.13) and (2.14) for $u(x) = x^\lambda \log^k x$ if $\lambda \notin \mathbb{Z}_{<0}$. Moreover we have (2.12) and (2.15), namely,

$$I_0^{\mu'} \circ I_0^\mu(u) = I_0^{\mu+\mu'}(u) \text{ and } I_0^0(u) = u \tag{2.18}$$

for $x^\lambda \log^k x$ if $\lambda \notin \mathbb{Z}_{<0}$ and $\lambda + \mu \notin \mathbb{Z}_{<0}$.

We examine $I_0^\mu(u)$ for the solution u to a differential equation with a singularity at the origin. If the origin is a regular singular point of the equation, the solutions

$$u(x) = \sum_{j=0}^{k} \phi_j(x) x^\lambda \log^j x \tag{2.19}$$

with $\lambda \in \mathbb{C}$ and suitable holomorphic functions $\phi_j(x)$ at the origin span a space of the local solutions. We may assume that c is also a singular point and $x \in (0,c)$ are regular point. Then $u(x) \in C_0^m[0,c)$ if $\operatorname{Re}\lambda > m$. For any given $m \in \mathbb{Z}_{>0}$, $u(x)$ in (2.19) is a finite sum of the functions $C_{i,j}x^{\lambda+i}\log^j x$ with $i \in \mathbb{Z}_{\geq 0}$ and a function in $C_0^m[0,c)$ and therefore if $\operatorname{Re}\lambda \notin \mathbb{Z}$, $I_0^\mu(u)$ is naturally defined. As a consequence, $I_0^\mu(x^\tau u(x))$ is defined when $\operatorname{Re}\tau$ and $\operatorname{Re}\mu$ are sufficiently large and $I_0^\mu(x^\tau u(x))$ extends holomorphically for $(\tau,\mu) \in \{\tau \in \mathbb{C} \mid \tau + \lambda \notin \mathbb{Z}_{<0}\} \times \mathbb{C}$ and satisfies (2.13) and (2.14). Moreover if $\lambda + \mu \notin \mathbb{Z}_{<0}$, we have (2.18). Thus we have the following proposition.

PROPOSITION 2.3. *Suppose $u(x)$ is of the form* (2.19) *with convergent series $\phi_j(x)$ and $\lambda \in \mathbb{C} \setminus \mathbb{Z}_{<0}$, Then $I_0^\mu(u)$ is naturally defined for any $\mu \in \mathbb{C}$ and μ is a holomorphic parameter of $I_0^\mu(u)$ and we have* (2.13) *and* (2.14). *Moreover we have* (2.18) *if $\lambda + \mu \notin \mathbb{Z}_{<0}$.*

If the origin is an unramified irregular singular point of the equation, the local solutions $u(x)$ with the asymptotic expansion

$$\begin{gathered} u(x) \sim \sum_{j=0}^{k} \phi_j(x) x^\lambda \log^j x \cdot \exp\left(-\tfrac{C_1}{x^{m_1}} - \tfrac{C_2}{x^{m_2}} - \cdots - \tfrac{C_K}{x^{m_K}}\right) \text{ for } x \to e^{i\theta_0}0, \\ \phi_j \in \mathbb{C}[\![x]\!],\ C_j \in \mathbb{C} \text{ and } m_j \in \mathbb{Z} \text{ with } m_1 > m_2 > \cdots > m_K > 0 \end{gathered} \tag{2.20}$$

span the space of the local solutions. Here the condition $x \to e^{i\theta_0}0$ means that

$$V_{\theta_0,\epsilon,r} := \{se^{i\theta} \in \mathbb{C} \mid 0 < s < r,\ |\theta - \theta_0| \leq \epsilon\} \ni x \to 0 \tag{2.21}$$

for suitable $\epsilon > 0$ and $r > 0$. Namely if $\phi_j(x) = a_{j,0} + a_{j,1}x + a_{j,2} + \cdots$, then

$$\lim_{r\to+0} \sup_{x\in V_{\theta_0,\epsilon,r}} |E_N(x)x^{-N}| = 0 \text{ for } N \in \mathbb{Z}_{\geq 0}, \text{ where}$$

$$E_N(x) = u(x)x^{-\lambda}\exp\left(\tfrac{C_1}{x^{m_1}} + \tfrac{C_2}{x^{m_2}} + \cdots + \tfrac{C_K}{x^{m_K}}\right) - \sum_{\nu=0}^{N}\sum_{j=0}^{k} a_{j,\nu}v^i \log^j x.$$

We define $\exp(-\frac{C_1}{x^{m_1}} - \frac{C_2}{x^{m_2}} - \cdots - \frac{C_K}{x^{m_K}}) = 1$ when $K = 0$. We note that $\partial^m u$ satisfies a similar asymptotic which is obtained by applying ∂^m to the right hand side of (2.20).

Moreover we note that if the origin is a ramified singular point, it changes into an unramified singular point under a new coordinate $y = x^{\frac{1}{q}}$ with a suitable $q \in \mathbb{Z}_{>0}$.

THEOREM 2.4. *Put* $\theta_0 = \operatorname{Arg} L'(\alpha)$ *for the path* (2.2) *of the integration* (2.1) *to define* $I_0^\mu u$. *Suppose* $u(x)$ *has the asymptotic* (2.20) *with the condition*

$$K = 0 \text{ and } \lambda \in \mathbb{C} \setminus \mathbb{Z}_{<0} \tag{2.22}$$

or

$$\operatorname{Re} C_1 e^{-im_1\theta_0} > 0. \tag{2.23}$$

Then $I_0^\mu(u)$ *is defined for any* $\mu \in \mathbb{C}$ *and* μ *is a holomorphic parameter of* $I_0^\mu(u)$ *and we have* (2.13) *and* (2.14).

If the condition (2.22) *together with* $\lambda+\mu \notin \mathbb{Z}_{<0}$ *is valid or the condition* (2.23) *is valid,* (2.18) *holds for any* $\mu' \in \mathbb{C}$.

PROOF. We may assume that $x = L(t)$ with $t \in [t_0, t_1]$ and $L(s) = \frac{s-\alpha}{t_1-\alpha}L(t_1)$ for $s \in [\alpha, t_1]$. Then under the new coordinate $y = e^{-i\theta_0}x$ we may assume $L|_{[\alpha,t_0]}$ equals $[0, c]$ with $c > 0$. Then the theorem is clear by the argument we have done. □

REMARK 2.5. i) Suppose $K = 0$ in Theorem 2.4. Put $\tilde{u} = x^{\tilde{\lambda}-\lambda}u$. It follows from (2.17) that $\frac{1}{\Gamma(\tilde{\lambda}+1)}I_0^\mu \tilde{u}$ is well defined and holomorphic for $(\tilde{\lambda}, \mu) \in \mathbb{C}^2$. In this interpretation $\frac{1}{\Gamma(\lambda+1)}I_0^\mu u$ is well-defined even if $\lambda \in \mathbb{Z}_{<0}$.

ii) Since

$$I_a^{-n}(u) = \partial^n I_0^0(u) = \partial^n u,$$

we sometimes denote I_a^μ by $\partial^{-\mu}$.

DEFINITION 2.6. We define a Riemann-Liouville integral $\tilde{I}^\mu_{e^{i\theta_0}\infty}$ of $u(x)$ by the complex integral

$$(\tilde{I}^\mu_{e^{i\theta_0}\infty}u)(x) := \frac{1}{\Gamma(\mu)}\int_x^\infty u(t)(t-x)^{\mu-1}dt = \frac{1}{\Gamma(\mu)}\int_x^{e^{i\theta_0}\infty} u(t)(t-x)^{\mu-1}dt$$

along a path

$$L : [\alpha, \infty) \ni t \mapsto L(t) \in \mathbb{C},\ L(\alpha) = x,\ \lim_{t\to\infty}|L(t)| = \infty,\ \lim_{t\to\infty}\operatorname{Arg} L(t) = \theta_0,$$
$$\alpha = t_0 < t_1 < t_2 < \cdots < t_m,\ L|_{[t_{j-1},t_j]} \in C^1[t_{j-1}, t_j] \quad (j = 1, \ldots, m),$$
$$L|_{[t_m,\infty)} \in C^1[t_m, \infty),\quad L(s) \neq L(t) \quad (\alpha \leq s < t).$$

Here $u(x)$ is holomorphic along L. For simplicity, we may denote $I^\mu_{e^{i\theta_0}\infty}$ by I_a^∞.

Suppose $u(x)$ has an asymptotic

$$u(x) \sim \sum_{j=0}^{k} \phi_j(\tfrac{1}{x})x^{-\lambda}\log^j x \cdot \exp(-C_1x^{m_1} - \cdots - C_kx^{m_K}) \text{ for } x \to e^{i\theta_0}\infty. \tag{2.24}$$

Here $m_j \in \mathbb{Z}$, $\phi_j \in \mathbb{C}[\![x]\!]$ and $m_1 > m_2 > \cdots > m_K > 0$ and the condition $x \to e^{i\theta_0}\infty$ is defined by $\frac{1}{x} \to e^{-i\theta_0}0$. Since

$$\begin{aligned}\int_x^\infty u(t)(t-x)^{\mu-1}dt &= \int_0^{\frac{1}{x}} u(\tfrac{1}{s})(\tfrac{1}{s} - \tfrac{1}{\frac{1}{x}})^{\mu-1}\tfrac{ds}{s^2} \\ &= x^{\mu-1}\int_0^{\frac{1}{x}} s^{-\mu-1}u(\tfrac{1}{s})(\tfrac{1}{x} - s)^{\mu-1}ds,\end{aligned}$$

we have

$$(\tilde{I}^\mu_{e^{i\theta_0}\infty}u)(x) = x^{\mu-1}\cdot\Big(I^\mu_{e^{-i\theta_0}0}x^{-\mu-1}u(\tfrac{1}{x})\Big)(\tfrac{1}{x}). \tag{2.25}$$

In particular

$$\tilde{I}^\mu_\infty(x^{-\lambda}) = x^{\mu-1}\big((I^\mu_0x^{\lambda-\mu-1})\,\big|_{x\mapsto\frac{1}{x}}\big) = x^{\mu-1}\frac{\Gamma(\lambda-\mu)}{\Gamma(\lambda)}x^{1-\lambda} = \frac{\Gamma(\lambda-\mu)}{\Gamma(\lambda)}x^{-\lambda+\mu}.$$

Hence if

$$K = 0 \text{ and } \lambda - \mu \notin \mathbb{Z}_{\le 0} \tag{2.26}$$

or

$$\operatorname{Re} C_1e^{im\theta_0} > 0 \tag{2.27}$$

is valid, $I^\mu_{e^{i\theta_0}\infty}u$ is defined and satisfies (2.13) and (2.14). Moreover we have (2.18) if the condition $\lambda + \mu + \mu' \notin \mathbb{Z}_{\le 0}$ and (2.26) are valid or the condition (2.27) is valid. In the same way we have

$$(I^\mu_cu)(x) = x^{\mu-1}\cdot\Big(\tilde{I}^\mu_{\frac{1}{c}}x^{-\mu-1}u(\tfrac{1}{x})\Big)(\tfrac{1}{x}) \tag{2.28}$$

for $x > 1$, which is useful to examine the asymptotic of $(I^\mu_cu)(x)$ for $x \to \infty$.

REMARK 2.7 (Cf. [**O1**, p.5]). Let $Pu = 0$ be a differential equation with $P \in W[x]$ and $u_1, \ldots, u_n$ be linearly independent solutions to the equation in a neighborhood of a point $x_0 = \epsilon e^{i\operatorname{Arg}x}$. Here ϵ with $0 < \epsilon \ll 1$ and n is the order of P and $x \neq 0$ and we suppose that tx is not the singular point of the equation for $t \in (0,1)$. Then we define

$$(\hat{I}^\mu_0u)(x) = \frac{1}{\Gamma(\mu)(1-e^{2\pi i\mu})}\int^{+x,+0,-x,-0} u(t)(t-x)^{\mu-1}dt$$

by the integration along the Pochhammer cycle $L_1+C_1+L_2+C_2+L_3+C_3+L_4+C_4$

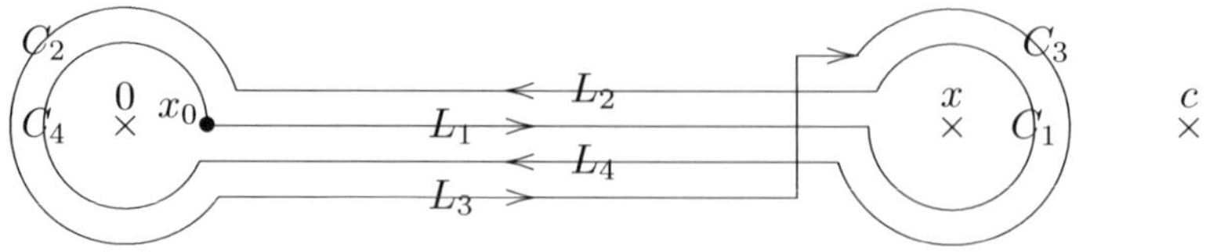

We can define for any $\mu \in \mathbb{C}$ by the analytic continuation to $\mu \in \mathbb{Z}_{<0}$.

The analytic continuation along the path $[0, 2\pi] \ni t \to x_0 e^{i\theta t}$ induces transformations $u(x) \mapsto u^1(x) := u(e^{2\pi i}x)$ and $\mathbf{u} \mapsto \mathbf{u}A$. Here $A \in GL(n, \mathbb{C})$ is called the local monodromy matrix. This integration is used by [**DR**] to define a transformation of solutions to Fuchsian systems.

If $(I_0^\mu \mathbf{u})(x) = \int_0^x \mathbf{u}(t)(t-x)^{\mu-1}dt$ is naturally defined, then

$$\hat{I}_0^\mu u = I_0^\mu u - I_0^\mu u^1 \text{ and } \hat{I}_0^\mu \mathbf{u} = (I_0^\mu \mathbf{u})(1-A).$$

For example,

$$\hat{I}_0^\mu x^\lambda = \frac{(1-e^{2\pi i\lambda})\Gamma(\lambda+1)}{\Gamma(\lambda+\mu+1)} x^{\lambda+\mu}.$$

Suppose $u(x)$ is a solution to the equation $Pu = 0$ with $P \in W[x]$. We may assume that the path has no intersection with the disc $D_\epsilon := \{x \in \mathbb{C} \mid |x| < \epsilon\}$. Then the above integration is defined even if there are several singular points of the equation in D_ϵ, which will be useful to examine the confluence of singular points. The monodromy around a circle enclosing several singular points is discussed in [**O4**].

3. Asymtotic of Riemann-Liouville integral

In this section we examine the asymptotic of holomorphic functions at singular points under the Riemann-Liouville transform so that the result can be applied to solutions to linear differential equations.

THEOREM 3.1. *Assume a holomorphic function $u(x)$ has an asymptotic*

$$u(x) \sim \sum_{j=0}^{k} \phi_j(x) x^\lambda \log^j x \cdot \exp\left(-\tfrac{C_1}{x^{m_1}} - \tfrac{C_2}{x^{m_2}} - \cdots - \tfrac{C_K}{x^{m_K}}\right) \text{ for } x \to e^{i\theta_0}0 \tag{3.1}$$

with $\phi_j \in \mathbb{C}[\![x]\!]$, $\theta_0 \in \mathbb{R}$, k, $K \in \mathbb{Z}_{\geq 0}$, λ, $C_j \in \mathbb{C}$, $m_j \in \mathbb{Z}$ and $m_1 > m_2 > \cdots > m_K > 0$. Here $x \to e^{i\theta_0}0$ means $e^{-i\theta_0}x \to +0$.

To define $v(x) = (I_0^\mu u)(x)$ by (2.1) *we moreover assume the condition* (2.22)

$$K = 0 \text{ and } \lambda \notin \mathbb{Z}_{<0}$$

or the condition (2.23)

$$\operatorname{Re} C_1 e^{-m_1\theta_0} > 0 \qquad (\theta_0 := \operatorname{Arg} L'(\alpha)).$$

i) *Then $v(x)$ has the asymptotic*

$$v(x) \sim \sum_{j=0}^{k} \psi_j(x) x^{\lambda_0} \log^j x \cdot \exp\Bigl(-\tfrac{C_1}{x^{m_1}} - \tfrac{C_2}{x^{m_2}} - \cdots - \tfrac{C_K}{x^{m_K}}\Bigr) \text{ for } x \to e^{i\theta_0}0. \tag{3.2}$$

Here $\psi_j \in \mathbb{C}[\![x]\!]$ and $\lambda_0 \in \mathbb{C}$. Fix $k_0 \in \mathbb{Z}$ satisfying $\phi_j(0) = 0$ for $k_0 < j \leq k$ and $-1 \leq k_0 \leq k$. Then $\psi_j(0) = 0$ for $k_0 < j \leq k$.

CASE I. Assume (2.22). Then

$$\lambda_0 = \lambda + \mu, \quad \psi_{k_0}(0) = \frac{\Gamma(\lambda+1)}{\Gamma(\lambda+\mu+1)}\phi_{k_0}(0). \tag{3.3}$$

Moreover if $k = 0$ and $\phi_0(x) = a_0 + a_1 x + a_2 x + \cdots$, then

$$\psi_0 = \frac{\Gamma(\lambda+1)}{\Gamma(\lambda+\mu+1)}\Bigl(a_0 + \frac{(\lambda+1)_1 a_1}{(\lambda+\mu+1)_1}x + \frac{(\lambda+1)_2 a_2}{(\lambda+\mu+1)_2}x^2 + \cdots\Bigr). \tag{3.4}$$

If ϕ_j are convergent power series and $u(x)$ equals the right hand side of (3.1), then ψ_j are convergent power series and $v(x)$ equals the right hand side of (3.2).

CASE II. Assume (2.23). Then

$$\lambda_0 = \lambda + (m_1 + 1)\mu \text{ and } \psi_{k_0}(0) = (m_1 C_1)^{-m_1\mu}\phi_{k_0}(0). \tag{3.5}$$

Let $c \in \mathbb{C}$ and L be a piecewise C^1-path which satisfies

$$L : [\alpha, \beta] \ni t \to \mathbb{C} \text{ with } L(\alpha) = 0 \text{ and } L(\beta) = c.$$

For $x = L(t)$ with $\alpha \le t < \beta$ we define $I_0^\mu(u)(x)$ by the path $L|_{[\alpha,t]}$. We examine $u(x)$ and $(I_0^\mu u)(x)$ when $x = L(t)$ is close to c, namely, t is close to β. We define $\operatorname{Arg}\bigl(L(s_2) - L(s_1)\bigr)$ with $\alpha \le s_1 < s_2 \le \beta$, $\theta_0 = \operatorname{Arg} L'(\alpha)$, $\theta_1 = \operatorname{Arg} L'(\beta)$ and $(c - x)^\lambda$ so that they are compatible.

We note that $L(s_1) \ne L(s_2)$ but $L(\alpha)$ may equal $L(\beta)$. Assume

$$u(x) = \bigl(a' + o(1)\bigr)(c - x)^{\lambda'} \exp\Bigl(\frac{C_1'}{(c-x)^{m_1'}} + \cdots + \frac{C_{K'}}{(c-x)^{m'_{K'}}}\Bigr) \text{ for } x \to c - e^{i\theta_1}0.$$

Here a', $C_j' \in \mathbb{C}$, $m_j \in \mathbb{Z}$ and $m_0' > m_1' > \cdots > m_{K'}' > 0$.

CASE III. Assume $K' = 0$ and $\operatorname{Re}(\lambda' + \mu) < 0$. If $L(\alpha) = L(\beta)$ and $K = 0$, we moreover assume $\operatorname{Re}\lambda > \operatorname{Re}\lambda'$. Then

$$v(x) = \frac{\Gamma(-\lambda' - \mu)}{\Gamma(-\lambda')}\bigl(a' + o(1)\bigr)(c - x)^{\lambda'+\mu} \text{ for } x \to c - e^{i\theta_1}0.$$

CASE IV. Assume $\operatorname{Re} C_1' e^{-im_1'\theta_1} > 0$. Then

$$v(x) = (m_1' C_1')^{-m_1'\mu}\bigl(a' + o(1)\bigr)(c - x)^{\lambda+(m_0'+1)\mu} \exp\Bigl(\frac{C_1'}{(c-x)^{m_1'}} + \cdots + \frac{C_{K'}}{(c-x)^{m'_{K'}}}\Bigr).$$

For $s_0 \in (\alpha, \beta)$ we put $x_0 = L(s_0)$ and define $v_0(x) = (I_{x_0}^\mu u)(x)$ by the path $L|_{[x_0,t]}$ with $x = L(t)$. Then the asymptotic of $v_0(x)$ is same as that of $v(x)$ given in Case III and Case IV.

CASE V. Assume $K' = 0$ and $\operatorname{Re}(\lambda' + \mu) > 0$ or $\operatorname{Re} C_1 e^{-im_1'\theta_1} < 0$. If $L(\alpha) = L(\beta)$ and $K = 0$, we moreover assume $\operatorname{Re}(\lambda + \mu) > 0$. Then

$$v(x) = \frac{1}{\Gamma(\mu)}\int_{e^{i\theta_0}0}^{c-e^{i\theta_1}0} u(t)(c - t)^{\mu-1}dt + o(1) \text{ for } x \to c - e^{i\theta_1}0.$$

REMARK 3.2. Under the assumption in Theorem 3.1 Case III, we have $(\check{I}_c u)(x) = \frac{\Gamma(\lambda'+1)}{\Gamma(\lambda'+\mu+1)}(a' + o(c - x))(c - x)^{\lambda'+\mu}$ and therefore

$$I_0^\mu(u) = \frac{\sin(\lambda' + \mu)}{\sin\lambda'}\check{I}_c^\mu(u) + o\bigl((c - x)^{\lambda'+\mu}\bigr) \text{ for } x \to c - e^{i\theta_1}0.$$

Note that the connection coefficient between the Riemann-Liouville transforms of a function is written by trigonometric functions. This is commonly true because we can construct local solutions by successive applications of the transforms. See the examples (6.9) and (6.18).

PROOF OF THEOREM 3.1. Since $I_0^\mu(u) = I_0^{\mu+m}(\partial^m u)$ with $m \in \mathbb{Z}_{\ge 0}$, we may assume $\operatorname{Re}\mu > 0$ to prove the theorem.

First suppose $K = 0$ and $\lambda \notin \mathbb{Z}$. Then the statement ii) follows from the value $T_{a,b}^\mu(u)(1)$ in [**O1**, Lemma 12.2] and the statement i) follows from (2.16) and the fact that $I_0^\mu(u) \in C_0^m(0, 1)$ if $u \in C_0^m(0, 1)$.

The statement i) in the theorem follows from [**O6**, Lemma 5.1]. We will show the statement Case IV in the theorem. The proof of other statements are easier.

Put $C_1' = C$, $m_1' = m$, $m_j = m_j'$ and $K' = K$ for simplicity. Suppose $\operatorname{Re} C > 0$. Let $\phi \in \mathrm{C}[0,1]$. Then

$$\begin{aligned}
I(x) &:= \int_0^x (1-t)^\lambda e^{\frac{C}{(1-t)^m}+\frac{C_2}{(1-t)^{m_2}}+\cdots+\frac{C_K}{(1-t)^{m_K}}}(x-t)^{\mu-1}\phi(t)dt \\
&= x^\mu \int_0^1 (1-xs)^\lambda e^{\frac{C}{(1-xs)^m}+\frac{C_2}{(1-xs)^{m_2}}+\cdots}(1-s)^{\mu-1}\phi(xs)ds \quad (t=xs) \\
&= x^\mu \int_0^1 (1-x+xs_1)^\lambda e^{\frac{C}{(1-x+xs_1)^m}+\frac{C_2}{(1-x+xs_1)^{m_2}}+\cdots} s_1^{\mu-1}\phi(x-xs_1)ds_1 \\
&\quad (s_1 = 1-s) \\
&= x^\mu(1-x)^\lambda \int_0^1 (1+\frac{xs_1}{1-x})^\lambda e^{\frac{C}{(1-x)^m(1+\frac{xs_1}{1-x})^m}+\frac{C_2}{(1-x)^{m_2}(1+\frac{xs_1}{1-x})^{m_2}}+\cdots} \\
&\quad \cdot s_1^{\mu-1}\phi(x-xs_1)ds_1 \\
&= x^\mu(1-x)^{\lambda+\mu}\int_0^{\frac{1}{1-x}} (1+xs_2)^\lambda e^{\frac{C}{(1-x)^m(1+xs_2)^m}+\frac{C_2}{(1-x)^{m_2}(1+xs_2)^{m_2}}+\cdots} \\
&\quad \cdot s_2^{\mu-1}\phi\big(x-x(1-x)s_2\big)ds_2 \quad (s_1 = (1-x)s_2) \\
&= x^\mu(1-x)^{\lambda+\mu}e^{\frac{C}{(1-x)^m}+\frac{C_2}{(1-x)^{m_2}}+\cdots}\int_0^{\frac{1}{1-x}}(1+xs_2)^\lambda s_2^{\mu-1} \\
&\quad \cdot e^{\frac{C}{(1-x)^m}(\frac{1}{(1+xs_2)^m}-1)+\frac{C_2}{(1-x)^{m_2}}(\frac{1}{(1+xs_2)^{m_2}}-1)+\cdots}\phi\big(x-x(1-x)s_2\big)ds_2 \\
&\quad (s_2 = (1-x)^m s_3) \\
&= x^\mu(1-x)^{\lambda+(m+1)\mu}e^{\frac{C}{(1-x)^m}+\frac{C_2}{(1-x)^{m_2}}+\cdots} \\
&\quad \cdot \int_0^{\frac{1}{(1-x)^{m+1}}}\big(1+x(1-x)^m s_3\big)^\lambda s_3^{\mu-1}e^{-f(x,(1-x)^m s_3)}\phi\big(x-x(1-x)^{m+1}s_3\big)ds_3
\end{aligned}$$

with

$$f(x,t) := \frac{C}{(1-x)^m}\left(1-\frac{1}{(1+xt)^m}\right)+\frac{C_2}{(1-x)^{m_2}}\left(1-\frac{1}{(1+xt)^{m_2}}\right)+\cdots.$$

Suppose $1 \le s_2 \le \frac{1}{1-x}$, $\frac{1}{2} < x < 1$ and $m > 0$. Then $xs_2 > \frac{1}{2}$ and

$$1-(\tfrac{2}{3})^m < 1-\frac{1}{(1+xs_2)^m} < 1.$$

Hence putting $M = \max_{0\le t\le 1}|\phi(t)|$ and $A = \frac{1}{1-x}$, we have

$$\begin{aligned}
&\lim_{x\to 1-0}\Big|\int_1^{\frac{1}{1-x}}(1+xs_2)^\lambda e^{\frac{C}{(1-x)^m}(\frac{1}{(1+xs_2)^m}-1)+\frac{C_2}{(1-x)^{m_2}}(\frac{1}{(1+xs_2)^{m_2}}-1)+\cdots} \\
&\quad \cdot\left(\frac{s_2}{(1-x)^m}\right)^{\mu-1}v\big(x-x(1-x)s_2\big)\frac{ds_2}{(1-x)^m}\Big| \\
&\le \lim_{A\to+\infty}(A-1)(1+A)^{|\lambda|}A^{(m+1)|\mu|+m}Me^{-\operatorname{Re}C(1-\frac{2^m}{3^m})A^m+|C_1|A^{m-1}+\cdots} = 0.
\end{aligned}$$

Suppose $0 \le s_2 = (1-x)^m s_3 \le 1$, $\frac{1}{2} < x < 1$, $0 < m_2 < m$ and $t > 0$. Then

$$\frac{(1+xs_2)^m - 1}{(1-x)^m(1+xs_2)^m} \ge \frac{mxs_2}{2^{m+1}(1-x)^m} = \frac{mxs_3}{2^{m+1}} \ge \frac{ms_3}{2^{m+2}},$$

$$\frac{\frac{1}{(1-x)^{m_2}}\big(1 - \frac{1}{(1+t)^{m_2}}\big)}{\frac{1}{(1-x)^m}\big(1 - \frac{1}{(1+t)^m}\big)} = (1-x)^{m-m_2}\frac{(1+t)^m - (1+t)^{m-m_2}}{(1+t)^m - 1}$$
$$< (1-x)^{m-m_2} \xrightarrow{x\to 1-0} 0.$$

Hence there exists $\delta > 0$ such that

$$|e^{-f(x,(1-x)^m s_3)}| \le e^{-\operatorname{Re} C \frac{ms_3}{2^{m+3}}} \text{ for } 0 \le s_3 \le \tfrac{1}{(1-x)^m} \text{ and } 1-\delta < x < 1.$$

Since

$$\lim_{x\to 1-0}\Big(\frac{C}{(1-x)^m}\Big(1 - \frac{1}{(1+x(1-x)^m s_3)^m}\Big) + \frac{C_2}{(1-x)^{m_2}}\Big(1 - \frac{1}{(1+x(1-x)^m s_3)^{m_2}}\Big) + \cdots\Big) = mCs_3,$$

Lebesgue's dominated convergence theorem proves

$$\lim_{x\to 1-0}\int_0^{\frac{1}{(1-x)^{m+1}}} \big(1 + x(1-x)^m s_3\big)^\lambda s_3^{\mu-1} e^{-f(x,(1-x)^m s_3)}\phi\big(x - x(1-x)^{m+1}s_3\big)ds_3$$
$$= \lim_{x\to 1-0}\int_0^{\frac{1}{(1-x)^{m}}} \big(1 + x(1-x)^m s_3\big)^\lambda s_3^{\mu-1} e^{-f(x,(1-x)^m s_3)}\phi\big(x - x(1-x)^{m+1}s_3\big)ds_3$$
$$= \int_0^\infty s_3^{\mu-1}e^{-mCs_3}\phi(1)ds_3 = (mC)^{-\mu}\int_0^\infty s^{\mu-1}e^{-s}\phi(1)ds = (mC)^{-\mu}\Gamma(\mu)\phi(1),$$

which implies

$$\frac{I(x)}{\Gamma(\mu)} = (mC)^{-\mu}\big(\phi(1) + o(1)\big)(1-x)^{\lambda+(m+1)\mu}e^{\frac{C}{(1-x)^m} + \frac{C_2}{(1-x)^{m_2}} + \cdots} \quad \text{for } x \to 1-0.$$

In the general case, we may assume that the path L is of the form (2.2) and (2.3). We may moreover assume that $x = L(t)$ with $t \in [t_{m-1}, t_m)$. Since we may assume $|(I_a^\mu u)(x) - (I^\mu_{L(t_{m-1}u)}(x)|$ is bounded for $x \to c - e^{i\theta_1}0$, we have only to examine the asymptotic $(I^\mu_{L(t_{m-1})}u)(x)$ when $x \to c - e^{i\theta_1}0$. Hence we have the statement Case IV from the estimate we have just proved. □

REMARK 3.3. i) The condition $\operatorname{Re}(\lambda' + \mu) < 0$ or $\operatorname{Re} C_1 e^{-im_0'\theta_1} > 0$ is essential for the claim ii) in the theorem, which assures that $\lim_{x\to c - e^{i\theta_1}0}|u(x)| = \infty$.

ii) If $\operatorname{Re}(\lambda' + \mu) > 0$ and $\lambda \notin \mathbb{Z}_{<0}$,

$$\big(I_0^\mu x^\lambda(1-x)^{\lambda'}\big)(1) = \frac{1}{\Gamma(\mu)}\int_0^1 x^\lambda(1-x)^{\lambda'+\mu-1}dx = \frac{\Gamma(\lambda+1)\Gamma(\lambda'+\mu)}{\Gamma(\mu)\Gamma(\lambda+\lambda'+\mu+1)}.$$

EXAMPLE 3.4. Let $u(x)$ and $v(x)$ be the functions given in Example 2.2.

When $C = 0$, the series expansion of the function (3.4) in the theorem is obtained by that of the holomorphic function $(1-x)^{\lambda'}\exp(-\frac{C'}{1-x})$ at the origin. Hence if $C = C' = 0$, the series expansion

$$(1-x)^{\lambda'} = \sum_{n=0}^\infty \frac{(-\lambda')_n}{n!}x^n$$

shows

$$v(x) = I_0^\mu\big(x^\lambda(1-x)^{\lambda'}\big) = \frac{\Gamma(\lambda+1)}{\Gamma(\lambda+\mu+1)}x^{\lambda+\mu}\sum_{n=0}^\infty \frac{(\lambda+1)_n(-\lambda')_n}{(\lambda+\mu+1)_n n!}x^n$$
$$= \frac{\Gamma(\lambda+1)}{\Gamma(\lambda+\mu+1)}x^{\lambda+\mu}F(\lambda+1,-\lambda';\lambda+\mu+1;x).$$

If $\operatorname{Re} C > 0$,

$$\lim_{x\to+0} v(x)x^{-\lambda-2\mu}e^{\frac{C}{x}} = C^{-\mu}.$$

If $C' = 0$ and $\operatorname{Re}(\lambda'+\mu) < 0$,

$$\lim_{x\to 1-0} v(x)(1-x)^{-\lambda'-\mu} = \frac{\Gamma(-\lambda'-\mu)}{\Gamma(-\lambda')}.$$

If $\operatorname{Re} C' > 0$,

$$\lim_{x\to 1-0} v(x)(1-x)^{-\lambda'-2\mu}e^{-\frac{C'}{1-x}} = (C')^{-\mu}.$$

EXAMPLE 3.5. Here we give another example of the claim i) in the theorem. Assume $\operatorname{Re} C > 0$ and $\lambda' = C' = 0$ in Example 2.2. Then $v(x)$ is

$$\begin{aligned}
(I_0^\mu x^\lambda e^{-\frac{C}{x}})(x) &= \frac{1}{\Gamma(\mu)}\int_0^x t^{\lambda+1}e^{-\frac{C}{t}}(x-t)^{\mu-1}\tfrac{dt}{t}\\
&= \frac{1}{\Gamma(\mu)}\int_1^\infty x^{\lambda+1}s_1^{-\lambda-1}e^{-C\frac{s_1}{x}}x^{\mu-1}(\tfrac{s_1-1}{s_1})^{\mu-1}\tfrac{ds_1}{s_1} \qquad (t=\tfrac{x}{s_1})\\
&= \frac{1}{\Gamma(\mu)}x^{\lambda+\mu}\int_1^\infty s_1^{-\lambda-\mu-1}(s_1-1)^{\mu-1}e^{-\frac{Cs_1}{x}}ds_1\\
&= \frac{1}{\Gamma(\mu)}x^{\lambda+\mu}e^{-\frac{c}{x}}\int_0^\infty s_1^\mu(s_1+1)^{-\lambda-\mu-1}e^{-\frac{c}{x}s_1}\tfrac{ds_1}{s_1}\\
&= \frac{1}{\Gamma(\mu)}x^{\lambda+2\mu}e^{-\frac{C}{x}}\int_0^\infty s_2^\mu(1+xs_2)^{-\lambda-\mu-1}e^{-Cs_2}\tfrac{ds_2}{s_2} \qquad (s_1=xs_2)\\
&= \frac{1}{\Gamma(\mu)}x^{\lambda+2\mu}e^{-\frac{C}{x}}\int_0^\infty \sum_{n=0}^\infty(-x)^n s_2^{\mu+n}\frac{(\lambda+\mu+1)_n}{n!}e^{-Cs_2}\tfrac{ds_2}{s_2}\\
&= \frac{C^{-\mu}}{\Gamma(\mu)}x^{\lambda+2\mu}e^{-\frac{c}{x}}\int_0^\infty\sum_{n=0}^\infty(-\tfrac{x}{C})^n\frac{(\lambda+\mu+1)_n}{n!}s^{\mu+n}e^{-s}\tfrac{ds}{s}\\
&\qquad \left(s=Cs_2,\ \int_0^\infty s^{\mu+n}e^{-s}\tfrac{ds}{s}=\Gamma(\mu+n)\right)\\
&\sim C^{-\mu}x^{\lambda+2\mu}e^{-\frac{c}{x}}\Big(\sum_{n=0}^\infty\frac{(\mu)_n(\lambda+\mu+1)_n}{n!}(-\tfrac{x}{C})^n\Big) \qquad (x\to+0)\\
&= C^{-\mu}x^{\lambda+2\mu}e^{-\frac{c}{x}}{}_2F_0(\mu,\lambda+\mu+1;-\tfrac{x}{C}).
\end{aligned}$$

EXAMPLE 3.6. We give an example related the singular point ∞.

$$\begin{aligned}
\tilde I_\infty^\mu(x^\lambda) &= x^{\mu-1}\big(I_0^\mu x^{-\mu-1-\lambda}\,|_{x\mapsto\frac{1}{x}}\big) = x^{\mu-1}\big(\tfrac{\Gamma(-\lambda-\mu)}{\Gamma(-\lambda)}x^{-1-\lambda}\,|_{x\mapsto\frac{1}{x}}\big)\\
&= \frac{\Gamma(-\lambda-\mu)}{\Gamma(-\lambda)}x^{\lambda+\mu} \qquad (x\to+\infty,\ \lambda+\mu\notin\mathbb{Z}_{\geq 0}).
\end{aligned} \tag{3.6}$$

We give other examples on the asymptotic when $x \to \infty$. Owing to (2.25), we have

$$\begin{aligned}\tilde{I}^{\mu}_{\infty}(x^{\lambda}e^{-x}) &= \frac{1}{\Gamma(\mu)}\int_x^{\infty} t^{\lambda+1}e^{-t}(t-x)^{\mu-1}\tfrac{dt}{t}\\ &= \frac{1}{\Gamma(\mu)}\int_1^{\infty} x^{\lambda+\mu}t^{\lambda}e^{-xt}(t-1)^{\mu-1}dt\\ &= \frac{x^{\lambda+\mu}e^{-x}}{\Gamma(\mu)}\int_0^{\infty}(t+1)^{\lambda}e^{-xt}t^{\mu-1}dt\\ &= \frac{x^{\lambda}e^{-x}}{\Gamma(\mu)}\int_0^{\infty}(1+\tfrac{t}{x})^{\lambda}e^{-t}t^{\mu-1}dt\\ &= \frac{x^{\lambda}e^{-x}}{\Gamma(\mu)}\int_0^{\infty}\sum_{n=0}^{\infty}\frac{(-\lambda)_n}{n!}(-\tfrac{t}{x})^n e^{-t}t^{\mu-1}dt,\end{aligned}$$

$$\begin{aligned}\tilde{I}^{\mu}_{\infty}(x^{\lambda}e^{-x}) &\sim x^{\lambda}e^{-x}\sum_{n=0}^{\infty}\frac{(-\lambda)_n(\mu)_n}{n!}(-x)^{-n}\\ &= x^{\lambda}e^{-x}{}_2F_0(-\lambda,\mu;-\tfrac{1}{x}) \quad (x\to+\infty).\end{aligned} \tag{3.7}$$

This is also obtained from Example 3.5 and the relation (2.28).

If $\operatorname{Re} C_1 > 0$, then

$$\begin{aligned}&\tilde{I}^{\mu}_{+\infty}(x^{\lambda}\exp(-C_1x^{m_1}-\cdots)) = x^{\mu-1}\big(I^{\mu}_{+0}(x^{-\mu-1}x^{-\lambda}\exp(-\tfrac{C_1}{x^{m_1}}-\cdots)\,|_{x\to\frac{1}{x}}\big)\\ &= x^{\mu-1}\big((m_1C_1)^{-\mu}\big(1+o(1)\big)x^{-\lambda-\mu-1+(m_1+1)\mu}\exp(-\tfrac{C_1}{x^{m_1}}-\cdots)\ \ (x\to+0)\,|_{x\to\frac{1}{x}}\big)\\ &= \big(1+o(1)\big)(m_1C_1)^{-\mu}x^{\lambda-(m_1-1)\mu}\exp(-C_1x^{m_1}-\cdots)\ \ (x\to+\infty).\end{aligned}$$

In general we have the following theorem by Theorem 3.1 and the equalities (2.25) and (2.28).

THEOREM 3.7. *Assume $u(x)$ has an asymptotic*

$$u(x) \sim \sum_{j=0}^{k}\phi_j(\tfrac{1}{x})x^{-\lambda}\log^j\tfrac{1}{x}\cdot\exp(-C_1x^{m_1}-C_2x^{m_2}-\cdots-C_Kx^{m_K}) \quad \text{for } x\to e^{i\theta_0}\infty \tag{3.8}$$

with $\phi_j \in \mathbb{C}[\![x]\!]$, λ, $C_j \in \mathbb{C}$, $\theta_0 \in \mathbb{R}$, $m_j \in \mathbb{Z}$ and $m_1 > m_2 > \cdots > m_K > 0$.

i) *To define $v(x) = (\tilde{I}^{\mu}_{e^{i\theta_0}\infty}u)(x)$ we assume (2.26) or (2.27). Then $v(x)$ has the asymptotic*

$$v(x) \sim \sum_{j=0}^{k}\psi_j(\tfrac{1}{x})x^{-\lambda_0}\log^j\tfrac{1}{x}\cdot\exp\Big(-C_1x^{m_1}-C_2x^{m_2}-\cdots-C_Kx^{m_K}\Big) \quad \text{for } x\to e^{i\theta_0}\infty. \tag{3.9}$$

Here $\psi_j \in \mathbb{C}[\![x]\!]$ and $\lambda_0 \in \mathbb{C}$. Fix $k_0 \in \mathbb{Z}$ satisfying $\phi_j(0) = 0$ for $k_0 < j \le k$ and $-1 \le k_0 \le k$. Then $\psi_j(0) = 0$ for $k_0 < j \le k$.

CASE I. Assume (2.22). Then

$$\lambda_0 = \lambda-\mu, \quad \psi_{k_0}(0) = \frac{\Gamma(\lambda-\mu)}{\Gamma(\lambda)}\phi_{k_0}(0). \tag{3.10}$$

Moreover if $k = 0$ and $\phi_0(x) = a_0 + a_1x + a_2x + \cdots$, then

$$\psi_0(\tfrac{1}{x}) = \frac{\Gamma(\lambda-\mu)}{\Gamma(\lambda)}\Big(a_0 + \frac{(\lambda-\mu)_1 a_1}{(\lambda)_1}x^{-1} + \frac{(\lambda-\mu)_2 a_2}{(\lambda)_2}x^{-2} + \cdots\Big). \tag{3.11}$$

If ϕ_j are convergent power series and $u(x)$ equals the right hand side of (3.8), then ψ_j are convergent power series and $v(x)$ equals the right hand side of (3.9).

CASE II. Assume (2.23). Then

$$\lambda_0 = \lambda - (m_1 - 1)\mu, \quad \psi_{k_0}(0) = (m_1C_1)^{-m_1\mu}\phi_{k_0}(0). \tag{3.12}$$

ii) *Let $c \in \mathbb{C}$ and L be a piecewise C^1-path which satisfies*

$$L : [\alpha, \infty] \ni t \to \mathbb{C} \text{ with } L(\alpha) = c \text{ and } \lim_{t\to\infty} \operatorname{Arg} L'(t) = \theta_0.$$

For $x = L(t)$ with $t \in (\alpha, \infty)$, we assume that $w(x) = (I_c^\mu u)(x)$ is defined by the path $L|_{[\alpha,t]}$.

CASE III. Assume $K = 0$ and $\operatorname{Re}\lambda < 1$. Then

$$w(x) = \frac{\Gamma(1-\lambda)}{\Gamma(1-\lambda+\mu)}\big(a + o(1)\big)x^{-\lambda+\mu} \text{ for } x \to e^{i\theta_0}\infty.$$

CASE IV. Assume $\operatorname{Re} C_1e^{-im_1\theta_0} < 0$. Then

$$w(x) = (-m_1C_1)^{-m_1\mu}\big(a + o(1)\big)x^{-\lambda-(m_1-1)\mu}\exp\big(-C_1x^{m_1} - \cdots - C_Kx^{m_K}\big)$$
$$\text{for } x \to e^{i\theta_0}\infty.$$

REMARK 3.8. Suppose $k = 0$ and $m_1 = K = 1$ in Theorem 3.7. Owing to Example 3.6, we have

$$\phi_0(x) = a_0 + a_1x + \cdots, \quad \psi_0(x) = b_0 + b_1x + \cdots$$

$$b_j = \sum_{\nu=0}^{j} a_{j-\nu}\frac{(\mu)_\nu(\lambda+\mu+1+j-\nu)_\nu}{(-C)^\nu\nu!} = \sum_{\nu=0}^{j} a_{j-\nu}\frac{(\mu)_\nu(-\lambda-\mu-j-1)_\nu}{C^\nu\nu!}.$$

4. Middle convolution of a differential operator

In this section we study the differential equation satisfied by the Riemann-Liouville transform of a solution to a linear differential equation on $P^1_{\mathbb{C}}$. The resulting equation is the middle convolution of the original equation.

For $P \in W[x]$ we put

$$\partial^K P = \partial^K \sum a_{i,j}x^i\partial^j = \sum_{i,j} c_{i,j}\partial^i\vartheta^j \tag{4.1}$$

Here we choose $K \in \mathbb{Z}_{\geq 0}$ so that $K \geq i - j$ for any (i, j) with $a_{i,j} \neq 0$. Then for $\mu \in \mathbb{C}$ we define the middle convolution $\operatorname{mc}_\mu(P)$ by

$$\operatorname{mc}_\mu(P) = \partial^{-N}\sum_{i,j} c_{i,j}\partial^i(\vartheta-\mu)^j \in W[x]. \tag{4.2}$$

Here $N \in \mathbb{Z}_{\geq 0}$ is taken to be maximal so that $\sum_{i,j} c_{i,j}\partial^i(\vartheta-\mu)^j \in \partial^N W[x]$ for $P \neq 0$. It is clear that $\operatorname{mc}_\mu(P)$ does not depend on the choice of K. Note that

$$\operatorname{mc}_{\mu'} \circ \operatorname{mc}_\mu(P) = \operatorname{mc}_{\mu+\mu'}(P) \text{ and } \operatorname{mc}_{-\mu} \circ \operatorname{mc}_\mu(P) = P \tag{4.3}$$

if $P \notin \partial W[x]$.

REMARK 4.1. For $P \in W(x)$ we define $\mathrm{mc}_\mu(P) = \mathrm{mc}_\mu(\phi P)$ with $\phi \in \mathbb{C}(x)$ so that $\phi \neq 0$, $\phi P \in W[x]$ and $\deg_x \phi P$ is minimal. In [**O1**, 1.1] we define the map R of $W(x)$ to $W[x]$ by $\mathrm{R}P = \phi P$.

THEOREM 4.2. *Let $P \in W[x]$ and $u(x)$ be a solution to $Pu = 0$ with the asymptotic expansion* (2.20)*. If $m_j \in \mathbb{Z}_{>0}$ and*

$$\lambda \notin \mathbb{Z} \text{ and } \lambda + \mu \notin \mathbb{Z} \tag{4.4}$$

or

$$\mathrm{Re}\, C_1 e^{-im_1\theta_0} > 0 \tag{4.5}$$

in (2.20)*, then*

$$\mathrm{mc}_\mu(P)(I_0^\mu u) = 0. \tag{4.6}$$

PROOF. Let $Q \in W[x]$. Then Qu has an asymptotic expansion of the form (2.20) by changing λ and $\hat\phi(x)$ by λ_0 and $\hat\phi_0(x)$. Here $\lambda_0 = \lambda + \mu \notin \mathbb{Z}$ if (4.4) is not valid. Under the notation (4.1) and (4.2)

$$0 = I_0^\mu(\partial^K Pu) = I_0^\mu(\sum c_{i,j}\partial^i\vartheta^j u) = \sum c_{i,j}\partial^i(\vartheta - \mu)I_0^\mu(u) = \partial^N \mathrm{mc}_\mu(P)I_0^\mu(u).$$

Put $u_0 = \mathrm{mc}_\mu(P)I_0^\mu(u)$. The equation $\partial^N u_0 = 0$ means $u_0 \in \mathbb{C}[x]$. On the other hand, Theorem 3.1 implies

$$u_0 \sim e^{-\frac{C_1}{x^{m_1}} - \cdots - \frac{C_K}{x^{m_K}}} \sum x^{\lambda_0}\phi_j(x)\log^j x \quad (x \to e^{i\theta_0}0)$$

and therefore $u_0 = 0$. □

REMARK 4.3. i) In the above theorem the condition $Pu = 0$ implies that $\mathrm{mc}_\mu(P)I_0^\mu(u)$ is a polynomial or 0 even without the condition $\lambda + \mu \notin \mathbb{Z}$, which follows from the above proof.

ii) Note that $\partial(x^2\partial + 1)u = 0$ with $u(x) = 1 + e^{\frac{1}{x}}$. Then $(x^2\partial + 1)u \neq 0$ and $u(x) \sim e^{\frac{1}{x}}$ for $x \to +0$ and $u(x) \sim 1$ for $x \to -0$.

We define

$$W_0[x]\partial^\mu := W[x] \otimes \mathbb{C}[\partial^{-1}] \otimes \partial^\mu,$$
$$\partial^{\mu'}\partial^{\mu'} = \partial^{\mu+\mu'} \text{ and } \partial^\mu x = x\partial^\mu + \mu\partial^{\mu-1}.$$

Then

$$W_0[x]\partial^{\mu'} \times W_0[x]\partial^\mu \ni (P', P) \mapsto P'P \in W[x]_0\partial^{\mu+\mu'}$$

and the map

$$\mathrm{Ad}(\partial^\mu) : W_0[x] \ni P \mapsto \partial^\mu P \partial^{-\mu} \in W_0[x]$$

defines an endomorphism of the ring $W_0[x]$. Then $\mathrm{mc}_\mu(P) = \partial^{N'-\mu}P\partial^\mu$. Here N' is a minimal integer so that $\partial^{N'-\mu}P\partial^\mu \in W[x]$.

REMARK 4.4. Replacing $\partial^{-\mu}$ by I_a^μ, $W_0[x]\partial^{-\mu}$ acts on a local singular solution $u(x)$ of a differential equation. Then the following calculation is valid.

$$Pu = 0 \Rightarrow P\partial^\mu\partial^{-\mu}u = 0 \Rightarrow (\partial^{N'-\mu}P\partial^\mu)(\partial^{-\mu}u) = 0 \Rightarrow \mathrm{mc}_\mu(P)I_a^\mu u = 0. \tag{4.7}$$

EXAMPLE 4.5. We calculate the equation $Pv=0$ satisfied by $v(x)=(I_0^\mu u)(x)$ with $u(x)=x^\lambda e^{-\frac{C}{x}}$, $\operatorname{Re}C>0$ and $x>0$, which is the function in Example 2.2 with $\lambda'=C'=0$. Namely, $v(x)=\frac{1}{\Gamma(\mu)}\int_0^x t^\lambda e^{-\frac{C}{t}}(x-t)^{\mu-1}dt$ satisfies $Pv=0$. Since

$$\frac{u'(x)}{u(x)}=\frac{\lambda}{x}+\frac{C}{x^2}=\frac{\lambda x+C}{x^2},$$

we have

$$\begin{aligned}
P&=\mathrm{mc}_\mu\operatorname{Ad}\bigl(x^\lambda e^{-\frac{C}{x}}\bigr)\partial\\
&=\mathrm{mc}_\mu\bigl(x^2\partial-(\lambda x+C)\bigr)\\
&=\mathrm{mc}_\mu\bigl(\partial x^2\partial-\partial(\lambda x+C)\bigr)\\
&=\mathrm{mc}_\mu\bigl(\partial x(x\partial-\lambda)-C\partial\bigr)\\
&=\mathrm{mc}_\mu\bigl((\vartheta+1)(\vartheta-\lambda)-C\partial\bigr)\\
&=(\vartheta+1-\mu)(\vartheta-\lambda-\mu)-C\partial\\
&=\vartheta^2-(\lambda+2\mu-1)\vartheta+(\mu+\lambda)(\mu-1)-C\partial\\
&=x^2\partial^2+x\partial-(\lambda-2\mu-1)x\partial+(\mu+\lambda)(\mu-1)-C\partial\\
&=x^2\partial^2-\bigl(C+(2\mu-\lambda-2)x\bigr)\partial+(\mu+\lambda)(\mu-1).
\end{aligned}$$

REMARK 4.6. To calculate mc_μ the following formula is useful

$$\begin{aligned}
x^n\partial^n&=\vartheta(\vartheta-1)\cdots(\vartheta-n+1),\\
\partial^n x^n&=(\vartheta+1)(\vartheta+2)\cdots(\vartheta+n),\\
\partial^m\vartheta^n&=(\vartheta+m)^n\partial^m,\\
\vartheta^2&=x^2\partial^2+x\partial,\ \vartheta^3=x^3\partial^3+3x^2\partial^2+x\partial,\\
\vartheta^4&=x^4\partial^4+6x^3\partial^3+7x^2\partial^2+x\partial.
\end{aligned}\tag{4.8}$$

This is easily checked by the action to the function x^λ which is an eigenvector to the operator ϑ. Calculations of $\mathrm{mc}_\mu(P)$, $\operatorname{Ad}(\phi)(P)$ and other related operations in $W(x)$ are realized by a computer program in [**O7**].

5. Differential equation on $P^1_{\mathbb{C}}$

In this section we review the Riemann scheme and the spectral type and the middle convolution mc_μ of a linear differential equation

$$Pu=0\tag{5.1}$$

on $P^1_{\mathbb{C}}$. See [**O6**] for the details.

Hereafter in this paper we assume that any singularity of the equation (5.1) is a regular singularity or an unramified irregular singularity. We will define a generalized Riemann scheme, which we denote by GRS in this paper, and a spectral type as in the case of Fuchsian differential equations defined by [**O1**].

Put

$$\lambda(x)=\lambda_0+\lambda_1x^{r_1}+\cdots+\lambda_mx^{r_m}\quad(0<r_1<\cdots<r_m).$$

Here $\lambda_j\in\mathbb{C}$. We define the *characteristic function* $e_\lambda(x)$ with the *exponent* $\lambda(x)$ by

$$e_\lambda(x):=x^{-\lambda_0}\exp\bigl(-\lambda_1\tfrac{x^{r_1}}{r_1}-\cdots-\lambda_m\tfrac{x^{r_m}}{r_m}\bigr)$$

and put $\check{e}_\lambda(x) := e_\lambda(\frac{1}{x}) = x^{\lambda_0}\exp\bigl(-\frac{\lambda_1}{r_1x^{r_1}} - \cdots - \frac{\lambda_m}{r_mx^{r_m}}\bigr)$. Then

$$\bigl(\vartheta+\lambda(x)\bigr)e_\lambda(x)=0,\quad \mathrm{Ad}\bigl(e_\lambda(x)\bigr)\vartheta=\vartheta+\lambda(x),\quad e_\lambda(x)e_\lambda(x)=e_{\lambda+\lambda}(x),$$
$$\bigl(\vartheta-\lambda(\tfrac{1}{x})\bigr)\check{e}_\lambda(x)=0,\quad \mathrm{Ad}\bigl(\check{e}_\lambda(x)\bigr)\vartheta=\vartheta-\lambda(\tfrac{1}{x}).$$

DEFINITION 5.1. Let $x = c$ be a singularity of the equation (5.1). For a polynomial $\lambda \in \mathbb{C}[x]$ and a positive integer m the equation has a *generalized characteristic exponent* $[\lambda]_{(m)}$ if the equation has formal solutions

$$u_\nu(y) = \check{e}_{\lambda+\nu}(y) + \psi_\nu(y)\check{e}_{\lambda+m}(y) \quad (\nu = 0, \dots, m-1)$$

with $\psi_\nu \in \mathbb{C}[\![x]\!]_{(\infty)}$ and

$$\mathbb{C}[\![x]\!]_{(\infty)} = \bigoplus_{j=0}^{\infty} \mathbb{C}[\![x]\!]\log^j x.$$

When $m = 1$, $[\lambda]_{(1)}$ is called a characteristic exponent of the equation and may be simply denoted by λ.

Here y is given by

$$y = \begin{cases} x - c & (c \neq \infty), \\ \frac{1}{x} & (c = \infty) \end{cases} \tag{5.2}$$

with $c = c_j$.

A *generalized Riemann scheme*, GRS in short, is the table

$$\begin{Bmatrix} x = c_0 = \infty & \cdots & x = c_p \\ [\lambda_{0,1}]_{(m_{0,1})} & \cdots & [\lambda_{p,1}]_{(m_{p,1})} \\ \vdots & \vdots & \vdots \\ [\lambda_{0,n_1}]_{(m_{0,n_1})} & \cdots & [\lambda_{p,n_p}]_{(m_{p,n_p})} \end{Bmatrix}. \tag{5.3}$$

Here

$$n = m_{j,1} + \cdots + m_{j,n_j} \qquad (j = 0, \dots, p)$$

are $(p+1)$ tuples of partitions of n. The Riemann scheme corresponding to (5.3) is given by putting

$$[\lambda]_{(m)} := \begin{pmatrix} \lambda \\ \lambda+1 \\ \vdots \\ \lambda+m-1 \end{pmatrix} \text{ and } [\lambda]_m := \begin{pmatrix} \lambda \\ \lambda \\ \vdots \\ \lambda \end{pmatrix} \in \mathbb{C}[x]^m. \tag{5.4}$$

Suppose

$$\lambda_{j,\nu'} - \lambda_{j,\nu} \notin \{0, 1, \dots, m_{j,\nu} - 1\} \quad (1 \le \nu < \nu' \le n_j,\ j = 0, \dots, p). \tag{5.5}$$

Then we define that P has GRS (5.3) if $[\lambda_{j,\nu}]_{(m_{j,\nu})}$ $(\nu = 1, \dots, n_j)$ are generalized characteristic exponents at $x = c_j$ for $j = 0, \dots, p$. (See the definition of GRS in [**O1**] when (5.5) is not valid.)

REMARK 5.2. Suppose

$$\deg(\lambda_{j,\nu} - \lambda_{j,\nu'}) > 0 \text{ or } \lambda_{j,\nu} - \lambda_{j,\nu'} \notin \mathbb{Z} \quad (1 \le \nu < \nu' \le n_j,\ j = 0, \dots, p). \tag{5.6}$$

Then P has GRS (5.3) if and only if P has the Riemann scheme corresponding to (5.3) and (5.1) has linearly independent solutions of the form $\psi(y)\check{e}_\lambda(y)$ with $\psi(x) \in \mathbb{C}[\![x]\!]$, namely, they have not any $\log y$ term.

REMARK 5.3. In **[O1]** we use the notation

$$\begin{Bmatrix} x=c & (r_1) & \cdots & (r_m) \\ \lambda_0 & \lambda_1 & \cdots & \lambda_m \end{Bmatrix} \quad \text{for} \quad \begin{Bmatrix} x=c \\ \lambda_0+\lambda_1 x^{r_1}+\cdots+\lambda_m x^{r_m} \end{Bmatrix},$$

which implies the existence of a solution $u(x) \sim x^{-\lambda_0}\exp(-\frac{\lambda_1}{r_1}x^{r_1}-\cdots-\frac{\lambda_m}{r_m}x^{r_m})$ for $x\to\infty$ when $c=\infty$.

Let $\{[\lambda_{j,1}]_{(m_{j,1})},\ldots,[\lambda_{j,n_j}]_{(m_{j,n_j})}\}$ be the set of generalized exponents of P at the singular point c_j. Then $n=m_{j,1}+\cdots+m_{j,n_j}$ be a partition of $n=\operatorname{ord}P$. For $r\in\mathbb{Z}_{\ge 0}$ we define equivalence relations $\underset{j,r}{\sim}$ between the elements of $I_n:=\{1,\ldots,n\}$ as follows. For $i\in I_n$, we put $\nu_{j,i}\in\{1,\ldots,n_j\}$ by

$$m_{j,1}+\cdots+m_{j,\nu_{j,i}-1}<i\le m_{j,1}+\cdots+m_{j,\nu_{j,i}}$$

and define

$$i\underset{j,r}{\sim}i' \iff \begin{cases}\nu_{j,i}=\nu_{j,i'} & (r=0),\\ \deg\big(\lambda_{j,\nu_{j,i}}(x)-\lambda_{j,\nu_{j,i'}}(x)\big)<r & (r\ge 1).\end{cases} \tag{5.7}$$

Let $n_{j,r}$ be the number of equivalence classes under $\underset{j,r}{\sim}$ and let R_j be the Poincaré rank of P at the singular point c_j. Then

$$n_{j,0}=n_j\ge n_{j,1}\ge\cdots\ge n_{j,R_j}\ge n_{j,R_j+1}=1.$$

By a suitable permutation of the indices $\nu\in\{1,\ldots,n_j\}$ of $m_{j,\nu}$ we may assume

$$i\le i''\le i',\ i\underset{j,r}{\sim}i'\ \Rightarrow\ i\underset{j,r}{\sim}i''.$$

Let

$$n=m_{j,1}^{(r)}+\cdots+m_{j,n_{j,r}}^{(r)} \tag{5.8}$$

be the corresponding partition of n such that $m_{j,\nu}^{(0)}=m_{j,\nu}$ and for $\nu=1,\ldots,n_{j,r}$

$$I_{j,\nu}^{(r)}:=\{\nu\in\mathbb{Z}_{\ge0}\mid m_{j,1}^{(r)}+\cdots+m_{j,\nu-1}^{(r)}<\nu\le m_{j,1}^{(r)}+\cdots+m_{j,\nu}^{(r)}\} \tag{5.9}$$

give the equivalence classes under $\underset{j,r}{\sim}$. Note that $\{I_{j,\nu}^{(r)}\mid\nu=1,\ldots,n_{j,r}\}$ is a refinement of $\{I_{j,\nu}^{(r+1)}\mid\nu=1,\ldots,n_{j,r+1}\}$. Then we define that the $(R_0+\cdots+R_p+p+1)$ tuples of partitions $\mathbf{m}=\big(\mathbf{m}_j^{(r)}\big)_{\substack{r=0,\ldots,R_j\\ j=0,\ldots,p}}=\big(m_{j,\nu}^{(r)}\big)_{\substack{\nu=1,\ldots,n_{j,r}\\ r=0,\ldots,R_j\\ j=0,\ldots,p}}$ of n is the *spectral type* of P and that of GRS (5.3). Then the number of full parameters of GRS (5.3) with the spectral type $\big(m_{j,\nu}^{(r)}\big)_{\substack{\nu=1,\ldots,n_{j,r}\\ r=0,\ldots,R_j\\ j=0,\ldots,p}}$ equals $R=\sum_{j=0}^{p}\sum_{r=0}^{R_j}n_{j,r}-1$. Here we note that we always impose the Fuchs-Hukuhara relation on GRS.

As in the case of Fuchsian differential equation, this spectral type is expressed by writing the numbers $m_{j,\nu}^{(r)}$. The numbers are separated by "," indicating different singular points and by "|" indicating different levels of the equivalence relations:

$$m_{0,1}^{(0)}m_{0,2}^{(0)}\cdots m_{0,n_{0,0}}^{(0)}|\cdots|m_{0,1}^{(R_1)}\cdots m_{0,n_{0,R_1}}^{(R_1)},m_{1,1}^{(0)}\cdots,\cdots m_{p,n_{p,R_p}}^{(R_p)}$$

The *index of the rigidity* of GRS (5.3) is defined by that of the tuples of the partitions $\big(m_{j,\nu}^{(r)}\big)$ (cf. [**O3**, **O1**]):

$$\operatorname{idx}\{\lambda_{\mathbf{m}}\} := \operatorname{idx}\mathbf{m} = \operatorname{idx}\big(m_{j,\nu}^{(r)}\big) = 2n^2 - \sum_{j=0}^{p}\sum_{r=0}^{R_j}\Big(n^2 - \sum_{\nu=1}^{n_{j,r}}\big(m_{j,\nu}^{(r)}\big)^2\Big)$$

$$(5.10)\qquad = 2n^2 - \sum_{j=0}^{p}\Big(n^2 - \sum_{\nu=1}^{n_j} m_{j,\nu}^2\Big)$$

$$- \sum_{j=0}^{p}\sum_{\nu=1}^{n_j}\sum_{\nu'=1}^{n_j} m_{j,\nu}m_{j,\nu'}\deg\big(\lambda_{j,\nu}(x) - \lambda_{j,\nu'}(x)\big).$$

Put $\operatorname{ord}\mathbf{m} = m_{j,1} + \cdots + m_{j,n_j} = n$. As in the Fuchsian case (cf. [**O1**, Definition 4.17]), *the Fuchs-Hukuhara relation* is given by

$$(5.11)\qquad \sum_{j=0}^{p}\sum_{\nu=1}^{n_j} m_{j,\nu}\lambda_{j,\nu}(0) = \operatorname{ord}\mathbf{m} - \tfrac{1}{2}\operatorname{idx}\mathbf{m}.$$

The generalized Riemann scheme of $\operatorname{mc}_\mu(P)$ is given by the following theorem.

THEOREM 5.4 ([**O1**, Theorem 5.2], [**Hi**, Theorem 3.2]). *Suppose $P \in W[x]$ has the generalized Riemann scheme $\big\{[\lambda_{j,\nu}]_{(m_{j,\nu})}\big\}_{\substack{\nu=1,\dots,n_j\\ j=0,\dots,p}}$ given in* (5.3) *and it is irreducible in $W(x)$. We may assume $\lambda_{j,1} = 0$ for $j = 1, \dots, p$ and $\mu = \lambda_{0,1} - 1$. Here some $m_{j,1}$ are allowed to be zero. If $\{\lambda_{j,\nu}\}$ and μ are generic (cf.* [**O1**, **Hi**]*) under this assumption,* $\operatorname{mc}_\mu(P)$ *has GRS $\big\{[\lambda'_{j,\nu}]_{(m'_{j,\nu})}\big\}_{\substack{\nu=1,\dots,n_j\\ j=0,\dots,p}}$ given by*

$$d(\mathbf{m}) := 2n - \sum_{j=0}^{p}\sum_{r=0}^{R_j}\big(n - m_{j,1}^{(r)}\big),$$

$$\begin{aligned}
m'_{j,\nu} &= m_{j,\nu} - \delta_{\nu,1}\cdot d(\mathbf{m}) && (1 \le \nu \le n_j,\ 0 \le j \le p),\\
\lambda'_{j,0} &= \delta_{j,0}\cdot(1-\mu) && (j = 0,\dots,p),\\
\lambda'_{j,\nu} &= \lambda_{j,\nu} + \big((-1)^{\delta_{j,0}} + \deg\lambda_{j,\nu}\big)\cdot\mu && (1 \le \nu \le n_j,\ 0 \le j \le p)
\end{aligned}$$

and the index of rigidity and the irreducibility of P are kept under mc_μ.

REMARK 5.5. The above equality $\lambda'_{j,\nu} = \lambda_{j,\nu} + \big((-1)^{\delta_{j,0}} + \deg\lambda_{j,\nu}\big)\cdot\mu$ was wrongly written as $\lambda'_{j,\nu} = \lambda_{j,\nu} + (-1)^{\delta_{j,0}}\big(1 + \deg\lambda_{j,\nu}\big)\cdot\mu$ in [**O6**, Theorem 5.3], which should be corrected as above. Similarly the equalities in [**O6**, Remark 5.2] should be

$$I_\mu x^\lambda = \frac{\Gamma(\lambda+1)}{\Gamma(\lambda+\mu+1)}x^{\lambda+\mu} \quad (x > 0),$$

$$\frac{1}{\Gamma(\mu)}\int_x^{\infty e^{i\theta_0}} u(t)(t-x)^{\mu-1}dt = \Big(\sum_{\nu=0}^{k} c'_\nu x^{-\nu} + o(x^{-k})\Big)x^{-\lambda-(m_0-1)\mu}\exp\Big(-\sum_{j=1}^{K} C_j x^{m_j}\Big)$$
$$(V'_{\theta_0,\theta_1,L} \ni x \to \infty).$$

REMARK 5.6. We can construct a differential equation with GRS (5.3) in the following way.

We may assume $m_{j,1} \ge m_{j,\nu}$ for $\nu = 1, \dots, n_j$ and $j = 0, \dots, p$. Suppose the equation $Pu = 0$ has this GRS. We apply suitable additions to this equation so that the resulting GRS satisfies $\lambda_{j,1} = 0$ for $j = 1, \ldots, p$ and $\deg \lambda_{0,1} = 0$. Then we apply $\mathrm{mc}_{\lambda_{1,0}-1}$ to this resulting equation. If the GRS is *rigid*, namely, the index of rigidity equals 2, then the order of the equation decreases by the application of this middle convolution (cf. [**O6**, Remark 5.4 ii)]). Repeating this procedure, we finally get the trivial equation $u' = 0$. Inverting this procedure, we can construct the differential equation $Pu = 0$ with a given rigid GRS together with an integral representation of solutions to the equation.

There exist 345 Fuchsian spectral types with order up to 8, which are listed in [**O1**, 13.2.3]. Hence the rigid spectral types allowing unramified singularities are easily obtained up to order 8. For example, the spectral type $421, 43, 43, 52$ in the list corresponds to the following 5 spectral types with irregular singularities

$$421|43,43,52 \quad 421,43|43,52 \quad 412|52,43,43 \quad 412|52,43|43 \quad 421|43|43,52.$$

The following is an example of the reduction of spectral types.

$$\begin{aligned}\underline{4}12|\underline{5}2,\underline{4}3,\underline{4}3 &\xrightarrow{-3} 112|22,13,13 \rightsquigarrow \underline{2}11|\underline{2}2,\underline{3}1,\underline{3}1 \xrightarrow{-2} 11|2,11,11 \\ &\rightsquigarrow \underline{1}1,\underline{1}1,\underline{1}1\,(\text{Gauss}) \xrightarrow{-1} 1,1,1\,(\text{trivial})\end{aligned}$$

The arrow $\to$ represents a middle convolution and $\rightsquigarrow$ represents an addition. This reduction gives a representation of a solution by a triple integral by means of the inverse of the above procedure. The number -3 above the first arrow indicates $-d(\mathbf{m}) = 3+2+3+3-2\cdot 7$. In general there are several way of the reduction of a given GRS. Then we have different integral representations of the local solution. See the last example in this paper.

6. Examples

In this last section we give simple examples to explain our results. The arguments in this section will work in the general case.

6.1. Gauss hypergeometric family. Gauss hypergeometric equation and its confluence are most fundamental and instructive. We examine them according to their spectral types.

CASE 1. $\underline{11,11,11}$

We may assume that the singular points of the equation with the spectral type $11, 11, 11$ are 0, 1 and ∞ by a fractional linear transformation. Then the corresponding GRS is

$$H_2 : \begin{Bmatrix} x=0 & 1 & \infty \\ \lambda_{0,1} & \lambda_{1,1} & \lambda_{2,1} \\ \lambda_{0,2} & \lambda_{1,2} & \lambda_{2,2} \end{Bmatrix} \quad \text{(the Fuchs-Hukuhara relation: } \textstyle\sum \lambda_{i,j} = 1)$$

and the equation with this GRS is uniquely determined since the spectral type is rigid. Here $\lambda_{j,\nu}$ are complex parameters. Note that there are symmetries given by $\lambda_{i,\nu} \leftrightarrow \lambda_{i,3-\nu}$ for $i \in \{0,1,2\}$ and $\nu \in \{1,2\}$. Another symmetry $(\lambda_{0,1}, \lambda_{0,2}) \leftrightarrow (\lambda_{1,1}, \lambda_{1,2})$ is induced from a fractional linear transformation. There are six symmetries of this type.

Let $u_{j,\lambda_{j,\nu}}(x)$ be the local solution characterized by

$$u_{j,\lambda_{j,\nu}}(x) = f_j(x)^{\lambda_{j,\nu}}\phi_{j,\nu}(x), \quad f_j(x) = \begin{cases} x & (j=0) \\ 1-x & (j=1) \end{cases},$$

$\phi_{j,\nu}(x)$ are holomorphic at $x=j$ and $\phi_{j,\nu}(j)=1$ for $j=0,1$ and $\nu=1,2$. Then we have

$$u_{0,\lambda_{0,2}}(x) = \sum_{\nu=1}^{2} c(0:\lambda_{0,2} \rightsquigarrow 1:\lambda_{1,\nu})u_{1,\lambda_{1,\nu}}(x) \quad (x\in(0,1))$$

Here the *connection coefficients* $c(0:\lambda_{0,2} \rightsquigarrow 1:\lambda_{1,\nu})$ holomorphically depend on $\lambda_{j,\nu}$ when $\lambda_{j,1}-\lambda_{j,2} \notin \mathbb{Z}$ for $j=0,1$. They are determined by Theorem 3.1 as follows.

A local solution corresponding to the exponent $\lambda_{0,2}$ is given by

$$v(x) = x^{\tilde\lambda}(1-x)^{\tilde\lambda'}I_0^{\mu}\big(x^{\lambda}(1-x)^{\lambda'}\big).$$

Corresponding transformation of the GRS

$$\begin{aligned}(6.1)\quad &\begin{Bmatrix} x=0 & 1 & \infty \\ \underline{\lambda} & \lambda' & -\lambda-\lambda' \end{Bmatrix} \xrightarrow{\partial^{-\mu}} \begin{Bmatrix} x=0 & 1 & \infty \\ 0 & 0 & 1-\mu \\ \underline{\lambda+\mu} & \lambda'+\mu & -\lambda-\lambda'-\mu \end{Bmatrix} \xrightarrow{\times x^{\tilde\lambda}(1-x)^{\tilde\lambda'}} \\ &\begin{Bmatrix} x=0 & 1 & \infty \\ \tilde\lambda=\lambda_{0,1} & \tilde\lambda'=\lambda_{1,1} & 1-\mu-\tilde\lambda-\tilde\lambda'=\lambda_{2,1} \\ \underline{\lambda+\mu+\tilde\lambda}=\lambda_{0,2} & \lambda'+\mu+\tilde\lambda'=\lambda_{1,2} & -\lambda-\lambda'-\mu-\tilde\lambda-\tilde\lambda'=\lambda_{2,2} \end{Bmatrix}\end{aligned}$$

follows from Theorem 5.4 (cf. [**O1**, (1.8)]).

Theorem 3.1 Case I and Case III imply

$$(6.2)\qquad v(x) = \frac{\Gamma(\lambda+1)}{\Gamma(\lambda+\mu+1)}u_{0,\lambda_{0,2}}(x) = Cu_{1,\lambda_{1,1}}(x) + \frac{\Gamma(-\lambda'-\mu)}{\Gamma(-\lambda')}u_{1,\lambda_{1,2}}(x)$$

with $C\in\mathbb{C}$ and therefore we have

$$\begin{aligned} c(0:\lambda_{0,2} \rightsquigarrow 1:\lambda_{1,2}) &= \frac{\Gamma(\lambda+\mu+1)\cdot\Gamma(-\lambda'-\mu)}{\Gamma(\lambda+1)\cdot\Gamma(-\lambda')} \\ &= \frac{\Gamma(\lambda_{0,2}-\lambda_{0,1}+1)\cdot\Gamma(\lambda_{1,1}-\lambda_{1,2})}{\Gamma(\lambda_{0,2}+\lambda_{1,1}+\lambda_{2,1})\cdot\Gamma(\lambda_{0,2}+\lambda_{1,1}+\lambda_{2,2})}. \end{aligned}$$

By the symmetry $\lambda_{1,1}\leftrightarrow\lambda_{1,2}$ we have

$$(6.3)\qquad u_{0,\lambda_{0,2}} = \sum_{\nu=1}^{2}\frac{\Gamma(\lambda_{0,2}-\lambda_{0,1}+1)\cdot\Gamma(\lambda_{1,3-\nu}-\lambda_{1,\nu})}{\Gamma(\lambda_{0,2}+\lambda_{1,3-\nu}+\lambda_{2,1})\cdot\Gamma(\lambda_{0,2}+\lambda_{1,3-\nu}+\lambda_{2,1})}u_{1,\lambda_{1,\nu}}.$$

Here we note that $C\in\mathbb{C}$ is given by Remark 3.3 ii), which corresponds to Theorem 3.1 Case V. Applying symmetries given by fractional linear transformation, we have other connection relations, which are given in [**O2**, §7 and (9.15)] in a unified form.

REMARK 6.1. i) Here the poles of $\Gamma(\lambda_{0,2}-\lambda_{0,1}+1)$ correspond to the existence of logarithmic term of the local solution $u_{0,\lambda_{0,2}}$ and the poles of $\Gamma(\lambda_{1,1}-\lambda_{1,2})$ correspond to the existence of logarithmic term of the local solution at $x=1$ with the exponent $\lambda_{1,2}$. The poles of the numerator correspond to the fact that the local solution $u_{0,\lambda_{0,2}}$ satisfies a differential equation of the first order whose characteristic exponent at $x=1$ equals $\lambda_{1,1}$. This fact holds in general for the connection coefficients of rigid Fuchsian differential equations (cf. [**O1**, Chapter 12]).

ii) The symmetries induced by $\lambda_{j,1} \leftrightarrow \lambda_{j,2}$ for $j = 1, 2$ and $x \mapsto \frac{x}{x-1}$ give 8 different Riemann-Liouville integral expression of the solution corresponding to the exponent $\lambda_{0,2}$.

iii) The condition for the irreducibility is give by

$$\lambda_{0,1} + \lambda_{1,i} + \lambda_{2,j} \notin \mathbb{Z} \quad (i = 1,\, 2 \text{ and } j = 1,\, 2).$$

Equation (3.4) in Theorem 3.1 gives the series expansion of $I_0^\mu\bigl(x^\lambda(1-x)^{\lambda'}\bigr)$ in Example 3.4, which shows

$$F(\alpha,\beta,\gamma;x) = \frac{\Gamma(\beta)}{\Gamma(\gamma)} x^{\gamma-1} I_0^{\gamma-\beta}\bigl(x^{1-\beta}(1-x)^{-\alpha}\bigr).$$

Then the GRS of $F(\alpha,\beta,\gamma;x)$ equals

$$\begin{Bmatrix} x=0 & 1 & \infty & \\ \underline{0} & 0 & \alpha & ;x \\ 1-\gamma & \gamma-\alpha-\beta & \beta & \end{Bmatrix}.$$

Owing to Theorem 4.2, we get Gauss hypergeometric equation. See [**O1**, Example 1.8 i)] for the calculation and related results. Moreover [**WW**, Chapter XIV] gives fundamental results related to Gauss hypergeometric function and [**O2**] gives an elementary approach to Gauss hypergeometric function without an integration.

CASE 2. $\underline{11|11,11}$

We may assume that the singular points of the equation with the spectral type $11|11,11$ are ∞ and 0 and moreover ∞ is an irregular singular point. Then the corresponding GRS equals

$$\begin{Bmatrix} x=0 & \infty \\ \lambda_{0,1} & \lambda_{1,1}+\alpha x \\ \lambda_{0,2} & \lambda_{1,2}+\beta x \end{Bmatrix} \quad \Bigl(\alpha \neq \beta,\ \sum_{j,\nu} \lambda_{j,\nu} = 1\Bigr).$$

We may moreover assume GRS equals

$$\begin{Bmatrix} x=0 & \infty \\ \frac12 - m & k - \frac{x}{2} \\ \frac12 + m & \underline{-k+\frac{x}{2}} \end{Bmatrix} \tag{6.4}$$

by the transformations $x \mapsto cx$ and $\mathrm{Ad}(x^\lambda e^{\lambda' x})$ with suitable c, λ, λ', m, $k \in \mathbb{C}$. We can construct the equation with this GRS and its solution by the operations, namely, additions and a middle convolution by the transformation of GRS

$$\begin{Bmatrix} x=0 & \infty \\ 0 & 0 \end{Bmatrix} \xrightarrow{x^\lambda e^{-x}} \begin{Bmatrix} x=0 & \infty \\ \lambda & x-\lambda \end{Bmatrix} \xrightarrow{\partial^{-\mu}} \begin{Bmatrix} x=0 & \infty \\ 0 & 1-\mu \\ \lambda+\mu & \underline{-\lambda+x} \end{Bmatrix}$$

$$\xrightarrow{\times(x^{-\frac{\lambda+\mu-1}{2}} e^{\frac{x}{2}})} \begin{Bmatrix} x=0 & \infty \\ \frac12 - \frac{\lambda+\mu}{2} & \frac{\lambda-\mu+1}{2} - \frac{x}{2} \\ \frac12 + \frac{\lambda+\mu}{2} & \underline{-\frac{\lambda-\mu+1}{2} + \frac{x}{2}} \end{Bmatrix},$$

$$\lambda = m+k-\tfrac12,\ \mu = m-k-\tfrac12,\ m = \tfrac{\lambda+\mu+1}{2},\ k = \tfrac{\lambda-\mu}{2}.$$

Then the local solution $v(x)$ at ∞ corresponding to the exponent $\frac{x}{2} - k$ has the asymptotic $v(x) \sim x^k e^{-\frac{x}{2}}$ for $x \to +\infty$ is the Whittaker function $W_{k,m}(x)$. In fact (cf. [**WW**, CHAPTER XVI]),

$$\begin{aligned}
W_{k,m}(x) &:= \frac{x^k e^{-\frac{x}{2}}}{\Gamma(m+\frac{1}{2}-k)} \int_0^\infty t^{m-k-\frac{1}{2}} e^{-t} (1+\tfrac{t}{x})^{m+k-\frac{1}{2}} dt \\
&= \frac{x^{-m+\frac{1}{2}} e^{-\frac{x}{2}}}{\Gamma(m+\frac{1}{2}-k)} \int_0^\infty t^{m-k-\frac{1}{2}} e^{-t} (x+t)^{m+k-\frac{1}{2}} dt \\
&= \frac{x^{-m+\frac{1}{2}} e^{\frac{x}{2}}}{\Gamma(m+\frac{1}{2}-k)} \int_x^\infty (s-x)^{m-k-\frac{1}{2}} e^{-s} s^{m+k-\frac{1}{2}} ds \\
&= x^{-m+\frac{1}{2}} e^{\frac{x}{2}} \tilde{I}_{+\infty}^{m-k+\frac{1}{2}} (x^{m+k-\frac{1}{2}} e^{-x}) \\
&\approx x^k e^{-\frac{x}{2}} \quad (x \to +\infty).
\end{aligned}$$

Here $\approx$ means the ratio of the both sides of the equation takes the value 1 at the limit and this asymptotic follows from Theorem 3.7 Case I. More precisely, (3.7) shows

$$W_{k,m}(x) \sim x^k e^{-\frac{x}{2}} {}_2F_0(\tfrac{1}{2}-m-k, \tfrac{1}{2}+m-k; -\tfrac{1}{x}) \quad \text{for } x \to +\infty. \tag{6.5}$$

It is clear that $\int_x^\infty$ in the above last integration may be replaced by $\int_x^{e^{i\theta_0}\infty}$ if $|\theta_0| < \frac{\pi}{2}$. Put $\theta_0 = \operatorname{Arg} x$. Suppose $|x| > 1$ and $\frac{\pi}{2} < |\theta_0| < \frac{3\pi}{2}$. Put $L = L_1 + C + L_2$. Here L_1 is the linear path from $e^{i\theta_0}\infty$ to $e^{i\theta_0}$, C is the path $[0,1] \ni t \mapsto e^{i(1-t)\theta_0}$ and L_2 is the path from 1 to $+\infty$. We define $\tilde{I}_{+\infty}^{\mu}$ by a part of the path L.

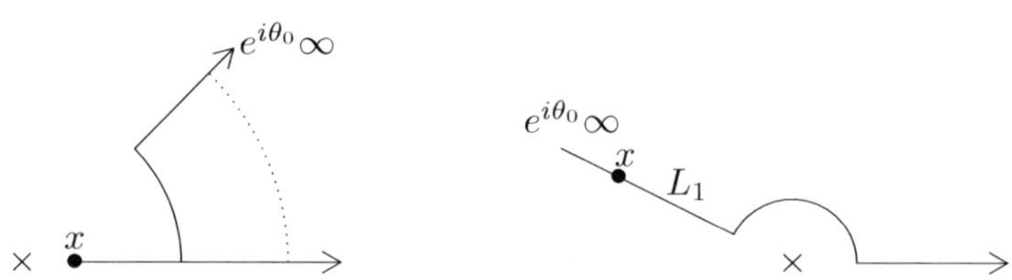

Then Theorem 3.7 Case IV implies the asymptotic $W_{k,m}(x) \approx x^k e^{-\frac{x}{2}}$ for $x \to e^{i\theta_0}\infty$. Thus we have

$$W_{k,m}(x) \approx x^k e^{-\frac{x}{2}} \text{ for } x \to e^{i\theta_0}\infty \text{ with } |\theta_0| < \tfrac{3\pi}{2} \text{ and } |\theta_0| \neq \tfrac{\pi}{2}. \tag{6.6}$$

Here it is known that the condition $|\theta_0| \neq \frac{\pi}{2}$ is not necessary (cf. [**Hu2**]).

Moreover Theorem 3.1 Case III implies

$$W_{k,m}(x) \approx \frac{\Gamma(-\lambda-\mu)}{\Gamma(-\lambda)} x^{\frac{1}{2}+m} = \frac{\Gamma(-2m)}{\Gamma(\frac{1}{2}-m-k)} x^{\frac{1}{2}+m} \quad (\operatorname{Re}(\lambda+\mu) < 0,\ x \to +0).$$

Since

$$\begin{aligned}
&\mathrm{Ad}(x^{\frac{1}{2}-m}e^{\frac{x}{2}})\mathrm{mc}_{m+k+\frac{1}{2}}\,\mathrm{Ad}(x^{m+k-\frac{1}{2}}e^{-x})\partial\\
&\quad= \mathrm{Ad}(x^{\frac{1}{2}-m}e^{\frac{x}{2}})\mathrm{mc}_{m+k+\frac{1}{2}}\big(\partial - \tfrac{m+k-\frac{1}{2}}{x} + 1\big)\\
&\quad= \mathrm{Ad}(x^{\frac{1}{2}-m}e^{\frac{x}{2}})\mathrm{mc}_{m-k+\frac{1}{2}}\big(\partial(x\partial - (m+k-\tfrac{1}{2}) + x)\big)\\
&\quad= \mathrm{Ad}(x^{\frac{1}{2}-m}e^{\frac{x}{2}})\mathrm{mc}_{m-k+\frac{1}{2}}\big(\partial x(\partial+1) - (m+k-\tfrac{1}{2})\partial)\big)\\
&\quad= \mathrm{Ad}(x^{\frac{1}{2}-m}e^{\frac{x}{2}})\big((\partial x - (m-k+\tfrac{1}{2}))(\partial+1) - (m+k-\tfrac{1}{2})\partial\big)\\
&\quad= \mathrm{Ad}(x^{\frac{1}{2}-m}e^{\frac{x}{2}})\big(x\partial^2 + (x-2m+1)\partial + k - m + \tfrac{1}{2}\big)\\
&\quad= x(\partial - \tfrac{\frac{1}{2}-m}{x} - \tfrac{1}{2})^2 + (x-2m+1)(\partial - \tfrac{\frac{1}{2}-m}{x} - \tfrac{1}{2}) + k - m + \tfrac{1}{2}\\
&\quad= x\partial^2 - \tfrac{x}{4} + k + \tfrac{\frac{1}{4}-m^2}{x},
\end{aligned}$$

$W_{k,m}(x)$ is a solution to the Whittaker equation

$$u'' = \Big(\frac{1}{4} - \frac{k}{x} - \frac{\frac{1}{4}-m^2}{x^2}\Big)u. \tag{6.7}$$

Another local solution $M_{k,m}(\mathrm{x})$ given in [**WW**, CHAPTER XVI] corresponding to the exponent $m+\frac{1}{2}$ at 0 is obtained by the integral

$$\begin{aligned}
\int_0^1 t^{m-k+\frac{1}{2}}(1-t)^{m+k-\frac{1}{2}}e^{xt}dt &= \int_0^1 (1-s)^{m-k-\frac{1}{2}}s^{m+k-\frac{1}{2}}e^{x(1-s)}ds\\
&= e^x\int_0^1 (1-s)^{m-k-\frac{1}{2}}s^{m+k+\frac{1}{2}}e^{-xs}\tfrac{ds}{s}\\
&= e^x\int_0^x (1-\tfrac{s}{x})^{m-k-\frac{1}{2}}(\tfrac{s}{x})^{m+k+\frac{1}{2}}e^{-s}\tfrac{ds}{s}\\
&= x^{-2m}e^x\int_0^x (x-s)^{m-k-\frac{1}{2}}s^{m+k-\frac{1}{2}}e^{-s}ds\\
&= x^{-2m}e^x\Gamma(m-k+\tfrac{1}{2})I_0^{m-k+\frac{1}{2}}(x^{m+k-\frac{1}{2}}e^{-x})
\end{aligned}$$

and then

$$\begin{aligned}
M_{k,m}(x) &:= \frac{x^{m+\frac12}e^{-\frac{x}{2}}\Gamma(2m+1)}{\Gamma(\frac12+m-k)\Gamma(\frac12+m+k)}\int_0^1 t^{m-k+\frac12}(1-t)^{m+k-\frac12}e^{xt}dt\\
&= \frac{x^{-m+\frac12}e^{\frac{x}{2}}\Gamma(2m+1)}{\Gamma(m+k+\frac12)}I_0^{m-k+\frac12}(x^{m+k-\frac12}e^{-x})\\
&= \frac{x^{-m+\frac12}e^{\frac{x}{2}}\Gamma(2m+1)}{\Gamma(m+k+\frac12)}I_0^{m-k+\frac12}\sum_{n=0}^{\infty}x^{m+k-\frac12+n}\frac{(-1)^n x^n}{n!}\\
&= \frac{x^{-m+\frac12}e^{\frac{x}{2}}\Gamma(2m+1)}{\Gamma(m+k+\frac12)}\sum_{n=0}^{\infty}\frac{\Gamma(m+k+\frac12+n)(-1)^n}{\Gamma(2m+n+1)n!}x^{2m+n}\\
&= x^{m+\frac12}e^{\frac{x}{2}}\sum_{n=0}^{\infty}\frac{(m+k+\frac12)_n}{(2m+1)_n n!}(-x)^n\\
&= x^{m+\frac12}e^{\frac{x}{2}}\,{}_1F_1(m+k+\tfrac12;2m+1;-x).
\end{aligned}$$

By the symmetry $m\leftrightarrow -m$ of GRS (6.4), $M_{k,-m}(x)$ is a solution to (6.7) and

$$W_{k,m}(x)=W_{k,-m}(x)=\frac{\Gamma(-2m)}{\Gamma(\frac12-m-k)}M_{k,m}(x)+\frac{\Gamma(2m)}{\Gamma(\frac12+m-k)}M_{k,-m}(x). \tag{6.8}$$

Hence we have

(6.9)

$$\begin{aligned}
\tilde I_{+\infty}^{m-k+\frac12}(x^{m+k-\frac12}e^{-x}) &= \frac{\Gamma(-2m)\Gamma(2m+1)}{\Gamma(\frac12-m-k)\Gamma(\frac12+m+k)}I_0^{m-k+\frac12}(x^{m+k-\frac12}e^{x})\\
&\quad+\frac{\Gamma(2m)\Gamma(-2m+1)}{\Gamma(\frac12+m-k)\Gamma(\frac12-m+k)}I_0^{-m-k+\frac12}(x^{-m+k-\frac12}e^{-x})\\
&= \frac{\sin(m+k-\frac12)\pi}{\sin 2m\pi}I_0^{m-k+\frac12}(x^{m+k-\frac12}e^{-x})\\
&\quad+\frac{\sin(m-k+\frac12)\pi}{\sin 2m\pi}I_0^{-m-k+\frac12}(x^{-m+k-\frac12}e^{-x}).
\end{aligned}$$

Note that GRS (6.4) has the coordinate symmetry

$$\begin{Bmatrix} x=0 & \infty & \\ \frac12-m & k-\frac{x}{2} & ;x\\ \underline{\frac12+m} & \underline{-k+\frac{x}{2}} & \end{Bmatrix}=\begin{Bmatrix} x=0 & \infty & \\ \frac12-m & k+\frac{x}{2} & ;-x\\ \underline{\frac12+m} & \underline{-k-\frac{x}{2}} & \end{Bmatrix} \tag{6.10}$$

corresponding to $(x,k)\mapsto(-x,-k)$. Hence

$$M_{-k,m}(e^{\pm\pi i}x)=e^{\pm(m+\frac12)\pi i}M_{k,m}(x) \tag{6.11}$$

and $W_{-k,m}(e^{\pm\pi i}x)$ are solutions to (6.7). Note that

$$W_{-k,m}(x)\approx x^{-k}e^{-\frac{x}{2}}\ \text{for}\ x\to e^{i\theta_0}\infty\ \text{with}\ |\theta_0|<\tfrac32\pi.$$

Then we have

(6.12)

$$\begin{aligned} W_{-k,m}(e^{\pm\pi i}x) &= \frac{\Gamma(-2m)}{\Gamma(\frac{1}{2}-m+k)} M_{-k,m}(e^{\pm\pi i}x) + \frac{\Gamma(2m)}{\Gamma(\frac{1}{2}+m+k)} M_{-k,-m}(e^{\pm\pi i}x) \\ &= \frac{\Gamma(-2m)e^{\pm(\frac{1}{2}+m)\pi i}}{\Gamma(\frac{1}{2}-m+k)} M_{k,m}(x) + \frac{\Gamma(2m)e^{\pm(\frac{1}{2}-m)\pi i}}{\Gamma(\frac{1}{2}+m+k)} M_{k,-m}(x), \\ &\approx e^{\mp k\pi i} x^{-k} e^{\frac{x}{2}} \text{ for } x \to e^{i(\theta_0 \mp \pi)}\infty \text{ with } |\theta_0| < \tfrac{3}{2}\pi. \end{aligned}$$

Theorem 3.7 Case IV shows

$$\begin{aligned} M_{k,m}(x) &= \frac{x^{-m+\frac{1}{2}} e^{\frac{x}{2}} \Gamma(2m+1)}{\Gamma(m+k+\frac{1}{2})} I_0^{m-k+\frac{1}{2}} (x^{m+k-\frac{1}{2}} e^{-x}) \\ &\approx \frac{\Gamma(2m+1)}{\Gamma(m+k+\frac{1}{2})} e^{\mp(m-k+\frac{1}{2})\pi i} x^k e^{-\frac{x}{2}} \text{ for } x \to e^{\mp\pi i}\infty \end{aligned} \tag{6.13}$$

and

$$M_{k,m}(x) = \frac{\Gamma(2m+1)e^{\mp(m-k+\frac{1}{2})\pi i}}{\Gamma(m+k+\frac{1}{2})} W_{k,m}(x) + C_{\pm} W_{-k,m}(e^{\pm\pi i}x) \tag{6.14}$$

with $C_{\pm} \in \mathbb{C}$ and $|\operatorname{Arg} x \pm \frac{\pi}{2}| < \pi$, which is determined by the asymptotic of $M_{k,m}(x)$ for $x \to +\infty$. Namely, (6.11), (6.13) and (6.12) imply

$$M_{k,m}(x) = e^{\mp(m+\frac{1}{2})\pi i} M_{-k,m}(e^{\pm\pi i}x) \approx \frac{\Gamma(2m+1)}{\Gamma(m-k+\frac{1}{2})} x^{-k} e^{\frac{x}{2}} \quad (x \to +\infty),$$

$$C_{\pm} = \frac{\Gamma(2m+1)e^{\pm k\pi i}}{\Gamma(m-k+\frac{1}{2})}. \tag{6.15}$$

On the other hand, we can also get (6.14) by (6.8) and (6.12).

Equation (6.7) is irreducible if and only if

$$m \pm k + \tfrac{1}{2} \notin \mathbb{Z}. \tag{6.16}$$

This is proved as follows. First we note that the reducibility is a closed condition. If $m \pm k + \frac{1}{2} \in \mathbb{Z}$, the relations (6.8) and (6.12) imply the reducibility because of the vanishing of one of the connection coefficients.

If the equation is reducible, there exists a solution $u(x)$ satisfies an equation of the first order. Then $u(x) \sim Cx^{\frac{1}{2}-m}$ or $Cx^{\frac{1}{2}+m}$ with $C \neq 0$ for $x \to +0$ and $u(x) \sim C'x^k e^{-\frac{x}{2}}$ or $C'x^{-k}e^{\frac{x}{2}}$ with $C' \neq 0$ for $x \to +\infty$. For an example, suppose $u(x) \sim Cx^{\frac{1}{2}-m}$ for $x \to +0$ and $u(x) \sim C'x^k e^{-\frac{x}{2}}$ for $x \to +\infty$. Since $u(x)$ has no singularity in $\mathbb{C} \setminus \{0\}$, the Fuchs-Hukuhara relation for the differential equation of the first order implies $d := m + k - \frac{1}{2} \in \mathbb{Z}_{\geq 0}$ and $u(x) = C'W_{k,m}(x)$. Note that the equation satisfied by $u(x)$ may have apparent singularities. Then d is a number of zeros of $u(x)$ on $\mathbb{C} \setminus \{0\}$ and $x^{m-\frac{1}{2}} e^{\frac{x}{2}} u(x)$ corresponds a Laguerre polynomial of degree d under a certain normalization.

CASE 3. $\underline{11|11|11}$. There is only one singular point and we may assume that it is ∞.

$$(6.17)\quad \begin{Bmatrix} x=\infty \\ ax^2+bx \end{Bmatrix} \xrightarrow{\mathrm{mc}_{-\mu}} \begin{Bmatrix} x=\infty \\ 1+\mu \\ ax^2+bx-\mu \end{Bmatrix} \to \begin{Bmatrix} x=\infty \\ -\frac{x^2}{2}+m+\frac12 \\ \underline{\frac{x^2}{2}-m+\frac12} \end{Bmatrix},\ \mu=m-\tfrac12.$$

The above last transformation is obtained by a transformation $x\mapsto C_1x+C_0$ and $\mathrm{Ad}(e^{\frac{x^2}{4}})$ with $a=1$, $b=0$ and $\mu=m-\frac12$. Note that the normalized solution $u(x)$ corresponding to the exponent $\frac{x^2}{2}-m+\frac12$ has the asymptotic $u(x)\sim x^{-m-\frac12}e^{-\frac{x^2}{2}}$ for $x\to+\infty$. Then

$$\begin{aligned} u(x) &= e^{\frac{x^2}{4}}\tilde I_{+\infty}^{-\mu}(e^{-\frac{x^2}{2}}) = \frac{e^{\frac{x^2}{4}}}{\Gamma(\mu)}\int_x^\infty e^{-\frac{t^2}{2}}(t-x)^{-\mu-1}dt \\ &= \frac{e^{\frac{x^2}{4}}}{\Gamma(\mu)}\int_0^\infty e^{-\frac{(s+x)^2}{2}}s^{-\mu-1}dt \quad (s=t-x) \\ &= \frac{e^{-\frac{x^2}{4}}}{\Gamma(\mu)}\int_0^\infty e^{-\frac{t^2}{2}-tx}t^{-\mu-1}dt \quad (s\mapsto t) \\ &= D_\mu(x), \end{aligned}$$

$$D_\mu(x)\approx x^\mu e^{-\frac{x^2}{4}} \text{ for } x\to+\infty \quad \text{(cf. Theorem 3.7 Case IV.}$$

Here the parabolic cylinder function $D_\mu(x)$ is a solution to the Weber equation $Pu=0$, where

$$\begin{aligned} P &= \mathrm{Ad}(x^{\frac{x^2}{4}})\mathrm{mc}_{-\mu}\mathrm{Ad}(e^{-\frac{x^2}{2}})\partial \\ &= \mathrm{Ad}(x^{\frac{x^2}{4}})\mathrm{mc}_{-\mu}(\partial+x) \\ &= \mathrm{Ad}(x^{\frac{x^2}{4}})\mathrm{mc}_{-\mu}(\partial^2+x\partial+1) \\ &= \mathrm{Ad}(x^{\frac{x^2}{4}})(\partial^2+x\partial+\mu+1) \\ &= (\partial-\tfrac{x}{2})^2+x(\partial-\tfrac{x}{2})+\mu+1 \\ &= \partial^2+\tfrac12+\mu-\tfrac{x^2}{4}. \end{aligned}$$

Note that $D_\mu(x)=D_\mu(-x)$ and $D_{-\mu-1}(ix)$ is another solution to the equation.

It follows from GRS (6.17) that $\sqrt{x}D_\mu(\sqrt{2x})$ satisfies the Whittaker equation and the asymptotic for $x\to+\infty$ shows

$$D_\mu(x)=2^{\frac{\mu}{2}+\frac14}x^{-\frac12}W_{\frac{\mu}{2},-\frac14}(\tfrac{x^2}{2}).$$

The condition for the irreducibility is

$$\mu=m-\tfrac12\notin\mathbb{Z}.$$

Note that $e^{-\frac{x^2}{4}}D_\mu(x)$ is the Hermite polynomial of degree μ for $\mu\in\mathbb{Z}_{\geq0}$.

Versal unfolding (Gauss hypergeometric family)

The versal equation $Pu = 0$ of Gauss hypergeometric family is given in [**O1**, Example 2.5], which is

$$\begin{aligned}
P &= \mathrm{mc}_\mu \circ \mathrm{Ad}\Big(\exp\big(-\int \frac{\lambda_1 dx}{1-c_1x} - \int \frac{\lambda_2 x dx}{(1-c_1x)(1-c_2x)}\big)\Big)\partial \\
&= \mathrm{mc}_\mu\big(\partial + \frac{\lambda_1}{1-c_1x} + \frac{\lambda_2 x}{(1-c_1x)(1-c_2x)}\big) \\
&= \mathrm{mc}_\mu\big(\partial(1-c_1x)(1-c_2x)\partial + \partial(\lambda_1(1-c_2x)+\lambda_2x)\big) \\
&= \big((1-c_1x)\partial + c_1(\mu-1)\big)\big((1-c_2x)\partial + c_2\mu\big) + \lambda_1\partial + (\lambda_2-\lambda_1c_2)(x\partial+1-\mu) \\
&= (1-c_1x)(1-c_2x)\partial^2 \\
&\quad + \big((c_1+c_2)(\mu-1) + \lambda_1 + \big(2c_1c_2(1-\mu) + \lambda_2 - \lambda_1c_2\big)x\big)\partial \\
&\quad + (\mu-1)(c_1c_2\mu + \lambda_1c_2 - \lambda_2)
\end{aligned}$$

and a solution is given by

$$\begin{aligned}
&I_c^\mu \exp\Big(-\int \frac{\lambda_1 dx}{1-c_1x} - \int \frac{\lambda_2 x dx}{(1-c_1x)(1-c_2x)}\Big) \\
&= I_c^\mu\big((1-c_1x)^{\frac{\lambda_1}{c_1}+\frac{\lambda_2}{c_1(c_1-c_2)}}(1-c_2x)^{\frac{\lambda_2}{c_2(c_2-c_1)}}\big).
\end{aligned}$$

Here $c = \frac{1}{c_1}$ or $\frac{1}{c_2}$ or ∞. If $c_1c_2(c_1-c_2) \neq 0$, the corresponding GRS is

$$\begin{Bmatrix}
x = \frac{1}{c_1} & \frac{1}{c_2} & \infty \\
0 & 0 & 1-\mu \\
\frac{\lambda_1}{c_1} + \frac{\lambda_2}{c_1(c_1-c_2)} + \mu & \frac{\lambda_2}{c_2(c_2-c_1)} + \mu & -\frac{\lambda_1}{c_1} + \frac{\lambda_2}{c_1c_2} - \mu
\end{Bmatrix}.$$

Since

$$\big(\tfrac{\lambda_2}{c_2(c_2-c_1)} + \mu\big) + \big(-\tfrac{\lambda_1}{c_1} + \tfrac{\lambda_2}{c_1c_2} - \mu\big) = -\tfrac{\lambda_1}{c_1} + \tfrac{\lambda_2}{c_1(c_2-c_1)}$$

and

$$\lim_{c_2\to 0} c_2\big(\tfrac{\lambda_2}{c_2(c_2-c_1)} + \mu\big) = -\tfrac{\lambda_2}{c_1} \text{ and } \lim_{c_2\to 0}(1-c_2x)^{-\frac{K}{c_2}} = e^{Kx},$$

we have

$$\begin{Bmatrix}
x = \frac{1}{c_1} & \infty \\
0 & 1-\mu \\
\frac{\lambda_1}{c_1} + \frac{\lambda_2}{c_1^2} + \mu & -\frac{\lambda_2}{c_1}x - \frac{\lambda_1}{c_1} - \frac{\lambda_2}{c_1^2}
\end{Bmatrix} \quad (c_1 \neq c_2 = 0),$$

which is also obtained by

$$\tfrac{\lambda_1}{1-c_1x} + \tfrac{\lambda_2 x}{1-c_1x} = \tfrac{\lambda_1}{1-c_1x} + \tfrac{\lambda_2}{c_1(1-c_1x)} - \tfrac{\lambda_2}{c_1}.$$

6.2. Hypergeometric family. The spectral type of the generalized hypergeometric function ${}_nF_{n-1}(\alpha_1,\ldots,\alpha_n;\beta_1,\ldots,\beta_{n-1};x)$ is $\overbrace{1\cdots1}^{n}, (n-1)1, \overbrace{1\cdots1}^{n}$ and it is explained in [**O1**, 13.4].

We examine a confluence of the generalized hypergeometric function ${}_3F_2$ with the following spectral type:

<u>111|21, 111</u>

The corresponding GRS is realized by

$$\begin{Bmatrix} x=\infty & 0\\ \underline{x-\lambda} & \underline{\lambda}\end{Bmatrix} \xrightarrow{\mathrm{mc}_\mu} \begin{Bmatrix} x=\infty & 0\\ 1-\mu & 0\\ \underline{x-\lambda} & \underline{\lambda+\mu}\end{Bmatrix} \xrightarrow{\mathrm{Ad}(x^{\lambda'})} \begin{Bmatrix} x=\infty & 0\\ 1-\mu-\lambda' & \lambda'\\ \underline{x-\lambda-\lambda'} & \underline{\lambda+\mu+\lambda'}\end{Bmatrix}$$

$$\xrightarrow{\mathrm{mc}_{\mu'}} \begin{Bmatrix} x=\infty & x=0\\ 1-\mu'=1-\lambda_{1,1}+\lambda_{0,1} & 0\\ 1-\mu-\lambda'-\mu'=1-\lambda_{1,2}+\lambda_{0,1} & \lambda'+\mu'=\lambda_{0,2}-\lambda_{0,1}\\ \underline{x-\lambda-\lambda'}=x-\lambda_{1,3}+\lambda_{0,1} & \underline{\lambda+\mu+\lambda'+\mu'}=\lambda_{0,3}-\lambda_{0,1}\end{Bmatrix}$$

with $\quad \mu=\lambda_{1,2}-\lambda_{0,2},\ \ \mu'=\lambda_{1,1}-\lambda_{0,1},\ \ \lambda=\lambda_{0,3}-\lambda_{1,2},\ \ \lambda'=\lambda_{0,2}-\lambda_{1,1},$

$\lambda_{0,1}+\lambda_{0,2}+\lambda_{0,3}=\lambda_{1,1}+\lambda_{1,2}+\lambda_{1.3}.$

Solutions to the corresponding equation are given by

$$\begin{aligned}
u_\infty(x) &= \tilde I^{\mu'}_{+\infty} x^{\lambda'} \tilde I^{\mu}_{+\infty}(x^\lambda e^{-x})\\
&\sim \tilde I^{\mu'}_{\infty} x^{\lambda+\lambda'} \sum_{n=1}^\infty \frac{(-\lambda)_n(\mu)_n(-1)^n}{n!} x^{-n}e^{-x}\\
&\sim \Bigl(\sum_{m=0}^\infty\sum_{n=0}^\infty \frac{(-\lambda-\lambda'+n)_m(-\lambda)_n(\mu')_m(\mu)_n}{m!n!}(-x)^{-m-n}\Bigr)x^{\lambda+\lambda'}e^{-x}\\
&= \Bigl(\sum_{m=0}^\infty\sum_{n=0}^\infty \frac{(\lambda_{0,1}-\lambda_{1,3})_{m+n}(\lambda_{1,2}-\lambda_{0,3})_n(\lambda_{1,1}-\lambda_{0,1})_m(\lambda_{1,2}-\lambda_{0,2})_n}{(\lambda_{0,1}-\lambda_{1,3})_n m!n!}\\
&\qquad \times (-\tfrac{1}{x})^{m+n}\Bigr)x^{\lambda_{1,3}-\lambda_{0,1}}e^{-x} \approx x^{\lambda_{1,3}-\lambda_{0,1}}e^{-x} \quad (x\to+\infty),\\
u_\infty(x) &\approx \frac{\Gamma(-\lambda-\mu)\cdot\Gamma(-\lambda-\lambda'-\mu-\mu')}{\Gamma(-\lambda)\cdot\Gamma(-\lambda-\lambda'-\mu)} x^{\lambda+\lambda'+\mu+\mu'}\\
&= \frac{\Gamma(\lambda_{0,2}-\lambda_{0,3})\cdot\Gamma(\lambda_{0,1}-\lambda_{0,3})}{\Gamma(\lambda_{1,2}-\lambda_{0,3})\cdot\Gamma(\lambda_{1,1}-\lambda_{0,3})} x^{\lambda_{0,3}-\lambda_{0,1}} \quad (x\to+0\\
&\qquad \text{if } \mathrm{Re}(\lambda_{0,3}-\lambda_{0,2})<0 \text{ and } \mathrm{Re}(\lambda_{0,3}-\lambda_{0,1})<0),\\
u_0(x) &:= I^{\mu'}_0 x^{\lambda'} I^{\mu}_0(x^\lambda e^{-x}) = \frac{1}{\Gamma(\mu)\cdot\Gamma(\mu')}\int_0^x t^{\lambda'}(x-t)^{\mu'}\int_0^t s^\lambda e^{-s}(t-s)^\mu \tfrac{ds}{s}\tfrac{dt}{t}\\
&= I^{\mu'}_0 x^{\lambda'} \sum_{n=0}^\infty \frac{(-1)^n\Gamma(\lambda+1+n)}{\Gamma(\lambda+\mu+1+n)n!} x^{\lambda+\mu+n}\\
&= \sum_{n=0}^\infty \frac{(-1)^n\Gamma(\lambda+1+n)\cdot\Gamma(\lambda+\lambda'+\mu+1+n)}{\Gamma(\lambda+\mu+1+n)\cdot\Gamma(\lambda+\lambda'+\mu+\mu'+1+n)} x^{\lambda+\lambda'+\mu+\mu'+n}\\
&= \frac{\Gamma(\lambda_{0,3}-\lambda_{1,2}+1)\cdot\Gamma(\lambda_{0,3}-\lambda_{1,1}+1)}{\Gamma(\lambda_{0,3}-\lambda_{0,2}+1)\cdot\Gamma(\lambda_{0,3}-\lambda_{0,1}+1)} x^{\lambda_{0,3}-\lambda_{0,1}}\\
&\quad \times {}_2F_2(\lambda_{0,3}-\lambda_{1,2}+1,\lambda_{0,3}-\lambda_{1,1}+1;\lambda_{0,3}-\lambda_{0,2}+1,\lambda_{0,3}-\lambda_{0,1}+1;-x).
\end{aligned}$$

Here we remark that the connection coefficient of the solution $u_\infty(x)$ to the local solution $u_0(x)$ at the origin equals

$$\frac{\sin(\lambda_{0,2}-\lambda_{0,3})\pi\cdot\sin(\lambda_{0,1}-\lambda_{0,3})\pi}{\sin(\lambda_{1,2}-\lambda_{0,3})\pi\cdot\sin(\lambda_{1,1}-\lambda_{0,3})\pi}. \tag{6.18}$$

Applying $\mathrm{Ad}(x^{\lambda_{0,1}})$ to our GRS, it changes into

$$\begin{Bmatrix} x=\infty & x=0 \\ 1-\lambda_{1,1} & \lambda_{0,1} \\ 1-\lambda_{1,2} & \lambda_{0,2} \\ x-\lambda_{1,3} & \lambda_{0,3} \end{Bmatrix} \quad \text{with} \quad \lambda_{0,1}+\lambda_{0,2}+\lambda_{0,3}=\lambda_{1,1}+\lambda_{1,2}+\lambda_{1,3}. \tag{6.19}$$

Then we have obtained the normalized connection coefficients

$$c(\infty : x-\lambda_{1,3} \rightsquigarrow 0 : \lambda_{0,3}) = \frac{\Gamma(\lambda_{0,2}-\lambda_{0,3})\cdot\Gamma(\lambda_{0,1}-\lambda_{0,3})}{\Gamma(\lambda_{1,2}-\lambda_{0,3})\cdot\Gamma(\lambda_{1,1}-\lambda_{0,3})}. \tag{6.20}$$

The poles of the denominator of the connection coefficient implies that $u_{\infty,x-\lambda_{1,3}}$ satisfies a differential equation of the second order.

To get a solution corresponding to another exponent at ∞, we consider the procedure

$$\begin{Bmatrix} x=\infty & 0 \\ x-\lambda & \lambda \end{Bmatrix} \xrightarrow{\mathrm{mc}_\mu} \begin{Bmatrix} x=\infty & 0 \\ 1-\mu & 0 \\ \underline{x-\lambda} & \underline{\lambda+\mu} \end{Bmatrix} \xrightarrow{\mathrm{Ad}(x^{\lambda'}e^x)} \begin{Bmatrix} x=\infty & x=0 \\ -x+1-\mu-\lambda' & \lambda' \\ \underline{-\lambda-\lambda'} & \underline{\lambda+\mu+\lambda'} \end{Bmatrix}$$

$$\xrightarrow{\mathrm{mc}_{\mu'}} \begin{Bmatrix} x=\infty & x=0 \\ 1-\mu' & 0 \\ -x+1-\mu-\lambda'-\mu' & \lambda'+\mu' \\ \underline{-\lambda-\lambda'-\mu'} & \underline{\lambda+\mu+\lambda'+\mu'} \end{Bmatrix} \xrightarrow{\mathrm{Ad}(x^{\lambda_{0,1}})}$$

$$\xrightarrow{x\mapsto -x} \begin{Bmatrix} x=\infty & x=0 \\ 1-\mu'-\lambda_{0,1}=1-\lambda_{1,1} & \lambda_{0,1} \\ x+1-\mu-\lambda'-\mu'-\lambda_{0,1}=x-\lambda_{1,3} & \lambda'+\mu'+\lambda_{0,1}=\lambda_{0,2} \\ \underline{-\lambda-\lambda'-\mu'-\lambda_{0,1}}=1-\lambda_{1,2} & \underline{\lambda+\mu+\lambda'+\mu'+\lambda_{0,1}}=\lambda_{0,3} \end{Bmatrix}$$

and put

$$\begin{aligned}
\tilde u_\infty(x) &:= \tilde I^{\mu'}_{+\infty} x^{\lambda'} e^x \tilde I^{\mu}_{+\infty}(x^\lambda e^{-x}) \\
&\sim \tilde I^{\mu'}_{+\infty}\big(x^{\lambda+\lambda'}{}_2F_0(-\lambda,\mu;-\tfrac{1}{x})\big) \quad (x\to+\infty) \\
&\sim \frac{\Gamma(-\lambda-\lambda'-\mu')}{\Gamma(-\lambda-\lambda')} x^{\lambda+\lambda'+\mu'}{}_3F_1(-\lambda,\mu,-\lambda-\lambda'-\mu';-\lambda-\lambda';-\tfrac{1}{x}) \\
&\quad (x\to+\infty), \\
\tilde u_\infty(x) &\approx \frac{\Gamma(-\lambda-\mu)\cdot\Gamma(-\lambda-\lambda'-\mu-\mu')}{\Gamma(-\lambda)\cdot\Gamma(-\lambda-\lambda'-\mu)} x^{\lambda+\lambda'+\mu+\mu'} \quad (x\to+0,\ \mathrm{Re}(\lambda+\mu)<0, \\
&\quad \mathrm{Re}(\lambda+\lambda'+\mu+\mu')<0).
\end{aligned}$$

Thus we have

$$\begin{aligned}
&c(-\infty : 1-\lambda_{1,2} \rightsquigarrow 0 : \lambda_{0,3}) \\
&\quad = e^{\lambda_{0,3}\pi i}\frac{\Gamma(\lambda_{1,1}-\lambda_{1,2}+1)\cdot\Gamma(\lambda_{0,2}-\lambda_{0,3})\cdot\Gamma(\lambda_{0,1}-\lambda_{0,3})}{\Gamma(\lambda_{0,1}-\lambda_{1,2}+1)\cdot\Gamma(\lambda_{0,2}-\lambda_{1,2}+1)\cdot\Gamma(\lambda_{1,1}-\lambda_{0,3})}
\end{aligned} \tag{6.21}$$

and

$$\begin{aligned}
&{}_3F_1(-\lambda,-\lambda-\lambda'-\mu',\mu;-\lambda-\lambda';-\tfrac{1}{x}) \\
&\quad = {}_3F_1(\lambda_{0,1}-\lambda_{1,2}+1,\lambda_{0,2}-\lambda_{1,2}+1,\lambda_{0,3}-\lambda_{1,2}+1;\lambda_{1,2}-\lambda_{1,1}+1;-\tfrac{1}{x}).
\end{aligned}$$

Here the local solution $u_{-\infty,1-\lambda_{1,2}}$ at $-\infty$ is given by

$$(6.22)\qquad \begin{aligned} u_{-\infty,1-\lambda_{1,2}}(x) &:= \frac{\Gamma(-\lambda-\lambda')}{\Gamma(-\lambda-\lambda'-\mu)}(e^{\pi i}x)^{\lambda_{0,1}}\tilde{u}_\infty(e^{\pi i}x) \\ &\approx (e^{\pi i}x)^{\lambda_{1,2}-1} \text{ for } x\to e^{-\pi i}\infty \end{aligned}$$

which is uniquely characterized by this asymptotic and then

$$(6.23)\qquad u_{-\infty,1-\lambda_{1,2}} = \sum_{\nu=1}^{3} c(-\infty:1-\lambda_{1,2}\rightsquigarrow 0:\lambda_{0,\nu})u_{0,\lambda_{0,\nu}}.$$

If $\lambda_{0,\nu}-\lambda_{1,2}+1\in\mathbb{Z}_{\le 0}$, $u_{-\infty,1-\lambda_{1,2}}$ satisfies a differential equation of the first order for $\nu=1, 2, 3$. If $\lambda_{1,1}-\lambda_{0,\nu}\in\mathbb{Z}_{\le 0}$, $u_{-\infty,1-\lambda_{1,2}}$ satisfies a differential equation of the second order for $\nu=1, 2, 3$.

By the symmetry of GRS (6.19) with respect to the exponents $\lambda_{0,\nu}$ for $\nu=1, 2, 3$ and the exponents $1-\lambda_{1,\nu}$ for $\nu=1, 2$, we have normalized 3 solutions corresponding to the 3 exponents at each singular point and the connection relation between the singular points 0 and ∞.

The above procedure is valid for the spectral type

$$\underline{\overbrace{1\cdots 1}^{n}|(n-1)1,\overbrace{1\cdots 1}^{n}}.$$

Namely, applying $\mathrm{Ad}(x^{\lambda''})\mathrm{mc}_{\mu''}$ to (6.19) several times with various (λ'',μ''), we get

$$\begin{Bmatrix} x=\infty & x=0 \\ 1-\lambda_{1,\nu}\ (1\le\nu<n) & \lambda_{0,\nu}\ (1\le\nu\le n) \\ x-\lambda_{1,n} & \end{Bmatrix}\qquad \Bigl(\sum_{\nu=1}^{n}\lambda_{0,\nu}=\sum_{\nu=1}^{n}\lambda_{1,\nu}\Bigr),$$

which is symmetric with respect to the parameters in $\{\lambda_{0,\nu}\mid 1\le\nu\le n\}$ and the parameters in $\{\lambda_{1,\nu}\mid 1\le\nu<n\}$. The local solutions are given by

$$\begin{aligned} u_{0,\lambda_{0,n}} &= x^{\lambda_{0,\nu}}{}_{n-1}F_{n-1}(\lambda_{0,n}-\lambda_{1,\nu}\ (1\le\nu<n)\ ;\ \lambda_{0,n}-\lambda_{0,\nu}\ (1\le\nu<n)\ ;\ x), \\ u_{\infty,1-\lambda_{1,1}} &= (e^{\pi i}x)^{\lambda_{1,1}-1}{}_nF_{n-2}(\lambda_{0,\nu}-\lambda_{1,1}+1\ \ (1\le\nu\le n)\ ; \\ &\qquad \lambda_{1,1}-\lambda_{1,\nu}\ (1<\nu<n)\ ;\ -\tfrac{1}{e^{\pi i}x})\qquad (x\to e^{-\pi i}\infty), \\ u_{\infty,x-\lambda_{1,n}} &= x^{\lambda_{0,1}}\tilde{I}_\infty^{\lambda_{1,1}-\lambda_{0,1}}x^{\lambda_{0,2}-\lambda_{1,1}}\tilde{I}_\infty^{\lambda_{1,2}-\lambda_{0,2}}x^{\lambda_{0,3}-\lambda_{1,2}}\cdots \\ &\qquad \cdots\tilde{I}_\infty^{\lambda_{1,n-1}-\lambda_{0,n-1}}(x^{\lambda_{0,n}-\lambda_{1,n-1}}e^{-x}) \\ &\approx x^{\lambda_{1,n}}e^{-x}\qquad (x\to+\infty) \end{aligned}$$

and the connection coefficients are

$$\begin{aligned} c(-\infty:1-\lambda_{1,1}\rightsquigarrow 0:\lambda_{0,n}) &= e^{\lambda_{0,n}\pi i} \\ &\qquad\times\frac{\prod_{1<\nu<n}\Gamma(\lambda_{1,\nu}-\lambda_{1,1}+1)\cdot\prod_{1\le\nu<n}\Gamma(\lambda_{0,\nu}-\lambda_{0,n})}{\prod_{1<\nu<n}\Gamma(\lambda_{1,\nu}-\lambda_{0,n})\cdot\prod_{1\le\nu<n}\Gamma(\lambda_{0,\nu}-\lambda_{1,1}+1)}, \\ c(\infty:x-\lambda_{1,n}\rightsquigarrow 0:\lambda_{0,n}) &= \frac{\prod_{1\le\nu<n}\Gamma(\lambda_{0,\nu}-\lambda_{0,n})}{\prod_{1\le\nu<n}\Gamma(\lambda_{1,\nu}-\lambda_{0,n})}. \end{aligned}$$

6.3. Jordan Pochhammer family. The spectral type of Jordan-Pochhammer equation is

$$\underbrace{(p-1)1,(p-1)1,\cdots,(p-1)1}_{p+1\text{blocks}} \quad (p=2 \Rightarrow \text{Gauss})$$

Integral representation of the solution $u(x)$ of the corresponding versal equation is given by

$$u(x) = I_c^\mu(\phi), \tag{6.24}$$

$$\begin{aligned}\phi(x) &= \exp\Big(-\int \sum_{k=1}^{p} \frac{\lambda_k x^{k-1}}{(1-c_1x)(1-c_2x)\cdots(1-c_kx)}\Big)\\ &= \prod_{j=1}^{p}(1-c_jx)^{\sum_{k=j}^{p}\frac{\lambda_j}{c_j\prod_{j\le\nu\le k,\,\nu\ne j}(c_j-c_\nu)}} \qquad (c_i\ne c_j\ne 0 \text{ for } i\ne j).\end{aligned} \tag{6.25}$$

The GRS of the equation is

$$\begin{Bmatrix} x=\infty & x=\frac{1}{c_j}\ (j=1,\dots,p) \\ [1-\mu]_{(p-1)} & [0]_{(p-1)} \\ \displaystyle\sum_{k=1}^{p}\frac{(-1)^k\lambda_k}{\prod_{\nu=1}^{k}c_\nu}-\mu & \displaystyle\sum_{k=j}^{p}\frac{\lambda_k}{c_j\prod_{1\le\nu\le k,\nu\ne j}(c_j-c_\nu)}+\mu \end{Bmatrix} \quad (c_i\ne c_j\ne 0 \text{ for } i\ne j),$$

$$\begin{Bmatrix} x=\infty \\ [1-\mu]_{(p-1)} \\ \displaystyle\sum_{k=1}^{p}\lambda_k x^k+(p-1)\mu \end{Bmatrix} \quad (c_1=\cdots=c_p=0).$$

Suppose $\lambda_p = 1$ and $c_1 = \cdots = c_p = 0$. In view of Remark 4.6 the corresponding equation $Pu = 0$ is given by

$$\begin{aligned} P &= \mathrm{mc}_\mu\big(\partial+\sum_{k=1}^{p}\lambda_k x^{k-1}\big)\\ &= \mathrm{mc}_\mu\big(\partial^p+\sum_{k=1}^{p}\partial^{p-1}\lambda_k x^{k-1}\big)\\ &= \mathrm{mc}_\mu\big(\partial^p+\sum_{k=1}^{p}\lambda_k(\vartheta+p-1)(\vartheta+p-2)\cdots(\vartheta+p-k+1)\partial^{p-k}\big)\\ &= \partial^p+\sum_{k=1}^{p}\lambda_k(\vartheta-\mu+p-1)(\vartheta-\mu+p-2)\cdots(\vartheta-\mu+p-k+1)\partial^{p-k}\end{aligned}$$

and we have its solutions

$$\begin{aligned} u_m(x) &= \frac{1}{\Gamma(\mu)}\int_x^{e^{\frac{2m\pi}{p}i}\infty} e^{-\sum_{k=1}^{p}\lambda_k\frac{t^k}{k}}(t-x)^{\mu-1}dt\\ &\approx x^{-(p-1)\mu}e^{-\frac{x^p}{p}} \quad (x\to e^{\frac{2m\pi}{p}i}\infty)\end{aligned}$$

for $m = 0, \ldots, p-1$.

We consider the spectral type $21|21,21,21$.

$\underline{2}1|\underline{2}1,\underline{2}1,\underline{2}1 \xrightarrow{-2} 1|1,1,1$

$$\begin{Bmatrix} x=\infty & 0 & y \\ [1-\lambda_3]_{(2)} & [0]_{(2)} & [0]_{(2)} \\ \alpha x+\lambda_2 & \lambda_0 & \lambda_1 \end{Bmatrix} \qquad (\alpha\neq 0,\ \lambda_0+\lambda_1+\lambda_2=2\lambda_3)$$

$$\xrightarrow{\mathrm{mc}_{-\lambda_3}} \begin{Bmatrix} x=\infty & 0 & y \\ \alpha x+\lambda_2 & \lambda_0-\lambda_3 & \lambda_1-\lambda_3 \end{Bmatrix}.$$

We can consider another reduction.

$21|21,21,21 \rightsquigarrow \underline{2}1|\underline{2}1,\underline{2}1,\underline{1}2 \xrightarrow{-1} 11|11,11,2 \rightsquigarrow \underline{1}1|\underline{1}1,\underline{1}1 \xrightarrow{-1} 1,1,1$

$$\begin{Bmatrix} x=\infty & 0 & y \\ [1-\lambda_3]_{(2)} & [0]_{(2)} & [0]_{(2)} \\ \alpha x+\lambda_2 & \lambda_0 & \lambda_1 \end{Bmatrix} \qquad (\lambda_0+\lambda_1+\lambda_2=2\lambda_3)$$

$$\xrightarrow{\mathrm{Ad}\left((y-x)^{-\lambda_1}\right)} \begin{Bmatrix} x=\infty & 0 & y \\ [1+\lambda_1-\lambda_3]_{(2)} & [0]_{(2)} & [-\lambda_1]_{(2)} \\ \alpha x+\lambda_1+\lambda_2 & \lambda_0 & 0 \end{Bmatrix}$$

$$\xrightarrow{mc_{\lambda_1-\lambda_3}} \begin{Bmatrix} x=\infty & 0 & y \\ 1-\lambda_1+\lambda_3 & 0 & [-\lambda_3]_{(2)} \\ \alpha x+\lambda_1+\lambda_2 & \lambda_0+\lambda_1-\lambda_3 & \end{Bmatrix}$$

$$\xrightarrow{\mathrm{Ad}(y-x)^{\lambda_3}} \begin{Bmatrix} x=\infty & 0 \\ 1-\lambda_1 & 0 \\ \alpha x+\lambda_1+\lambda_2-\lambda_3 & \lambda_0+\lambda_1-\lambda_3 \end{Bmatrix}$$

$$\xrightarrow{\mathrm{mc}_{-\lambda_1}} \begin{Bmatrix} x=\infty & 0 \\ \alpha x+\lambda_1+\lambda_2-\lambda_3 & \lambda_0-\lambda_3 \end{Bmatrix}.$$

Then the normalized local solution u_{0,λ_0} corresponding to the exponent λ_0 is given by

$$\begin{aligned} u_{0,\lambda_0} &= \frac{\Gamma(\lambda_0+1)}{\Gamma(\lambda_0-\lambda_3+1)} I_0^{\lambda_3}\left(x^{\lambda_0-\lambda_3}(1-\tfrac{x}{y})^{\lambda_1-\lambda_3}e^{-\alpha x}\right) \\ &= \frac{\Gamma(\lambda_0+1)}{\Gamma(\lambda_0-\lambda_3+1)} I_0^{\lambda_3-\lambda_1}\Big((1-\tfrac{x}{y})^{-\lambda_3} I_0^{\lambda_1}\left(x^{\lambda_0-\lambda_3}e^{-\alpha x}\right)\Big). \end{aligned}$$

Here $|y|>|x|>0$.

REMARK 6.2. i) $I_0^{\lambda_1}\left(x^{\lambda_0-\lambda_3}e^{-\alpha x}\right)$ is expressed by using the Whittaker function $M_{k,m}(x)$.

ii) If $\alpha=0$, the equation is reducible.

iii) We remark that the function $u_{0,\lambda_0}(x,y)$ satisfies a confluent KZ equation with the variables x and y which is given in [**O5**, Example 7.6] with $(x_0,x_1,x_2)=(x,y,0)$ and $a=0$. We will discuss such equations in another paper. The equation is a confluence of Appell's F_1. Note that the coordinate transformation $(x,y)\mapsto(x,\frac{x}{y})$ is an automorphism of confluent KZ equations of this type (cf. [**O3**, §6]).

References

[DR] Michael Dettweiler and Stefan Reiter, *Middle convolution of Fuchsian systems and the construction of rigid differential systems*, J. Algebra **318** (2007), no. 1, 1–24, DOI 10.1016/j.jalgebra.2007.08.029. MR2363121

[Kz] Nicholas M. Katz, *Rigid local systems*, Annals of Mathematics Studies, vol. 139, Princeton University Press, Princeton, NJ, 1996, DOI 10.1515/9781400882595. MR1366651

[Ha] Yoshishige Haraoka, *On Oshima's middle convolution*, Josai Mathematical Monographs **12** (2020), 19–51.

[Hi] Kazuki Hiroe, *Linear differential equations on* $\mathbb{P}^1$ *and root systems*, J. Algebra **382** (2013), 1–38, DOI 10.1016/j.jalgebra.2013.02.012. MR3034471
5 (1937), 123–166.

[Hu1] Masuo Hukuhara, *Sur les points singuliers des équations différentielles ordinaires du premier ordre, II* (French), Proc. Imp. Acad. Tokyo **11** (1935), no. 9, 345–347. MR1568410

[Hu2] Masuo Hukuhara, *Sur les points singuliers des équations différentielles linéaires. III* (French), Mem. Fac. Sci. Kyūsyū Imp. Univ. A **2** (1942), 125–137. MR21644

[O1] Toshio Oshima, *Fractional calculus of Weyl algebra and Fuchsian differential equations*, MSJ Memoirs, vol. 28, Mathematical Society of Japan, Tokyo, 2012. MR2986408

[O2] Toshio Oshima, *An elementary approach to the Gauss hypergeometric function*, Josai Mathematical Monographs **6** (2013) 3–23. DOI 10.20566/1344777-06-3

[O3] Toshio Oshima, *Transformations of KZ type equations*, Microlocal analysis and singular perturbation theory, RIMS Kôkyûroku Bessatsu, B61, Res. Inst. Math. Sci. (RIMS), Kyoto, 2017, pp. 141–161. MR3699873

[O4] Toshio Oshima, *Semilocal monodromy of rigid local systems*, Formal and analytic solutions of diff. equations, Springer Proc. Math. Stat., vol. 256, Springer, Cham, 2018, pp. 189–199, DOI 10.1007/978-3-319-99148-1_10. MR3872194

[O5] Toshio Oshima, *Confluence and versal unfolding of Pfaffian systems*, Josai Mathematical Monographs **12** (2020), 117–151. DOI 10.20566/1344777-12-117

[O6] Toshio Oshima, *Versal unfolding of irregular singularities of a linear differential equation on the Riemann sphere*, Publ. Res. Inst. Math. Sci. **57** (2021), no. [3-4], 893–920, DOI 10.4171/prims/57-3-6. MR4322002

[O7] Toshio Oshima, A library of computer algebra `Risa/Asir`, 2008–2021. `http://www.ms.u-tokyo.ac.jp/~oshima/`

[WW] E. T. Whittaker and G. N. Watson, *A course of modern analysis*, Cambridge Mathematical Library, Cambridge University Press, Cambridge, 1996. An introduction to the general theory of infinite processes and of analytic functions; with an account of the principal transcendental functions; Reprint of the fourth (1927) edition, DOI 10.1017/CBO9780511608759. MR1424469

Center for Mathematical and Data Science, Hirakawacho 2-3-20, Chiyodaku, Tokyo, 102-0093, Japan

Email address: `oshima@ms.u-tokyo.ac.jp`

Contemporary Mathematics
Volume **782**, 2023
https://doi.org/10.1090/conm/782/15723

The Riemann-Hilbert approach to the generating function of the higher order Airy point processes

Mattia Cafasso and Sofia Tarricone

ABSTRACT. We prove a Tracy-Widom type formula for the generating function of occupancy numbers on several disjoint intervals of the higher order Airy point processes. The formula is related to a new vector-valued Painlevé II hierarchy we define, together with its Lax pair.

1. Introduction

Let us consider the higher order Airy functions

$$\operatorname{Ai}_n(x) := \frac{1}{\pi}\int_0^\infty \cos\left(\frac{y^{2n+1}}{2n+1} + xy\right) \mathrm{d}y, \qquad x \in \mathbb{R},\, n \in \mathbb{N}, \tag{1.1}$$

and the associated kernels

$$K_n(x,y) := \int_0^\infty \operatorname{Ai}_n(x+z)\operatorname{Ai}_n(y+z)\mathrm{d}z. \tag{1.2}$$

It is easy to prove, using standard arguments in the theory of point processes (Theorem 3 in [**14**]) that the kernels K_n, for any $n \geq 1$, define a determinantal point process whose correlation functions are given by the standard formula

$$\rho_{\ell;n}(x_1,\ldots,x_\ell) := \det\left(K_n(x_i,x_j)\right)_{i,j=1}^{\ell} \qquad \ell \geq 1,$$

see Appendix A in [**6**]. The importance of these point processes stems from applications to statistical physics and combinatorics. Indeed, they are associated to new universality classes generalizing the KPZ one (case $n = 1$). These universality classes describe both the limiting behavior of the momenta of non–interacting fermions trapped in an anharmonic potential [**13**] and the one of multicritical random partitions [**4, 10, 11**].

Let us denote with

$$\zeta_1^{(n)} > \zeta_2^{(n)} > \zeta_3^{(n)} > \ldots > \zeta_j^{(n)} > \ldots$$

2020 *Mathematics Subject Classification.* Primary 34M55; Secondary 60B20.

The authors acknowledge the support of the H2020-MSCA-RISE-2017 PROJECT No. 778010 IPaDEGAN and the International Research Project PIICQ, funded by CNRS. The second author was also supported by the Fonds de la Recherche Scientifique-FNRS under EOS project O013018F.

the (random) points in the process, fix a collection $\{A_j, j = 1, \ldots, k\}$ of intervals of the form

$$A_j = (x_j, x_{j-1}), \quad \text{with } +\infty =: x_0 > x_1 > \ldots > x_k > -\infty$$

and some real constants $\alpha_1, \ldots, \alpha_k$ such that $\alpha_j \in [0,1]$ for all j. We will denote

$$\#_{A_j}^{(n)} := \#\{\zeta_\ell^{(n)} | \zeta_\ell^{(n)} \in A_j,\ \ell \geq 0\}$$

the random variable counting the number of points contained in the interval A_j, often called the *occupancy number* of A_j. We are interested in studying the generating function of the occupancy numbers for the above mentioned A_j, i.e. the function

$$F_n(\vec{x}, \vec{\alpha}) = F_n(x_1, \ldots, x_k; \alpha_1, \ldots, \alpha_k) := \mathbb{E}\left[\prod_{j=1}^{k} (1-\alpha_j)^{\#_{A_j}^{(n)}}\right], \tag{1.3}$$

whose derivatives give the joint probability law of k given particles in the process. More precisely, given $1 \leq m_1 < \ldots < m_k$,
(1.4)

$$\mathbb{P}\left(\bigcap_{j=1}^{k} \left(\zeta_{m_j}^{(n)} < x_j\right)\right) = \sum \frac{(-1)^{j_1+\ldots j_k}}{j_1! j_2! \ldots j_k!} \frac{\partial^{j_1+j_2+\ldots+j_k}}{\partial \alpha_1^{j_1} \partial \alpha_2^{j_2} \ldots \partial \alpha_k^{j_k}} F_n(\vec{x}, \vec{\alpha})\Big|_{\vec{\alpha}=(1,\ldots,1)},$$

where the sum is taken over all indices $j_1, \ldots, j_k$ satisfying the conditions

$$j_1 < m_1,\ j_1 + j_2 < m_2, \cdots, \sum_{\ell=1}^{k} j_\ell < m_k.$$

(see for instance [**2**]). We also anticipate that $F_n(\vec{x}, \vec{\alpha})$ can be expressed as a Fredholm determinant, see (2.1) below and [**14**, Theorem 2]. This characterization is the first step needed in order to obtain the main result of this paper: a Tracy–Widom formula for $F_n(\vec{x}, \vec{\alpha})$, relating the latter to a vector–valued version of the Painlevé II hierarchy we are going to define. Our formula generalizes both the one obtained by Claeys and Doeraene [**7**] for the case $n = 1$ and arbitrary $k \geq 1$, and the one obtained by one of the authors, Claeys and Girotti for arbitrary $n \geq 1$ and $k = 1$ [**6**].

In order to state precisely our result, we need to introduce some (vector–valued) differential polynomials which will be used to define our hierarchy of equations. We will work with the ring

$$\mathcal{R} := \mathbb{C}[u_1, \ldots, u_k, Du_1, \ldots, Du_k, D^2u_1, \ldots D^2u_k, \ldots]$$

generated by k functions $u_j : \mathbb{R} \ni t \mapsto u_j(t)$, $j = 1, \ldots, k$ and its derivatives, and denote with D^{-1} the left–inverse of the derivation, such that $D^{-1}Dv = v$ for all v in $\mathrm{Im}(D)$. Given $\vec{v}, \vec{w} \in \mathcal{R}^k$, let us define

$$< \vec{v}, \vec{w} > := \vec{v}^\top \vec{w} \in \mathcal{R}, \quad \{\vec{v}, \vec{w}\} := \vec{v}\vec{w}^\top + \vec{w}\vec{v}^\top \in \mathrm{Mat}(k, \mathcal{R}),$$

$$[\vec{v}, \vec{w}] := \vec{v}\vec{w}^\top - \vec{w}\vec{v}^\top \in \mathrm{Mat}(k, \mathcal{R}).$$

We will also denote $\vec{u} := (u_1, \ldots, u_k)^\top \in \mathcal{R}^k$ and with $D(\mathcal{R}) = \mathrm{Im}(D)$ the set of total derivatives of any differential polynomials in $\mathcal{R}$. Analogously for $D(\mathrm{Mat}(k, \mathcal{R}))$.

DEFINITION 1.1. Suppose that $\vec{v} \in \mathcal{R}^k$ is such that

$$< \vec{u}, \vec{v} > \in D(\mathcal{R}) \quad \text{and} \quad \{\vec{u}, \vec{v}\} \in D(\mathrm{Mat}(k, \mathcal{R})).$$

We define

$$\mathcal{L}_+^{\vec{u}} \vec{v} := \mathrm{i}\, D\vec{v} - \mathrm{i}\left(D^{-1}\{\vec{u}, \vec{v}\}\right)\vec{u} - 2\mathrm{i}\left(D^{-1} < \vec{u}, \vec{v} >\right)\vec{u}.$$

Analogously, for any $\vec{v}$ such that $[\vec{u}, \vec{v}] \in D(\mathrm{Mat}(k, \mathcal{R}))$, we define

$$\mathcal{L}_-^{\vec{u}} \vec{v} := \mathrm{i}\, D\vec{v} + \mathrm{i}\left(D^{-1}[\vec{u}, \vec{v}]\right)\vec{u}.$$

THEOREM 1.2. *Let $F_n(\vec{x}, \vec{\alpha})$ defined as in (1.3) with $\alpha_j \neq \alpha_{j+1}$ for all $j \leq k-1$, and $\alpha_{k+1} \equiv 0$. Then*

$$F_n(\vec{x}, \vec{\alpha}) = \exp\left(-\int_0^\infty t < \vec{u}(t), \vec{u}(t) > \mathrm{d}t\right), \tag{1.5}$$

where $\vec{u}(t) = \vec{u}(n, \vec{x}+t, \vec{\alpha})$ satisfies the following (vector-valued) ordinary differential equation

$$\left(\mathcal{L}_+^{\vec{u}} \mathcal{L}_-^{\vec{u}}\right)^n \vec{u}(t) = -\,\mathrm{diag}\left(x_1 + t, \ldots, x_k + t\right)\vec{u}(t) \tag{1.6}$$

and have the following behavior at $+\infty$

$$\vec{u}(n, \vec{x} + t, \vec{\alpha}) = \left(\sqrt{\alpha_j - \alpha_{j+1}}\, \mathrm{Ai}_n(t + x_j)(1 + o(1))\right)_{j=1,\ldots,k}. \tag{1.7}$$

Moreover, if $\alpha_{j+1} < \alpha_j$ then $u_j(n, \vec{x}+t, \vec{\alpha})$ is real-valued for real t. If $\alpha_{j+1} > \alpha_j$, then $u_j(n, \vec{x} + t, \vec{\alpha})$ is purely imaginary for real t.

REMARK 1.3. We will call the collection of equations in (1.6) the *vector valued PII hierarchy*. Their formulation, from an algebraic point of view, is completely analogous to the one we previously introduced, in collaboration with Thomas Bothner [**5**], for the integro-differential Painlevé II hierarchy, see also [**12**]. Note, however, that the results contained in this article cannot be deduced from the ones in [**5**], because of the assumption of smoothness for the weight function w, see Section 1.3 in loc.cit.

REMARK 1.4. We write down explicitly the first two members of the hierarchy (1.6), using the shorthand notation $\dot{\vec{u}} = D\vec{u}$ to denote the derivative of $\vec{u}$. For $n = 1$, in order to write down equation (1.6), we compute (omitting the dependence on t)

$$\begin{aligned}\left(\mathcal{L}_+^{\vec{u}} \mathcal{L}_-^{\vec{u}}\right)\vec{u} = \mathcal{L}_+^{\vec{u}}\left(\mathcal{L}_-^{\vec{u}}\vec{u}\right) = \mathcal{L}_+^{\vec{u}}\left(\mathrm{i}\dot{\vec{u}}\right) &= -\ddot{\vec{u}} + D^{-1}D(\vec{u}\dot{\vec{u}}^{\top})\vec{u} + 2D^{-1}D\left(\frac{\vec{u}^{\top}\vec{u}}{2}\right)\vec{u} \\ &= -\ddot{\vec{u}} + 2\vec{u}\vec{u}^{\top}\vec{u}.\end{aligned}$$

Thus, equation (1.6) for $n = 1$ reads as

$$\ddot{\vec{u}} = 2\vec{u}\vec{u}^{\top}\vec{u} + \mathrm{diag}\left(x_1 + t, \ldots, x_k + t\right)\vec{u} \text{ i.e. } \begin{cases} \ddot{u}_1 = 2u_1 \sum_{j=1}^k u_j^2 + (t + x_1)u_1 \\ \ddot{u}_2 = 2u_2 \sum_{j=1}^k u_j^2 + (t + x_2)u_2 \\ \vdots \\ \ddot{u}_k = 2u_k \sum_{j=1}^k u_j^2 + (t + x_k)u_k \end{cases} \tag{1.8}$$

that coincide indeed with the coupled system of Painlevé II equations introduced in [**7**]. For $n = 2$, equation (1.6) is

$$\ddddot{\vec{u}} = 4\ddot{\vec{u}}\vec{u}^{\top}\vec{u} + 8\dot{\vec{u}}\dot{\vec{u}}^{\top}\vec{u} + 6\vec{u}\dot{\vec{u}}^{\top}\dot{\vec{u}} + 2u\ddot{\vec{u}}^{\top}\vec{u} - 6\vec{u}(\vec{u}^{\top}\vec{u})^2 - (t + \vec{x})\vec{u} \tag{1.9}$$

which is indeed a vector-valued version of the second member of the Painlevé II hierarchy, see e.g. equation (1.6) in [**6**].

REMARK 1.5. The composition of the two operators $\mathcal{L}_{\pm}^{\vec{u}}$ that appears in the definition of the hierarchy (1.6) should be intended as the vector-valued analogue of the recursion operator used in [**1**] to define the higher order Painlevé II equations. These operators are also very similar to the operators used in [**15**] to define the vector-valued Schrödinger hierarchy.

The paper is organised as follows: in Section 2 we prove that the generating function $F_n(\vec{x}+t,\vec{\alpha})$ is equal to the Fredholm determinant of an integrable operator of IIKS type [**9**]. As a byproduct, we formulate and use the Riemann-Hilbert problem 2.3 to compute the logarithmic derivative of $F_n(\vec{x}+t,\vec{\alpha})$ with respect to t, and this concludes Section 2. In Section 3 we associate to the Riemann-Hilbert problem 2.3 a Lax pair for the vector-valued Painlevé II hierarchy (1.6). Section 4 concludes, collecting all the previous results, the proof of Theorem 1.2.

2. $F_n(\vec{x},\vec{\alpha})$ and the associated Riemann-Hilbert problem

It is well known (see for instance [**14**]) that the generating function $F_n(\vec{x},\vec{\alpha})$ defined in (1.3) can be expressed as a Fredholm determinant. More precisely,

$$F_n(\vec{x},\vec{\alpha}) = \det\left(\mathrm{I} - \sum_{j=1}^{k} \alpha_j \mathbb{K}_{n|A_j}\right), \tag{2.1}$$

where $\mathbb{K}_n$ is the integral operator associated to the kernel (1.2) and, for any Borel subset $B \subseteq \mathbb{R}$, $\mathbb{K}_{n|B}$ indicates the restriction of $\mathbb{K}_n$ to B. For our purposes, it is convenient to recall a different representation of the kernel K_n written in (1.2) as a double contour integral. In what follows, let us denote

$$\psi_n(\lambda;t) := \frac{\lambda^{2n+1}}{2n+1} + \lambda t. \tag{2.2}$$

It is easy to show (see for instance [**5**]) that, for any real t,

$$\mathrm{Ai}_n(t) = \frac{1}{2\pi}\int_{\Gamma_+} \exp\Big(\mathrm{i}\psi_n(t;\lambda)\Big)\mathrm{d}\lambda = \frac{1}{2\pi}\int_{\Gamma_-} \exp\Big(-\mathrm{i}\psi_n(t;\lambda)\Big)\mathrm{d}\lambda, \tag{2.3}$$

where Γ_+ is any smooth contour oriented from $\infty e^{\mathrm{i}a}$ to $\infty e^{\mathrm{i}b}$ with $a \in \left(\frac{2n\pi}{2n+1},\pi\right)$ and $b \in (0,\frac{\pi}{2n+1})$ (see Fig. 1), and Γ_- its reflection with respect to the real axis. Actually, one can take (2.3) as an alternative definition of Ai_n (or, rather, of its analytical continuation). Then, combining (1.2) with (2.3), one proves [**5,6**]

LEMMA 2.1. *The kernel defined in* (1.2) *admits the double–contour integral representation*

$$K_n(x,y) = \frac{\mathrm{i}}{(2\pi)^2}\int_{\Gamma_+}\mathrm{d}\lambda \int_{\Gamma_-}\mathrm{d}\mu\, \frac{e^{\mathrm{i}(\psi_n(\lambda;x)-\psi_n(\mu;y))}}{\lambda-\mu}. \tag{2.4}$$

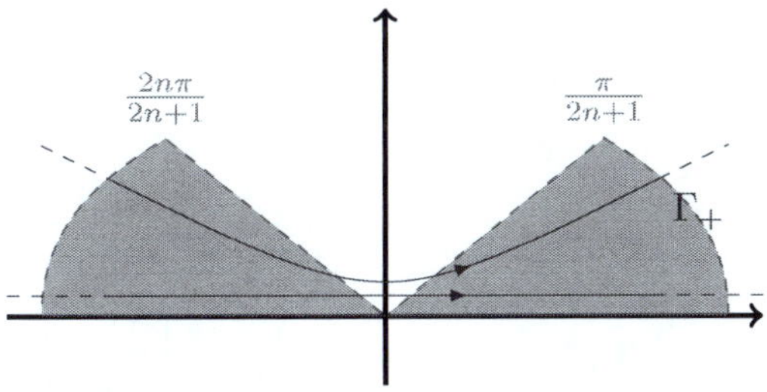

FIGURE 1. These are possible choices for the curve Γ_+ appearing the integral representation of the Airy function Ai_n.

We now define two vector–valued functions[1] $f_n, g_n : \Gamma \longrightarrow \mathbb{R}^{k+1}$, with $\Gamma := \Gamma_+ \cup \Gamma_-$:

$$
\begin{aligned}
f_n(\lambda) &:= \frac{1}{2\pi}\begin{pmatrix} \epsilon^{-\frac{i}{2}\psi_n(\lambda;0)}\chi_{\Gamma_-}(\lambda) \\ \hline \sqrt{\alpha_1-\alpha_2}\,\epsilon^{\frac{i}{2}\psi_n(\lambda;2t+2x_1)}\chi_{\Gamma_+}(\lambda) \\ \vdots \\ \sqrt{\alpha_k-\alpha_{k+1}}\,\epsilon^{\frac{i}{2}\psi_n(\lambda;2t+2x_k)}\chi_{\Gamma_+}(\lambda) \end{pmatrix}, \\
g_n(\lambda) &:= \begin{pmatrix} \epsilon^{\frac{i}{2}\psi_n(\lambda;0)}\chi_{\Gamma_+}(\lambda) \\ \hline \sqrt{\alpha_1-\alpha_2}\,\epsilon^{-\frac{i}{2}\psi_n(\lambda;2t+2x_1)}\chi_{\Gamma_-}(\lambda) \\ \vdots \\ \sqrt{\alpha_k-\alpha_{k+1}}\,\epsilon^{-\frac{i}{2}\psi_n(\lambda;2t+2x_k)}\chi_{\Gamma_-}(\lambda) \end{pmatrix}
\end{aligned}
\tag{2.5}
$$

and denote with $\mathbb{L}_n : L^2(\Gamma) \longrightarrow L^2(\Gamma)$ the associated integral operator with integrable (in the sense of [**9**]) kernel defined by the equation

$$
(\lambda-\mu)L_n(\lambda,\mu) = f_n^\top(\lambda)g_n(\mu). \tag{2.6}
$$

The following proposition is a generalization of Proposition 2.1 in [**6**] (see also [**3**] for the case $n=1$ and k arbitrary).

PROPOSITION 2.2. *The generating function $F_n(\vec{x}+t,\vec{\alpha})$ coincide with the Fredholm determinant of the integrable operator $\mathbb{L}_n$, i.e.*

$$
\det(\mathrm{I}-\mathbb{L}_n) = F_n(\vec{x}+t,\vec{\alpha}). \tag{2.7}
$$

[1]The notation we used in (2.5) reflects the fact that it will be convenient, in the sequel, to think about f_n, g_n as functions taking values in $\mathbb{R}\oplus\mathbb{R}^k$

PROOF. From the very definition of $\mathbb{L}_n$, using the natural polarization of $L^2(\Gamma) \simeq L^2(\Gamma_+) \oplus L^2(\Gamma_-)$, we can write in block form

$$\mathrm{I} - \mathbb{L}_n = \begin{pmatrix} \mathrm{I} & -\mathbb{F}_n \\ -\sum_{j=1}^{k} (\alpha_j - \alpha_{j+1}) \mathbb{G}_{n;x_j} & \mathrm{I} \end{pmatrix}$$

where $\mathbb{F}_n : L^2(\Gamma_+) \longrightarrow L^2(\Gamma_-)$, $\mathbb{G}_{n;x_j} : L^2(\Gamma_-) \longrightarrow L^2(\Gamma_+)$, $j = 1, \cdots, k$ and $\mathbb{F}_n, \mathbb{G}_{n;x_j}$ have kernels given by

$$(\mu - \lambda) F_n(\mu, \lambda) = \frac{1}{2\pi} \mathrm{e}^{\frac{i}{2}(\psi_n(\lambda;0) - \psi_n(\mu;0))} \chi_{\Gamma_-}(\mu) \chi_{\Gamma_+}(\lambda),$$

$$(\xi - \mu) G_{n;x_j}(\xi, \mu) = \frac{1}{2\pi} \mathrm{e}^{\frac{i}{2}(\psi_n(\xi;2t+2x_j) - \psi_n(\mu;2t+2x_j))} \chi_{\Gamma_+}(\xi) \chi_{\Gamma_-}(\mu).$$

We consider the two corresponding operators $\mathbb{F}_n, \mathbb{G}_{n;x_j}$ extended to the whole space $L^2(\Gamma) = L^2(\Gamma_+) \oplus L^2(\Gamma_-)$, acting trivially on the respective orthogonal component. We first notice that both the operators $\mathbb{F}_n, \mathbb{G}_{n;x_j}$ are Hilbert-Schmidt on the whole space. Indeed:

$$\|\mathbb{F}_n\|_2^2 = \frac{1}{(2\pi)^2} \int_{\Gamma_+} |\mathrm{d}\lambda| \int_{\Gamma_-} |\mathrm{d}\mu| \frac{\mathrm{e}^{-\Im(\psi_n(\lambda;0) - \psi_n(\mu;0))}}{|\mu - \lambda|^2} < +\infty \tag{2.8}$$

and also

$$\|\mathbb{G}_{n;x_j}\|_2^2 = \frac{1}{(2\pi)^2} \int_{\Gamma_-} |\mathrm{d}\mu| \int_{\Gamma_+} |\mathrm{d}\xi| \frac{\mathrm{e}^{-\Im(\psi_n(\xi;2t+2x_j) - \psi_n(\mu;2t+2x_j))}}{|\xi - \mu|^2} < +\infty. \tag{2.9}$$

Moreover, they are both trace-class, since they both can be obtained as composition of Hilbert-Schmidt operators. To see that, we consider a new contour $\Gamma_0 := \mathbb{R} + \epsilon$, not intersecting neither Γ_+ nor Γ_-. We start by the case of $\mathbb{G}_{n;x_j}$. We define the following two operators

$$\begin{aligned} &\mathbb{B}^{(1)}_{n;x_j} : L^2(\Gamma_-) \to L^2(\Gamma_0), \quad \text{with kernel} \quad B^{(1)}_{n;x_j}(\zeta, \mu) = \frac{\mathrm{e}^{-\frac{\mathrm{i}}{2}\psi_n(\mu;2x_j+2t)}}{(\zeta - \mu)} \\ &\mathbb{B}^{(2)}_{n;x_j} : L^2(\Gamma_0) \to L^2(\Gamma_+), \quad \text{with kernel} \quad B^{(2)}_{n;x_j}(\lambda, \zeta) = \frac{\mathrm{e}^{\frac{\mathrm{i}}{2}\psi_n(\lambda;2x_j+2t)}}{2\pi(\zeta - \lambda)}. \end{aligned} \tag{2.10}$$

Their composition gives $\mathbb{B}^{(2)}_{n;x_j} \circ \mathbb{B}^{(1)}_{n;x_j} = \mathbb{G}_{n;x_j}$ and, since they are both Hilbert-Schmidt, $\mathbb{G}_{n;x_j}$ is trace-class. The case $\mathbb{F}_n$ is treated similarly using the two operators

$$\begin{aligned} &\mathbb{C}^{(1)}_n : L^2(\Gamma_+) \to L^2(\Gamma_0) \quad \text{with kernel} \quad C^{(1)}_n(\lambda, \mu) = \frac{\mathrm{e}^{-\frac{\mathrm{i}}{2}\psi_n(\mu;0)}}{(\mu - \lambda)} \\ &\mathbb{C}^{(2)}_n : L^2(\Gamma_0) \to L^2(\Gamma_-) \quad \text{with kernel} \quad C^{(2)}_n(\zeta, \lambda) = \frac{\mathrm{e}^{\frac{\mathrm{i}}{2}\psi_n(\zeta;0)}}{2\pi(\lambda - \zeta)}. \end{aligned} \tag{2.11}$$

This means, in particular, that the Fredholm determinants $\det(\mathrm{I}-\mathbb{L}_n)$ and $\det(\mathrm{I}-\mathbb{G}_{n;x_j})$ are well defined, and moreover $\det(\mathrm{I}-\mathbb{G}_{n;x_j}) \equiv 1$. Hence,

$$\begin{aligned}\det(\mathrm{I}-\mathbb{L}_n) &= \det(\mathrm{I}-\mathbb{G}_{n;x_j})\det(\mathrm{I}-\mathbb{L}_n)\\ &= \det\left(\begin{pmatrix}\mathrm{I} & -\mathbb{F}_n\\ 0 & -\sum_{j=1}^{k}(\alpha_j-\alpha_{j+1})\mathbb{G}_{n;x_j}\circ\mathbb{F}_n\end{pmatrix}\right)\\ &= \det\left(\mathrm{I}-\sum_{j=1}^{k}(\alpha_j-\alpha_{j+1})\mathbb{G}_{n;x_j}\circ\mathbb{F}_n\right),\end{aligned}$$

where the equality between the first and the second line is easily proven using the block representation of $\mathbb{L}_n$ and $\mathbb{G}_{n;x_j}$ induced by the polarization $L^2(\Gamma) \simeq L^2(\Gamma_+)\oplus L^2(\Gamma_-)$.

Notice that each factor $\mathbb{G}_{n;x_j}\circ\mathbb{F}_n : L^2(\Gamma_+) \longrightarrow L^2(\Gamma_+)$ has kernel equal to

$$(G_{n;x_j}\circ F_n)(\xi,\lambda) = \frac{1}{(2\pi)^2}\mathrm{e}^{\frac{\mathrm{i}}{2}\left(\psi_n(\xi;2t+2x_j)+\psi_n(\lambda;0)\right)}\int_{\Gamma_-}\frac{\mathrm{e}^{-\mathrm{i}\psi_n(\mu;t+x_j)}}{(\xi-\mu)(\mu-\lambda)}\mathrm{d}\mu.$$

We now conjugate $\mathbb{G}_{n;x_j}\circ\mathbb{F}_n$ by the multiplication operator $\mathbb{P}_n$ with kernel $P_n(\lambda,\mu) = \mathrm{e}^{-\frac{\mathrm{i}}{2}\psi_n(\lambda,0)}\delta(\lambda-\mu)$ so to obtain

$$(P_n\circ G_{n;x_j}\circ F_n\circ P_n^{-1})(\xi,\lambda) = \frac{1}{(2\pi)^2}\mathrm{e}^{\mathrm{i}(t+x_j)\xi}\int_{\Gamma_-}\frac{\mathrm{e}^{\mathrm{i}(\psi_n(\lambda;0)-\psi_n(\mu,x_j+t)}}{(\xi-\mu)(\mu-\lambda)}\mathrm{d}\mu.$$

Next we observe that, in the double-contour integral representation (2.4) of K_n, one can deform the contour Γ_+ into $\mathbb{R}$. Using this property and conjugating once more with the standard Fourier transform (i.e. with the integral operator with kernel $\mathcal{F}(x,\xi) = \frac{1}{\sqrt{2\pi}}\mathrm{e}^{-\mathrm{i}x\xi}$) we finally obtain

$$\left(\mathcal{F}\circ P_n\circ G_{n;x_j}\circ F_n\circ P_n^{-1}\circ\mathcal{F}^{-1}\right)(x,y) \tag{2.12}$$

$$= \frac{1}{(2\pi)^2}\int_{\mathbb{R}}\frac{\mathrm{d}\xi}{\sqrt{2\pi}}\mathrm{e}^{\mathrm{i}(t+x_j-x)\xi}\int_{\Gamma_-}\mathrm{d}\mu\int_{\mathbb{R}}\frac{\mathrm{d}\lambda}{\sqrt{2\pi}}\frac{\mathrm{e}^{\mathrm{i}(\psi_n(\lambda;0)-\psi_n(\mu,x_j+t)}}{(\xi-\mu)(\mu-\lambda)}\mathrm{e}^{\mathrm{i}\lambda y} \tag{2.13}$$

$$= \begin{cases}\frac{\mathrm{i}}{(2\pi)^2}\int_{\Gamma_-}\mathrm{d}\mu\int_{\mathbb{R}}\mathrm{d}\lambda\frac{\mathrm{e}^{\mathrm{i}(\psi_n(\lambda;y)-\psi_n(\mu,x))}}{\lambda-\mu} & \text{if } x \geq t+x_j\\ 0 & \text{if } x < t+x_j\end{cases}, \tag{2.14}$$

where the latter equality is obtained deforming the outer contour $\mathbb{R}$ toward $+\infty\mathrm{e}^{\pm\pi\mathrm{i}}$ (depending on the sign of $t+x_j-x$) and taking a residue. Since the conjugations we performed do not change the value of the Fredholm determinant, we find

$$\det(\mathrm{I}-\mathbb{L}_n) = \det\left(\mathrm{I}-\sum_{j=1}^{k}(\alpha_j-\alpha_{j+1})\mathbb{K}_{n|[x_j+t,\infty)}\right),$$

and this latter is equal to $F_n(\vec{x}+t,\vec{\alpha})$ because of (2.1). □

As $\mathbb{L}_n$ is of integrable type, in the sense of [**9**], it is naturally associated to a Riemann-Hilbert problem we are going to define, and whose jumps are given by the $k+1$ dimensional matrix $J_Y(\lambda) := \mathbf{1}_{k+1} - 2\pi\mathrm{i} f_n(\lambda) g_n^\top(\lambda)$ for $\lambda\in\Gamma$.

RIEMANN-HILBERT PROBLEM 2.3. *Find a sectionally–analytic function* $Y(\bullet; n, \vec{x}+t, \vec{\alpha}) : \mathbb{C}/\Gamma \longrightarrow \mathrm{GL}(k+1, \mathbb{C})$ *such that*

a) Y *has continuous boundary values* $Y_\pm$ *as* $\lambda \in \Gamma$ *is approached from the left* $(+)$ *or right* $(-)$ *side, and they are related by*

$$Y_+(\lambda) = Y_-(\lambda) \left(\begin{array}{c|c} 1 & -\mathrm{i}\Theta_n^\top(-\lambda; \vec{x}+t, \vec{\alpha})\chi_{\Gamma_-}(\lambda) \\ \hline -\mathrm{i}\Theta_n(\lambda; \vec{x}+t, \vec{\alpha})\chi_{\Gamma_+}(\lambda) & \mathbf{1}_k \end{array} \right),$$

where $\Theta_n(\lambda; \vec{x}+t, \vec{\alpha})$ *is the (column) vector*

$$\Theta_n(\lambda; \vec{x}+t, \vec{\alpha}) := \left(\sqrt{\alpha_j - \alpha_{j+1}} \mathrm{e}^{\mathrm{i}\psi_n(\lambda; t+x_j)} \right)_{j=1}^k .$$

b) *There exists a matrix* $Y_1 = Y_1(n, \vec{x}+t, \vec{\alpha})$*, independent of* λ*, such that* Y *satisfies*

$$Y(\lambda) = \mathbf{1}_{k+1} + Y_1\lambda^{-1} + \mathcal{O}(\lambda^{-2}), \quad \lambda \to \infty. \tag{2.15}$$

We record in the following remark some symmetries that will be useful in the sequel.

REMARK 2.4. The jump matrix $J_Y(\lambda; n, \vec{x}+t, \vec{\alpha}) \equiv J_Y(\lambda)$ of the Riemann-Hilbert problem 2.3 satisfies the following two symmetries:

$$J_Y^{-\top}(-\lambda) = D_1 J_Y(\lambda) D_1 \qquad \overline{J_Y(\overline{\lambda})} = D_2 J_Y(-\lambda) D_2,$$

where $D_1 := \mathrm{diag}(1, -1 \ldots, -1)$ and $D_2 := \mathrm{diag}(1, c_1, \ldots, c_k)$ with $c_j := -\,\mathrm{sgn}(\alpha_j - \alpha_{j+1})$. Consequently, the unique solution Y to the Riemann-Hilbert problem satisfies the symmetry relations

$$D_1 Y^\top(-\lambda) D_1 = Y^{-1}(\lambda), \quad D_2\overline{Y(\overline{\lambda})}D_2 = Y(\lambda). \tag{2.16}$$

In particular, using the expansion of the function Y at $\lambda \to \infty$ together with the symmetries above, we obtain that

$$Y_1 = \left(\begin{array}{c|c} -\delta & \vec{u}^\top \\ \hline \vec{u} & \Delta \end{array} \right) \tag{2.17}$$

where the entry u_j, $j = 1, \ldots, k$ of $\vec{u} = \vec{u}(n, \vec{x}+t, \vec{\alpha})$ is real if $\alpha_j - \alpha_{j+1} > 0$ and purely imaginary if $\alpha_j - \alpha_{j+1} < 0$. Moreover, since $\det Y \equiv 1$, we also have that $\delta = \mathrm{Tr}(\Delta)$.

REMARK 2.5. The jump matrix $J_Y(\lambda; n, \vec{x}+t, \vec{\alpha}) \equiv J_Y(\lambda)$ can be factorized as

$$J_Y(\lambda) = \exp(M) J_Y(0) \exp(-M) \tag{2.18}$$

with $M \equiv M(\lambda; n, \vec{x}+t, \vec{\alpha}) := \mathrm{diag}(M_0, M_1, \ldots, M_k)$,

$$M_0 := -\frac{i}{k+1}\sum_{j=1}^k \psi_n(\lambda; t+x_j), \quad M_\ell := M_0 + i\psi_n(\lambda; x_\ell+t), \ \ \ell = 1, \ldots, k. \tag{2.19}$$

Note that the matrix $J_Y(0)$ does not depend on $\vec{x}, t$ or n.

PROPOSITION 2.6. *The unique solution Y of the Riemann–Hilbert problem 2.3 is related to the Fredholm determinant $F_n(\vec{x};\vec{\alpha})$ via the formula*

$$\frac{\partial}{\partial t}\ln F_n(\vec{x}+t,\vec{\alpha}) = \mathrm{i}(Y_1)_{1,1}. \tag{2.20}$$

PROOF. As we proved that $F_n(\vec{x}+t,\vec{\alpha})$ is the Fredholm determinant associated to the integrable kernel L_n defined in (2.6), the following general formula, proven in [**3**, Theorem 3.2], relates $F_n(\vec{x}+t,\vec{\alpha})$ to the (unique) solution of the Riemann–Hilbert problem 2.3:

$$\frac{\partial}{\partial t}\ln F(\vec{x}+t,\vec{\alpha}) = \int_\Gamma \mathrm{Tr}\left(Y_-^{-1}(\lambda)Y_-'(\lambda)\frac{\partial}{\partial t}J_Y(\lambda)J_Y^{-1}(\lambda)\right)\frac{d\lambda}{2\pi\mathrm{i}}, \tag{2.21}$$

where the symbol $'$ denotes the derivative w.r.t. the complex parameter λ. Thanks to the factorization of the jump matrix $J_Y(\lambda)$ given in (2.18) and the nature of the contour $\Gamma = \Gamma_+ \cup \Gamma_-$, the integral in the right hand side is computed as a formal residue at infinity. Indeed, we start noticing that

$$\begin{aligned}
&\int_{\Gamma_+\cup\Gamma_-}\mathrm{Tr}\left(Y_-^{-1}(\lambda)Y_-'(\lambda)\frac{\partial}{\partial t}J_Y(\lambda)J_Y^{-1}(\lambda)\right)\frac{\mathrm{d}\lambda}{2\pi\mathrm{i}} = \\
&\int_{\Gamma_+\cup\Gamma_-}\mathrm{Tr}\left(Y_-^{-1}(\lambda)Y_-'(\lambda)\left(\frac{\partial}{\partial t}M(\lambda) - J_Y(\lambda)\frac{\partial}{\partial t}M(\lambda)J_Y^{-1}(\lambda)\right)\right)\frac{\mathrm{d}\lambda}{2\pi\mathrm{i}} = \\
&\int_{\Gamma_+\cup\Gamma_-}\mathrm{Tr}\left(Y_-^{-1}(\lambda)Y_-'(\lambda)\frac{\partial}{\partial t}M(\lambda)\right)\frac{\mathrm{d}\lambda}{2\pi\mathrm{i}} - \int_{\Gamma_+\cup\Gamma_-}\mathrm{Tr}\left(Y_+^{-1}(\lambda)Y_+'(\lambda)\frac{\partial}{\partial t}M(\lambda)\right)\frac{\mathrm{d}\lambda}{2\pi\mathrm{i}} \\
&\underbrace{+\int_{\Gamma_+\cup\Gamma_-}\mathrm{Tr}\left(J_Y^{-1}(\lambda)J_Y'(\lambda)\frac{\partial}{\partial t}M(\lambda)\right)\frac{\mathrm{d}\lambda}{2\pi\mathrm{i}}}_{(\spadesuit)} = \\
&-\lim_{R\to+\infty}\int_{C_R}\mathrm{Tr}\left(Y^{-1}(\lambda)Y'(\lambda)\frac{\partial}{\partial t}M(\lambda)\right)\frac{\mathrm{d}\lambda}{2\pi\mathrm{i}}.
\end{aligned} \tag{2.22}$$

In the last passage we used that the term denoted with $(\spadesuit)$ is zero (as we will see in a moment) and we rewrote the two remaining terms by deforming the contour $\Gamma_+ \cup \Gamma_-$ into a circle C_R centered in zero and of increasing radius R. Indeed, $(\spadesuit)$ is zero because of the form of the jump matrix J_Y and the fact that $M(\lambda)$ is diagonal (we are computing the trace of a strictly lower/upper triangular matrix). Finally, using the asymptotic expansion of Y at infinity combined with

$$\frac{\partial M}{\partial t} = \frac{i\lambda}{k+1}\mathrm{diag}(-k,1,\dots,1),$$

we explicitly compute

$$-\lim_{R\to+\infty}\int_{C_R}\mathrm{Tr}\left(Y^{-1}(\lambda)Y'(\lambda)\frac{\partial}{\partial t}M(\lambda)\right)\frac{\mathrm{d}\lambda}{2\pi\mathrm{i}} = \mathrm{i}(Y_1)_{1,1}.$$

□

This result will be used in the last section to relate $F_n(\vec{x}+t,\vec{\alpha})$ to a distinguished solution of the vector–valued PII hierarchy, whose Lax pair is given in the following section.

3. A Lax pair associated to a vector valued PII hierarchy

We now introduce a new matrix $\Psi(\lambda) = \Psi(\lambda; n, \vec{x}+t, \vec{\alpha})$ solving a Riemann-Hilbert problem with constant jumps. More specifically, let

$$T^{(1)}(\lambda; n, \vec{x}+t) := \operatorname{diag}\left(1, \mathrm{e}^{\mathrm{i}\psi_n(\lambda; x_1+t)}, \cdots, \mathrm{e}^{\mathrm{i}\psi_n(\lambda; x_k+t)}\right), \tag{3.1}$$

$$T^{(2)}(\lambda; n, \vec{x}+t) := \mathrm{e}^{-\frac{\mathrm{i}}{k+1}\sum_{j=1}^{k}\psi_n(\lambda; x_j+t)}\mathbf{1}_{k+1} \tag{3.2}$$

and $T(\lambda) \equiv T(\lambda; n, \vec{x}+t) := T^{(1)}(\lambda; n, \vec{x}+t)T^{(2)}(\lambda; n, \vec{x}+t)$. We define

$$\Psi(\lambda; n, \vec{x}+t, \vec{\alpha}) := Y(\lambda; n, \vec{x}+t, \vec{\alpha})T(\lambda; n, \vec{x}+t). \tag{3.3}$$

It is easy to prove that this latter satisfies the following Riemann–Hilbert problem.

RIEMANN-HILBERT PROBLEM 3.1. *Find a sectionally–analytic function* $\Psi : \mathbb{C}/\Gamma \longrightarrow \mathrm{GL}(k+1, \mathbb{C})$ *such that*

a) Ψ *has continuous boundary values* $\Psi_\pm$ *as* $\lambda \in \Gamma$ *is approached from the left* $(+)$ *or right* $(-)$ *side, and they are related by*

$$\Psi_+(\lambda) = \Psi_-(\lambda)\left(\begin{array}{c|c} 1 & -\mathrm{i}\hat{\Theta}^\top\chi_{\Gamma_-}(\lambda) \\ \hline -\mathrm{i}\hat{\Theta}\chi_{\Gamma_+}(\lambda) & \mathbf{1}_k \end{array}\right), \tag{3.4}$$

where $\hat{\Theta} := \Theta(\lambda; n, \vec{x}+t, \vec{\alpha})_{|\lambda=0}$ *is the (column) vector*

$$\hat{\Theta} := \left(\sqrt{\alpha_j - \alpha_{j+1}}\right)_{j=1}^{k}. \tag{3.5}$$

b) Ψ *has the following asymptotic behavior*

$$\Psi(\lambda) = \left(\mathbf{1} + Y_1\lambda^{-1} + \mathcal{O}(\lambda^{-2})\right)T(\lambda), \quad \textit{as } \lambda \to \infty. \tag{3.6}$$

The properties of the function $\Psi(\lambda)$ listed above are then used to prove the following proposition.

PROPOSITION 3.2. *There exist two matrices* $A(\lambda) \equiv A(\lambda; n, \vec{x}+t, \vec{\alpha})$ *and* $B(\lambda) \equiv B(\lambda; n, \vec{x}+t, \vec{\alpha})$*, polynomial in* λ*, such that*

$$\begin{cases} \frac{\partial}{\partial\lambda}\Psi(\lambda) = A(\lambda)\Psi(\lambda), \\ \frac{\partial}{\partial t}\Psi(\lambda) = B(\lambda)\Psi(\lambda). \end{cases} \tag{3.7}$$

Moreover,

$$B(\lambda) = \left(\begin{array}{c|c} -\frac{\mathrm{i}k}{k+1}\lambda & -\mathrm{i}\vec{u}^\top \\ \hline \mathrm{i}\vec{u} & \frac{\mathrm{i}\lambda}{k+1}\mathbf{1}_k \end{array}\right) \tag{3.8}$$

and

$$A(\lambda) = \sum_{j=0}^{2n} A_j \lambda^{2n-j} + \hat{A}_{2n}, \tag{3.9}$$

with

$$A_0 = \frac{\mathrm{i}}{k+1} \operatorname{diag}(-k, 1, \dots, 1), \tag{3.10}$$

$$\hat{A}_{2n} = \frac{\mathrm{i}}{k+1} \operatorname{diag}\left(-kt - \sum_{j=1}^{k} x_j, t + kx_1 - \sum_{j \neq 1} x_j, \dots, t + kx_k - \sum_{j \neq k} x_j\right). \tag{3.11}$$

and $\vec{u}$ *as in* (2.17).

PROOF. For both equations the proof follows from applying Liouville theorem and using straightforward computations. For the first equation, we define

$$A(\lambda) := \frac{\partial}{\partial \lambda} \Psi(\lambda) \left(\Psi(\lambda)\right)^{-1}.$$

The matrix-valued function $A(\lambda)$ is analytic for every $\lambda \in \mathbb{C} \setminus \Gamma$. Moreover, for $\lambda \in \Gamma$ we have that $A_+(\lambda) = A_-(\lambda)$, thanks to the fact that $\Psi(\lambda)$ has constant jump condition along Γ. Thus $A(\lambda)$ is entire and behaves like a polynomial in λ of degree $2n$ at ∞. By Liouville theorem, we conclude that $A(\lambda)$ is a polynomial of degree $2n$ in λ. Using the asymptotic condition written in equation (2.15), we can then compute explicitely the leading coefficient and the constant coefficient of $A(\lambda)$ as in (3.10), (3.11). For the second equation, in an analogue way we define

$$B(\lambda) := \frac{\partial}{\partial t} \Psi(\lambda) \left(\Psi(\lambda)\right)^{-1}.$$

Using the same reasoning, we conclude that $B(\lambda)$ is a polynomial in λ of degree 1 and we compute its coefficients as in (3.8). □

Now, we show that the system (3.7) is the Lax pair for the vector-valued Painlevé II hierarchy (1.6). For any $j = 0, \dots, 2n$, we will denote with $a_j^{11}, a_j^{12}, a_j^{21}$, a_j^{22} the block-entries of A_j (and the same for $\hat{A}_{2n}$). Hence, a_j^{11} will be a scalar, a_j^{12}, a_j^{21} will be, respectively, a row and a column vector, and a_j^{22} a square matrix of size k. We will now study the compatibility condition

$$A(\lambda)B(\lambda) - B(\lambda)A(\lambda) = \frac{\partial B}{\partial \lambda}(\lambda) - \frac{\partial A}{\partial t}(\lambda). \tag{3.12}$$

The following two Lemmas are the analogue, for the vector–valued case, of Lemma 5.4 and 5.5 in [**5**] and the tecnique used in their proofs is inspired by the one used in [**15**]. The dependence of the variable $\vec{u}$ and $\{a_j^{ik}\}$ on $n, t, \vec{x}, \vec{\alpha}$ will not be made explicit in the following formulas. We will denote with a dot the derivative with respect to t, as this will be the only dynamical variable (the other variables will be considered as parameters).

LEMMA 3.3. *The compatibility condition of the Lax pair* (3.7) *is equivalent to the system of equations*

$$a_1^{12} = -\mathrm{i}\vec{u}^{\top}, \quad a_1^{21} = \mathrm{i}\vec{u}, \tag{3.13}$$

(3.14)
$$\begin{cases} \dot a_j^{11} = -\mathrm{i}(\vec u^\top a_j^{21} + a_j^{12}\vec u), & \dot a_j^{12} = -\mathrm{i}(a_{j+1}^{12} + \vec u^\top a_j^{22} - a_j^{11}\vec u^\top) \\ \dot a_j^{22} = \mathrm{i}(\vec u a_j^{12} + a_j^{21}\vec u^\top), & \dot a_j^{21} = \mathrm{i}(a_{j+1}^{21} + a_j^{11}\vec u - a_j^{22}\vec u) \end{cases}, \quad j = 1,\dots,2n-1$$

$$\begin{cases} \dot a_{2n}^{11} = -\mathrm{i}(\vec u^\top a_{2n}^{21} + a_{2n}^{12}\vec u), & \dot a_{2n}^{12} = -\mathrm{i}(\vec u^\top a_{2n}^{22} - a_{2n}^{11}\vec u^\top + \mathrm{i}\vec u^\top m_{t,\vec x}) \\ \dot a_{2n}^{22} = \mathrm{i}(\vec u a_{2n}^{12} + a_{2n}^{21}\vec u^\top), & \dot a_{2n}^{21} = \mathrm{i}(a_{2n}^{11}\vec u - a_{2n}^{22}\vec u - \mathrm{i}m_{t,\vec x}\vec u), \end{cases} \tag{3.15}$$

where $m_{t,\vec x} := \operatorname{diag}(x+t_1,\dots,x+t_k)$.

PROOF. The proof follows by direct computation, exploiting the polynomiality of the matrices $A(\lambda), B(\lambda)$ in λ. In particular, equation (3.13) corresponds to the term associated to the monomial λ^{2n} in the compatibility condition (3.12). Then system (3.14) corresponds, for every $j = 1,\dots,2n-1$, to the monomial λ^{2n-j} and, finally, (3.15) corresponds to λ^0. □

From equations (3.13), (3.14), and the first two equations of (3.15) we can derive some symmetries for the entries of A_j for $j = 1,\dots,2n$. In particular,

$$\begin{aligned}\frac{\partial}{\partial t}\operatorname{Tr}(a_j^{22}) &= \operatorname{Tr}(\dot a_j^{22}) = \mathrm{i}\operatorname{Tr}(\vec u a_j^{12} + a_j^{21}\vec u^\top) = \mathrm{i}\operatorname{Tr}(\vec u a_j^{12}) + i\operatorname{Tr}(a_j^{21}\vec u^\top) \\ &= \mathrm{i}a_j^{12}\vec u + i\vec u^\top a_j^{21} = -\dot a_j^{11}.\end{aligned}$$

Thus we conclude that

$$-\operatorname{Tr}(a_j^{22}) = a_j^{11}, \quad j = 1\dots,2n, \tag{3.16}$$

up to constant of integration. This constant is actually zero. This can be proven observing that $\lim_{t\to\infty} a_j^{22} = \lim_{t\to\infty} a_j^{11} = 0$. Indeed, both of them are polynomials in the entries of the matrix elements of Y_i, $i \geq 1$, and $Y_i \to 0$ for $t \to +\infty$, which is easily proven using the small norm theorem (see Section 4 below and, in particular, the proof of Proposition 4.1). Moreover, we can also deduce the following symmetries:

$$a_j^{21} = (-1)^j(a_j^{12})^\top, \quad a_j^{22} = (-1)^j(a_j^{22})^\top \quad \forall j = 1,\dots,2n. \tag{3.17}$$

This is proved by induction over j using the equations appearing in the Lemma 3.3.

LEMMA 3.4. *For every* $\ell = 2,\dots,2n$

$$a_\ell^{11} = -\mathrm{i}\sum_{j=1}^{\ell-1}\left(a_j^{11}a_{\ell-j}^{11} + a_j^{12}a_{\ell-j}^{21}\right) \quad \text{and} \quad a_\ell^{22} = \mathrm{i}\sum_{j=1}^{\ell-1}\left(a_j^{22}a_{\ell-j}^{22} + a_j^{21}a_{\ell-j}^{12}\right). \tag{3.18}$$

PROOF. Thanks to equation (3.16), we only need to prove the formula for a_ℓ^{22}. Consider the matrix

$$C := A^2 = \left(\begin{array}{c|c} -\frac{k^2}{(k+1)^2} & 0 \\ \hline 0^\top & -\dfrac{\mathbf{1}_k}{(k+1)^2} \end{array}\right)\lambda^{4n} + \sum_{\ell=1}^{4n}\lambda^{4n-\ell}C_\ell,$$

where $C_\ell = \sum_{j=0}^{\ell} A_j A_{\ell-j}$ for $\ell = 1,\dots,2n-1$ and $C_{2n} = \sum_{j=0}^{2n} A_j A_{2n-j} + A_0\hat{A}_{2n} + \hat{A}_{2n}A_0$ (these are the only coefficients we are going to use in the proof). We can write C_ℓ, for every value of ℓ, in the usual block-form

$$C_\ell := \left(\begin{array}{c|c} c_\ell^{11} & c_\ell^{12} \\ \hline c_\ell^{21} & c_\ell^{22} \end{array}\right),$$

and, in particular, each block-entry can be written in terms of the block entries of the matrix A

(3.19)

$$\begin{cases} c_\ell^{11} = \sum_{j=0}^{\ell}\left(a_j^{11}a_{\ell-j}^{11} + a_j^{12}a_{\ell-j}^{21}\right) + 2a_0^{11}\hat{a}_{2n}^{11}\delta_{\ell,2n} \\ c_\ell^{22} = \sum_{j=0}^{\ell}\left(a_j^{21}a_{\ell-j}^{12} + a_j^{22}a_{\ell-j}^{22}\right) + (a_0^{22}\hat{a}_{2n}^{22} + \hat{a}_{2n}^{22}a_0^{22})\delta_{\ell,2n} \\ c_\ell^{12} = \sum_{j=0}^{\ell}\left(a_j^{11}a_{\ell-j}^{12} + a_j^{12}a_{\ell-j}^{22}\right) \\ c_\ell^{21} = \sum_{j=0}^{\ell}\left(a_j^{21}a_{\ell-j}^{11} + a_j^{22}a_{\ell-j}^{21}\right) \end{cases} \qquad \ell = 1,\dots,2n.$$

From the compatibility condition (3.12), we deduce the following equation for the matrix C

$$\frac{\partial C}{\partial t}(\lambda) = B(\lambda)C(\lambda) - C(\lambda)B(\lambda) + \frac{\partial B}{\partial \lambda}(\lambda)A(\lambda) + A(\lambda)\frac{\partial B}{\partial \lambda}(\lambda). \tag{3.20}$$

In particular, by looking at the coefficients of the powers λ^m with $m = 4n,\dots,2n$ in equation (3.20), we obtain the following system of difference and differential equations for the block entries of each coefficient C_ℓ

$$\begin{cases} \dot{c}_\ell^{11} = -\mathrm{i}(\vec{u}^\top c_\ell^{21} + c_\ell^{12}\vec{u}) + 2(a_0^{11})^2\delta_{\ell,2n}, & \dot{c}_\ell^{12} = -\mathrm{i}(c_{\ell+1}^{12} + \vec{u}^\top c_\ell^{22} - c_\ell^{11}\vec{u}^\top) \\ \dot{c}_\ell^{22} = \mathrm{i}(\vec{u}c_\ell^{12} + c_\ell^{21}\vec{u}^\top) + 2(a_0^{22})^2\delta_{\ell,2n}, & \dot{c}_\ell^{21} = \mathrm{i}(c_{\ell+1}^{21} + c_\ell^{11}\vec{u} - c_\ell^{22}\vec{u}) \end{cases} \tag{3.21}$$

with $\ell = 1,\dots,2n$. Note that this is almost the same system satisfied by the matrix elements of A, see (3.14). In particular, the equations giving the entries $(1,2),(2,1)$ in (3.14) and (3.21) are exactly the same, while for the entries $(2,2),(1,1)$ the two sets of equations differ just for $\ell = 2n$.

Now, by using the equations above, we first prove that the entries of C_ℓ, for $\ell = 1,\dots,2n$ are multiple of the ones of A_ℓ, by induction over ℓ. Keeping in mind that the coefficient C_0 is explicitely written in the definition of the matrix C, we start by computing C_1:

(3.22)

$$c_1^{12} = \mathrm{i}\frac{1-k}{k+1}a_1^{12},\quad c_1^{21} = \mathrm{i}\frac{1-k}{k+1}a_1^{21},\quad c_1^{11} = 0 = \mathrm{i}\frac{1-k}{k+1}a_1^{11},\quad c_1^{22} = 0_k = \mathrm{i}\frac{1-k}{k+1}a_1^{22}.$$

We suppose now that the equation above holds for ℓ and we prove than it holds for $\ell+1$ too, by using the equations (3.21) together with equations (3.13), (3.14). In particular, from the third equation in the system (3.21) we recover $c_{\ell+1}^{12}$ as

$$c_{\ell+1}^{12} = \mathrm{i}\dot{c}_\ell^{12} - \vec{u}^\top c_\ell^{22} + c_\ell^{11}\vec{u}^\top = \mathrm{i}\frac{1-k}{k+1}\underbrace{(\mathrm{i}\dot{a}_{\ell+1}^{12} - \vec{u}^\top a_\ell^{22} + a_\ell^{11}\vec{u}^\top)}_{=a_{\ell+1}^{12}}, \tag{3.23}$$

using the induction hypothesis and the equation for $a_{\ell+1}^{12}$ in (3.14). The same procedure can be applied for the block entry $c_{\ell+1}^{21}$, to get the analogue result. Now, for the diagonal entries, we use instead the first two equations in the system (3.21) and obtain

$$\dot{c}^{22}_{\ell+1} = \mathrm{i}\left(\vec{u}c^{12}_{\ell+1} + c^{21}_{\ell+1}\vec{u}^\top\right) = \mathrm{i}\frac{1-k}{k+1}\underbrace{\mathrm{i}\left(\vec{u}a^{12}_{\ell+1} + a^{21}_{\ell+1}\vec{u}^\top\right)}_{=\dot{a}^{22}_{\ell+1}}. \tag{3.24}$$

Thus we conclude that $c^{22}_{\ell+1} = \mathrm{i}\frac{1-k}{k+1}a^{22}_{\ell+1}$ (from the equation above, this is true up to a constant of integration, that can be fixed to zero thanks to equations (3.19)). The same can be done for $c^{11}_{\ell+1}$. Thus the proportionality relation between the block entries of type $(1,2)$ or $(2,1)$ of A_ℓ and of C_ℓ is proved for $\ell = 1\dots,2n-1$, while for the diagonal block entries it holds for $\ell = 1\dots,2n-2$. For $\ell = 2n-1$ we have for the diagonal block entries

$$\dot{c}^{22}_{2n} = \mathrm{i}\frac{1-k}{k+1}\dot{a}^{22}_{2n} + 2(a^{22}_0)^2, \tag{3.25}$$

and an analogue relation for a^{11}_{2n}. Now, by using the first two equations in the system (3.19) for c^{22}_ℓ, we deduce the following chain

$$\mathrm{i}\frac{1-k}{k+1}a^{22}_\ell = c^{22}_\ell = \sum_{j=1}^{\ell-1}\left(a^{21}_j a^{12}_{\ell-j} + a^{22}_j a^{22}_{\ell-j}\right) + \underbrace{a^{21}_0 a^{12}_\ell + a^{22}_0 a^{22}_\ell + a^{21}_\ell a^{12}_0 + a^{22}_\ell a^{22}_0}_{=\mathrm{i}\frac{2}{k+1}a^{22}_\ell} \tag{3.26}$$

for $\ell = 1,\dots,2n-1$, from which the statement for a^{22}_ℓ is directly obtained. For the case $\ell = 2n$ the chain obtained by replacing the equation (3.19) for c^{22}_{2n} is the following one

$$\mathrm{i}\frac{1-k}{k+1}\dot{a}^{22}_{2n} + 2(a^{22}_0)^2 = \dot{c}^{22}_{2n} = \frac{d}{dt}\left(\sum_{j=0}^{\ell}\left(a^{21}_j a^{12}_{\ell-j} + a^{22}_j a^{22}_{\ell-j}\right)\right) + 2(a^{22}_0)^2. \tag{3.27}$$

Thus, by simplifying both sides the term $2\left(a^{22}_0\right)^2$ and then integrating, the formula for a^{22}_{2n} is obtained as for the previous values of ℓ. □

Combining the two lemmas 3.3 and 3.4, we are able to express all the coefficients of the Lax matrix $A(\lambda)$ in function of the vector $\vec{u}$ and its derivatives.

PROPOSITION 3.5. *The entries of the matrix $A(\lambda)$ are differential polynomials on u, given recursively by the formulas*

$$a^{21}_1 = \mathrm{i}\vec{u}, \quad a^{22}_1 = 0, \tag{3.28}$$

$$a^{21}_{j+1} = -\mathrm{i}\dot{a}^{21}_j - a^{11}_j\vec{u} + a^{22}_j\vec{u}, \quad a^{22}_{j+1} = \mathrm{i}\sum_{\ell=1}^{j}\left(a^{22}_\ell a^{22}_{j+1-\ell} + a^{21}_\ell a^{12}_{j+1-\ell}\right), \quad j = 1,\dots,2n \tag{3.29}$$

together with (3.16) *and* (3.17).

Moreover, using (3.29) together with (3.15) and the definition of the operators $\mathcal{L}^{\vec{u}}_\pm$, we prove that the vectors a^{21}_j, $j = 1,\dots,2n$ satisfy the recursion

$$a^{21}_{2j+1} = -\mathcal{L}^{\vec{u}}_+ a^{21}_{2j}, \quad a^{21}_{2j} = -\mathcal{L}^{\vec{u}}_- a^{21}_{2j-1}, \quad j = 1,\dots,n-1, \tag{3.30}$$

while a^{21}_{2n} satisfies the differential equation

$$-\mathcal{L}^{\vec{u}}_+ a^{21}_{2n} = -\mathrm{i}m_{t,\vec{x}}\vec{u}. \tag{3.31}$$

Hence, we proved recursively that $\vec{u}$, as defined in (2.17), satisfies the equation (1.6). Moreover, the entries of $\vec{u}$ are real/purely imaginary depending on the sign of $(\alpha_{j+1} - \alpha_j)$, as already observed in Remark 2.4. We are now left with the proofs of the equations (1.5) and (1.7).

4. The logarithmic derivative of $F_n(\vec{x}, \vec{\alpha})$

In this last section we finish the proof of Theorem 1.2 giving the relation between $F_n(\vec{x}, \vec{\alpha})$ and a particular solution $\vec{u}$ of the vector-valued Painlevé II hierarchy. The main ingredient is the first logarithmic derivative of the Fredholm determinant $F_n(\vec{x} + t, \vec{\alpha})$, computed at the end of Section 2.

PROPOSITION 4.1. *The distribution F_n defined in* (1.3) *satisfies the equation*

$$\frac{\partial^2}{\partial t^2} \ln F_n(\vec{x} + t, \vec{\alpha}) = - < \vec{u}(t), \vec{u}(t) > \tag{4.1}$$

and its integrated version

$$F_n(\vec{x}, \vec{\alpha}) = \exp\left(- \int_0^\infty t < \vec{u}(t), \vec{u}(t) > \mathrm{d}t \right), \tag{4.2}$$

where $\vec{u}(t) = \vec{u}(n, \vec{x}{+}t, \vec{\alpha})$ satisfies the following (vector-valued) ordinary differential equation

$$(\mathcal{L}_+^{\vec{u}} \mathcal{L}_-^{\vec{u}})^n \vec{u}(t) = - \operatorname{diag}(x_1 + t, \ldots, x_k + t) \vec{u}(t) \tag{4.3}$$

and have the following behavior at $+\infty$

$$\vec{u}(n, \vec{x} + t, \vec{\alpha}) = \left(\sqrt{\alpha_j - \alpha_{j+1}} \operatorname{Ai}_n(t + x_j)(1 + o(1)) \right)_{j=1,\ldots,k}. \tag{4.4}$$

PROOF. We start computing the λ^{-1}–term in the asymptotic expansion of $(\partial_t \Psi)\Psi^{-1}$ for $\lambda \to \infty$, where Ψ is the solution of the Riemann-Hilbert problem 3.1 with constant jump condition. The $(1,1)$–entry of this term, which is equal to zero because $B(\lambda)$ is polynomial in λ, leads to

$$\frac{\partial}{\partial t}(Y_1)_{11} - \mathrm{i}\vec{u}^{\top}\vec{u} = 0$$

and this equation, together with (2.20), gives (4.1).

We describe now the boundary behavior of $\vec{u}$ for $t \to +\infty$. We start by proving that the jump matrix $J_Y(\lambda)$ of the Riemann–Hilbert problem 2.3, for $t \to +\infty$, behaves like the identity matrix, so that the small norm theorem can be applied. We consider a rescaled complex variable w defined through the equation $\lambda = wt^{\frac{1}{2n}}$, so that the entries $(1, j+1)$ and $(j+1, 1)$ of $J_Y(\lambda)$ for $j = 1, \ldots, k$ are rewritten respectively as

(4.5)

$$\begin{aligned} &- \mathrm{i}\,(\Theta(\vec{x} + t, \vec{\alpha}, -\lambda))_j\, \chi_-(\lambda) = -\sqrt{\alpha_j - \alpha_{j+1}} \epsilon^{-\mathrm{i}t^{\frac{2n+1}{2n}} \left(\frac{1}{2n+1} w^{2n+1} + (1+\frac{x_j}{t})w\right)} \chi_{\Gamma_-}(wt^{\frac{1}{2n}}), \\ &- \mathrm{i}\,(\Theta(\vec{x} + t, \vec{\alpha}, \lambda))_j\, \chi_+(\lambda) = -\sqrt{\alpha_j - \alpha_{j+1}} \epsilon^{\mathrm{i}t^{\frac{2n+1}{2n}} \left(\frac{1}{2n+1} w^{2n+1} + (1+\frac{x_j}{t})w\right)} \chi_{\Gamma_+}(wt^{\frac{1}{2n}}). \end{aligned}$$

Note that the quantity $d_j := 1 + \frac{x_j}{t}$ is bounded in the regime $t \to \infty$ for any fixed x_j, as it converges to 1. Thus we can modify the curves Γ_- and Γ_+ into $\tilde{\Gamma}_-$ and

$\tilde{\Gamma}_+$ (as done in [6, Section 3]) so that

(4.6)

$$\Im\left(\frac{w^{2n+1}}{2n+1}+d_jw\right)<0,\quad \text{for}\quad w\in\tilde{\Gamma}_-\ \text{and}\ \Im\left(\frac{w^{2n+1}}{2n+1}+d_jw\right)>0,\quad \text{for}\quad w\in\tilde{\Gamma}_+.$$

In this way we obtain

$$\left\|J_Y(wt^{\frac{1}{2n}})-\mathbf{1}_{k+1}\right\|_\infty=\sqrt{\alpha_j-\alpha_{j+1}}\sup_{w\in\tilde{\Gamma}_\mp}\epsilon^{\pm t^{\frac{2n+1}{2n}}\Im\left(\frac{w^{2n+1}}{2n+1}+d_jw\right)}\to 0. \tag{4.7}$$

for $t\to+\infty$ and any fixed $x_j\in\mathbb{R}$. Moreover, also $\left\|J_Y(wt^{\frac{1}{2n}})-\mathbf{1}_{k+1}\right\|_2\to 0$ as

$$J_Y(wt^{\frac{1}{2n}})=\mathbf{1}_{k+1}+\mathcal{O}\left(w^{-2}\mathrm{e}^{-Ct^{\frac{2n+1}{2}}}\right)$$

for a suitable constant $C>0$ and uniformly on the contours. Thus, the rescaled function $X(w):=Y\big(wt^{\frac{1}{2n}}\big)$, thanks to the Riemann-Hilbert problem 2.3, satisfies the following conditions:

- it is analytic on $\mathbb{C}\setminus\tilde{\Gamma}_-\cup\tilde{\Gamma}_+$;
- it admits continuous boundary values $X_\pm$ while approaching from the left or from the right of the curves $\tilde{\Gamma}_-\cup\tilde{\Gamma}_+$ and they are related through $X_+(w)=X_-(w)J_Y\big(wt^{\frac{1}{2n}}\big)$ for all w along the curves;
- for $|w|\to\infty$ it behaves like the identity matrix, i.e. $X(w)\sim\mathbf{1}_{k+1}+\sum_{j\geq 1}\frac{X_j}{w^j}$.

Note that, in particular, $X_1=t^{-\frac{1}{2n}}Y_1$ (we will use it in a moment). Moreover, by applying the small norm theorem (see for instance Theorem 5.1.5 in [8]) we can conclude that $X(w)\sim\mathbf{1}_{k+1}$ for $t\to+\infty$ and fixed finite x_j and, in particular, $Y_j\to 0$ for $t\to+\infty$ (we already used this result in Section 3).

On the other hand, because of its properties described above, X satisfies the integral equation

$$X(w)=\mathbf{1}_{k+1}-\int_{\tilde{\Gamma}_+\cup\tilde{\Gamma}_-}X_-(v)\frac{f_n(vt^{\frac{1}{2n}})g_n^\top(vt^{\frac{1}{2n}})}{v-w}dv. \tag{4.8}$$

Expanding the right hand side for $w\to\infty$ we find

$$X_1=\int_{\tilde{\Gamma}_+\cup\tilde{\Gamma}_-}X_-(v)f_n(vt^{\frac{1}{2n}})g_n^\top(vt^{\frac{1}{2n}})dv, \tag{4.9}$$

and thus we can conclude that, for $t\to+\infty$ and fixed x_j,

$$\begin{aligned}u_j(t;\vec{x},\vec{\alpha})=(Y_1)_{1,j+1}=t^{\frac{1}{2n}}(X_1)_{1,j+1}&\sim\frac{\sqrt{\alpha_j-\alpha_{j+1}}}{2\pi\mathrm{i}}\int_{\Gamma_+}\epsilon^{\mathrm{i}\left(\frac{z^{2n+1}}{2n+1}+(t+x_j)z\right)}dz\\ &=\sqrt{\alpha_j-\alpha_{j+1}}\mathrm{Ai}_n(t+x_j)\end{aligned}$$

for any $j=1,\dots,k$, where in the last two equalities we used the small norm theorem and the integral representation of the n-th Airy function. Finally, the equation (4.2) is obtained integrating twice (4.1), and checking that one can set the integration constants equal to zero because of the asymptotic behavior of $\vec{u}$ as $t\to+\infty$. □

Acknowledgment

The authors are grateful to Thomas Bothner for the many useful discussions on the vector-valued Painlevé II hierarchy and its relation with the integro-differential one.

References

[1] H. Airault, *Rational solutions of Painlevé equations*, Stud. Appl. Math. **61** (1979), no. 1, 31–53, DOI 10.1002/sapm197961131. MR535866

[2] Jinho Baik, Percy Deift, and Eric Rains, *A Fredholm determinant identity and the convergence of moments for random Young tableaux*, Comm. Math. Phys. **223** (2001), no. 3, 627–672, DOI 10.1007/s002200100555. MR1866169

[3] M. Bertola and M. Cafasso, *The transition between the gap probabilities from the Pearcey to the airy process—a Riemann-Hilbert approach*, Int. Math. Res. Not. IMRN **7** (2012), 1519–1568, DOI 10.1093/imrn/rnr066. MR2913183

[4] Dan Betea, Jérémie Bouttier, and Harriet Walsh, *Multicritical random partitions* (English, with English and French summaries), Sém. Lothar. Combin. **85B** (2021), Art. 33, 12. MR4311914

[5] Thomas Bothner, Mattia Cafasso, and Sofia Tarricone, *Momenta spacing distributions in anharmonic oscillators and the higher order finite temperature Airy kernel*, Ann. Inst. Henri Poincaré Probab. Stat. **58** (2022), no. 3, 1505–1546, DOI 10.1214/21-aihp1211. MR4452641

[6] Mattia Cafasso, Tom Claeys, and Manuela Girotti, *Fredholm determinant solutions of the Painlevé II hierarchy and gap probabilities of determinantal point processes*, Int. Math. Res. Not. IMRN **4** (2021), 2437–2478, DOI 10.1093/imrn/rnz168. MR4218326

[7] Tom Claeys and Antoine Doeraene, *The generating function for the Airy point process and a system of coupled Painlevé II equations*, Stud. Appl. Math. **140** (2018), no. 4, 403–437, DOI 10.1111/sapm.12209. MR3798331

[8] Alexander R. Its, *Large N asymptotics in random matrices: the Riemann-Hilbert approach*, Random matrices, random processes and integrable systems, CRM Ser. Math. Phys., Springer, New York, 2011, pp. 351–413, DOI 10.1007/978-1-4419-9514-8_5. MR2858439

[9] A. R. Its, A. G. Izergin, V. E. Korepin, and N. A. Slavnov, *Differential equations for quantum correlation functions*, Proceedings of the Conference on Yang-Baxter Equations, Conformal Invariance and Integrability in Statistical Mechanics and Field Theory, Internat. J. Modern Phys. B **4** (1990), no. 5, 1003–1037, DOI 10.1142/S0217979290000504. MR1064758

[10] Taro Kimura and Ali Zahabi, *Unitary matrix models and random partitions: universality and multi-criticality*, J. High Energy Phys. **7** (2021), Paper No. 100, 46, DOI 10.1007/jhep07(2021)100. MR4316997

[11] Taro Kimura and Ali Zahabi, *Universal edge scaling in random partitions*, Lett. Math. Phys. **111** (2021), no. 2, Paper No. 48, 16, DOI 10.1007/s11005-021-01389-y. MR4244925

[12] Alexandre Krajenbrink, *From Painlevé to Zakharov-Shabat and beyond: Fredholm determinants and integro-differential hierarchies*, J. Phys. A **54** (2021), no. 3, Paper No. 035001, 51, DOI 10.1088/1751-8121/abd078. MR4209129

[13] P. Le Doussal, M. D. Majumdar, and G. Schehr, *Multicritical edge statistics for the momenta of fermions in non-harmonic traps*, Physical Review Letters, 2018.

[14] A. Soshnikov, *Determinantal random point fields* (Russian, with Russian summary), Uspekhi Mat. Nauk **55** (2000), no. 5(335), 107–160, DOI 10.1070/rm2000v055n05ABEH000321; English transl., Russian Math. Surveys **55** (2000), no. 5, 923–975. MR1799012

[15] Oliver H. Warren and John N. Elgin, *The vector nonlinear Schrödinger hierarchy*, Phys. D **228** (2007), no. 2, 166–171, DOI 10.1016/j.physd.2007.03.006. MR2331592

LAREMA, UMR 6093, UNIV ANGERS, CNRS, SFR MATH-STIC, FRANCE
Email address: cafasso@math.univ-angers.fr

INSTITUT DE RECHERCHE EN MATHÉMATIQUE ET PHYSIQUE, UCLOUVAIN, CHEMIN DU CYCLOTRON 2, B-1348 LOUVAIN-LA-NEUVE, BELGIUM
Email address: sofia.tarricone@uclouvain.be

Contemporary Mathematics
Volume **782**, 2023
https://doi.org/10.1090/conm/782/15724

A system of nonlinear difference equations for recurrence relation coefficients of a modified Jacobi weight

Yang Chen, Galina Filipuk, and Maria das Neves Rebocho

ABSTRACT. In this paper we study one modification of the Jacobi weight. We show that the coefficients in the three term recurrence relation for polynomials orthogonal with respect to this weight satisfy a complicated system of nonlinear difference equations.

1. Introduction

In recent years there has been a lot of interest to study various generalizatons of classical weights and to find nonlinear difference equations for the recurrence coefficients. In some cases it was also possible to show the connection of the recurrence coefficients to the solutions of the Painlevé equations (see numerous references in [**VA2018**]). In [**CFR2019**] the modifed Laguerre weight $w(x) = x^{\alpha}e^{-x}e^{-s/x}$, $\alpha > -1$, $s > 0$, $x > 0$ was studied and a relation to the solutions of the third Painlevé equation was established. In [**MLC2018**] another variation was studied: $w(x) = e^{-x^2-t/x^2}$ on $x \in \mathbb{R}$, $t > 0$. The Jacobi-type modifications of the weight were recently studied in [**MC2020, MC2021**]. The weights were respectively $w(x) = (1-x^2)^{\alpha}e^{-t/x^2}$, $x \in [-1,1]$, $\alpha > 0$, $t \geq 0$ and $w(x) = (1-x^2)^{\alpha}e^{-t/(1-x^2)}$, $x \in [-1,1]$, $\alpha > 0$, $t \geq 0$. In this paper we study the weight (1.2) and show that the discrete system for the recurrence coefficients is very cumbersome. It is quite unlikely that the theree-term recurrence coefficients have any connection to the Painlevé equations. The number of variables in the discrete system is large and an elimination of some of them would lead to even more complicated difference equations.

It is well-known [**C1978, S1975**] that a sequence $\{P_n(x)\}_{n\geq 0}$ of monic orthogonal polynomials of degree n in x satisfies a three-term recurrence relation. The three-term recurrence relation is given by

$$xP_n(x) = P_{n+1}(x) + \alpha_n P_n(x) + \beta_n P_{n-1}(x)\,, \quad n = 0, 1, \ldots \tag{1.1}$$

2020 *Mathematics Subject Classification.* Primary 33C47, 39A99.
Key words and phrases. Orthogonal polynomials, difference equations, modified weight.
The second author was partially supported by the OPUS 2017/25/B/ST1/00931 grant.
The third author acknowledges the FCT grants UID-B-MAT/00212/2020 and UID-P-MAT/00212/2020 at CMA-UBI. The work of the third author was partially supported by Dirección General de Investigación e Innovación, Consejería de Educación e Investigación of the Comunidad de Madrid (Spain), and Universidad de Alcalá under grant CM/JIN/2021-014, Proyectos de I+D para Jóvenes Investigadores de la Universidad de Alcalá 2021.

with initial conditions $P_0(x) = 1$, $P_{-1}(x) = 0$. The recurrence coefficients of $\{P_n\}_{n\geq 0}$, α_n and β_n, satisfy $\alpha_n \in \mathbb{R}$, $n = 0, 1, \ldots$, and $\beta_n > 0$, $n = 1, 2, \ldots$.

In recent years there has been a considerable interest to derive nonlinear difference equations for the recurrence coefficients of orthogonal polynomials (e.g., [**CG12015, CG22015**] and the references therein). For certain weights (see, for instance, [**CC2015, CI2010, M1995**]) the recurrence coefficients are related to the solutions of either discrete or differential Painlevé equation (as functions of a parameter in the weight).

In this paper we derive a system of nonlinear difference equations satisfied by the recurrence coefficients of a two parameter family of polynomials orthogonal with respect to the weight

$$w(x) = w(x; s, t) = e^{-s/x - t/x^2}(1 - x)^{\alpha}\,, \tag{1.2}$$

where $0 < x < 1$, $\alpha > 0$, $s, t > 0$. The orthogonality relation with respect to the weight function $w(x; s, t)$ defined in (1.2) is given by

$$\int_0^1 P_n(x)P_m(x)w(x; s, t)dx = h_n\delta_{n,m}\, n,\ m = 0, 1, \ldots. \tag{1.3}$$

The polynomials P_n and the constants h_n depend on s, t, but, to simplify the notation, we omit this dependence for brevity.

Note that the weight (1.2) can be represented as $w(x; s, t) = w_0(x)e^{-s/x-t/x^2}$, where $w_0(x) = (1 - x)^{\alpha}$ is a (particular case of) the Jacobi weight. Various deformations of the Jacobi weights have been extensively studied in, for example, [**BCE2010, CD2010**] (see other references therein).

For the purposes of the present paper we remark that (1.2) is a member of the so-called semi-classical class, as it satisfies a homogeneous differential equation with polynomial coefficients [**M1995**]. In this case, we have $Aw' = Cw$, where

$$A(x) = x^3(x - 1)\,,\quad C(x) = \alpha x^3 + sx^2 + (2t - s)x - 2t\,, \tag{1.4}$$

equivalently, we have the first order nonhomogeneous differential equation for the corresponding Stieltjes function (see [**M1991**])

$$AS' = CS + D\,, \tag{1.5}$$

with D being a polynomial depending on A, C.

The main tools to be used rely on results within the theory of semi-classical orthogonal polynomials, namely, differential systems [**M1995, M1991**] for the polynomials and the ladder operator approach [**CI1997**]. Let us emphasize that, in the account of the results from [**BPR2013**], the method for deriving the difference equations for α_n, β_n can be extended to the more general class of orthogonal polynomials whose Stieltjes function satisfies the Riccati equation of the form

$$A(x)S'(x) = B(x)S(x)^2 + C(x)S(x) + D(x), \tag{1.6}$$

where A, B, C, D are polynomials of degrees $4, 4, 3, 2$, respectively, and D depends on A, B and C. These are members of the so-called Laguerre-Hahn class two [**M1991**].

The reminder of the paper is organized as follows. In Section 2 we show the linear systems for orthogonal polynomials, to be used in the derivation of the difference equations for the recurrence coefficients. In Section 3 we show the difference equations involving α_n, β_n, together with some comments on the extension to the Laguerre-Hahn class.

2. Linear systems for orthogonal polynomials and compatibility conditions

In this section we briefly recall necessary formulas (see [**FR2016**] and the references therein). We shall use the notation $' = d/dx$.

The sequence of monic associated polynomials of the first kind, denoted by $\{P_n^{(1)}\}_{n\geq 0}$, satisfies a three term recurrence relation similar to (1.1):

$$P_n^{(1)}(x) = (x-\alpha_n)P_{n-1}^{(1)}(x) - \beta_n P_{n-2}^{(1)}(x)\,, \quad n = 1,2,\ldots \tag{2.1}$$

with $P_{-1}^{(1)}(x) = 0,\ P_0^{(1)}(x) = 1$.

The recurrence relations (1.1) and (2.1) can be written in the matrix form

$$Y_n = \mathcal{A}_n Y_{n-1}\,,\quad Y_n = \begin{bmatrix} P_{n+1} & P_n^{(1)} \\ P_n & P_{n-1}^{(1)} \end{bmatrix},\quad \mathcal{A}_n = \begin{bmatrix} x-\alpha_n & -\beta_n \\ 1 & 0 \end{bmatrix},\quad n \geq 1\,, \tag{2.2}$$

with initial conditions $Y_0 = \begin{bmatrix} x-\alpha_0 & -\beta_0 \\ 1 & 0 \end{bmatrix}$. The matrix $\mathcal{A}_n$ is known as the transfer matrix.

THEOREM 2.1 (see [**BPR2013**]). *Let S be a Stieltjes function, let $\{Y_n\}_{n\geq 0}$ be the corresponding sequence defined in (2.2), and let α_n, β_n be the coefficients in the recurrence relation. The following statements are equivalent:*

(a) *S satisfies the following Riccati equation with polynomial coefficients*

$$AS' = BS^2 + CS + D\,;$$

tem Y_n satisfies the matrix Sylvester equation

$$AY_n' = \mathcal{B}_n Y_n - Y_n \mathcal{C}\,,\quad n \geq 0\,, \tag{2.3}$$

where

$$\mathcal{B}_n = \begin{bmatrix} l_n & \Theta_n \\ -\Theta_{n-1}/\beta_n & l_{n-1} + (x-\alpha_n)\Theta_{n-1}/\beta_n \end{bmatrix},\quad \mathcal{C} = \begin{bmatrix} C/2 & -D \\ B & -C/2 \end{bmatrix}$$

with l_n, Θ_n polynomials of uniformly bounded degrees, satisfying the initial conditions

$$\Theta_{-1} = D\,,\ \Theta_0 = A + (x-\alpha_0)(C/2 - l_0) + B\,,\ l_{-1} = C/2\,,\ l_0 = -C/2 - (x-\alpha_0)D\,; \tag{2.4}$$

(b) *the transfer matrix $\mathcal{A}_n = \begin{bmatrix} x-\alpha_n & -\beta_n \\ 1 & 0 \end{bmatrix}$ satisfies the matrix Sylvester equation*

$$A\mathcal{A}_n' = \mathcal{B}_n \mathcal{A}_n - \mathcal{A}_n \mathcal{B}_{n-1}\,,\quad n \geq 1\,. \tag{2.5}$$

Furthermore, we have [**BPR2013**, Corollary 1]

$$\operatorname{tr} \mathcal{B}_n = 0\,,\quad n \geq 0\,, \tag{2.6}$$

$$\det \mathcal{B}_n = \det \mathcal{B}_0 + A\sum_{k=1}^{n} \frac{\Theta_{k-1}}{\beta_k}\,,\quad n \geq 1\,, \tag{2.7}$$

with $\det \mathcal{B}_0 = D(A+B) - (C/2)^2$.

From entries $(1,1)$ and $(1,2)$ of Eq. (2.5) we find two non-trivial equations, respectively:

$$A = (x-\alpha_n)(l_n - l_{n-1}) + \Theta_n - \beta_n \frac{\Theta_{n-2}}{\beta_{n-1}}, \tag{2.8}$$

$$l_n + (x-\alpha_n)\frac{\Theta_{n-1}}{\beta_n} = l_{n-2} + (x-\alpha_{n-1})\frac{\Theta_{n-2}}{\beta_{n-1}}. \tag{2.9}$$

Note that, after some basic computations, (2.8) implies (2.7) and (2.9) implies (2.6).

One of the key ingredients in the sequel is the differential system (2.3). This is the matrix form of the so-called structure relations [**Di1988**]

$$A\left(P_n^{(1)}\right)' = DP_{n+1} + (l_n + C/2)P_n^{(1)} + \Theta_n P_{n-1}^{(1)}, \quad n \geq 0, \tag{2.10}$$

$$AP'_{n+1} = (l_n - C/2)P_{n+1} - BP_n^{(1)} + \Theta_n P_n, \quad n \geq 0. \tag{2.11}$$

3. Main results

We will use the following notation:

$$P_n^{(1)}(x) = x^n - \eta_n x^{n-1} + \nu_n x^{n-2} + \mu_n^{(1)} x^{n-3} + \cdots + P_n^{(1)}(0), \tag{3.1}$$

$$\begin{aligned} P_{n+1}(x) = {} & x^{n+1} - (\eta_n + \alpha_0)x^n + (\nu_n + \alpha_0\eta_n - \beta_1)x^{n-1} \\ & + \mu_{n+1}x^{n-2} + \cdots + P_{n+1}(0), \end{aligned} \tag{3.2}$$

where

$$\eta_n = \sum_{k=1}^{n} \alpha_k, \quad \nu_n = \sum_{1\leq i<j\leq n} \alpha_i\alpha_j - \sum_{k=2}^{n} \beta_k, \quad n \geq 1. \tag{3.3}$$

In the lemma that follows we deduce the data to be used in the more general setting, the Laguerre-Hahn class two case. Such a data will then be applied, in Theorem 3.2, to deduce the system of difference equations for the recurrence coefficients of the orthogonal polynomials related to the weight (1.2).

LEMMA 3.1. *Let S be a Stieltjes function satisfying $AS' = BS^2 + CS + D$, let $\{P_n\}_{n\geq 0}$ be the corresponding sequence of monic orthogonal polynomials, satisfying the recurrence relation* (1.1),

$$P_{n+1}(x) = (x-\alpha_n)P_n(x) - \beta_n P_{n-1}(x), \quad n = 0, 1, 2, \ldots.$$

Set

$$A(x) = x^3(x-1), \quad B(x) = b_4x^4 + b_3x^3 + b_2x^2 + b_1x + b_0,$$
$$C(x) = c_3x^3 + c_2x^2 + c_1x + c_0, \quad D(x) = d_2x^2 + d_1x + d_0.$$

Then, the polynomials l_n, Θ_n satisfy $\deg(l_n) = 3, \deg(\Theta_n) = 2$. *Using the notation*

$$l_n(x) = \ell_{n,3}x^3 + \ell_{n,2}x^2 + \ell_{n,1}x + \ell_{n,0}, \quad \Theta_n(x) = \theta_{n,2}x^2 + \theta_{n,1}x + \theta_{n,0},$$

then we have

$$\begin{aligned}
\ell_{n,3} &= \frac{1}{2}(2n - c_3 - 2d_2),\\
\ell_{n,2} &= \frac{1}{2}(2\eta_n + 2d_2\alpha_0 - 2d_1 - c_2 - 2n),\\
\theta_{n,2} &= \frac{1}{2}(2d_1\alpha_0 + 2d_2\beta_1 - 2\eta_n + 2\eta_n^2 - 4\nu_n - c_1 - 2d_0 - 2\ell_{n,1}),\\
d_2 &= -(1 + b_4 + c_3),\\
d_1 &= 1 - b_3 - c_2 - 2\alpha_0 - 2b_4\alpha_0 - c_3\alpha_0,\\
d_0 &= 2\alpha_0 - b_2 - c_1 - 2b_3\alpha_0 - c_2\alpha_0 - 3\alpha_0^2 - 3b_4\alpha_0^2 - c_3\alpha_0^2 - 3\beta_1 - 2b_4\beta_1 - c_3\beta_1,\\
\ell_{n,0} &= \frac{1}{2}\Big(2b_1 + c_0 - 6\mu_{n+1} - 2b_4\mu_{n+1} + 2b_4\mu_n^{(1)} + 2b_2\alpha_0 - 2\alpha_0^2\\
&\quad +2b_3\alpha_0^2 + 2\alpha_0^3 + 2b_4\alpha_0^3 - 4\beta_1 + 2b_3\beta_1 + 6\alpha_0\beta_1 + 4b_4\alpha_0\beta_1\\
&\quad +2b_2\eta_{n-1} + c_1\eta_{n-1} - 2\ell_{n,1}\eta_{n-1} - 2\alpha_0\eta_{n-1} + 2b_3\alpha_0\eta_{n-1}\\
&\quad +2\alpha_0^2\eta_{n-1} + 2b_4\alpha_0^2\eta_{n-1} + 4\beta_1\eta_{n-1} + 2b_4\beta_1\eta_{n-1} - 2b_2\eta_n\\
&\quad -c_1\eta_n + 2\ell_{n,1}\eta_n + 2\alpha_0\eta_n - 2b_3\alpha_0\eta_n - 2\alpha_0^2\eta_n - b_4\alpha_0^2\eta_n\\
&\quad +2\beta_1\eta_n - 2\eta_{n-1}\eta_n + 2\eta_{n-1}\eta_n^2 - 2\theta_{n,1} + 4\nu_n - 6\alpha_0\nu_n\\
&\quad -2b_4\alpha_0\eta_n - 4\eta_{n-1}\nu_n - 2\eta_n\nu_n\Big),\\
\mu_n^{(1)} &= \frac{1}{4 + 2b_4 + c_3}\,(4\mu_{n+1} - b_1 - c_0 + 2b_4\mu_{n+1} + c_3\mu_{n+1} - 2b_2\alpha_0\\
&\quad -c_1\alpha_0 + 3\alpha_0^2 - 3b_3\alpha_0^2 - c_2\alpha_0^2 - 4\alpha_0^3 - 4b_4\alpha_0^3 - c_3\alpha_0^3 + 3\beta_1\\
&\quad -2b_3\beta_1 - c_2\beta_1 - 8\alpha_0\beta_1 - 6b_4\alpha_0\beta_1 - 2c_3\alpha_0\beta_1 - 4\beta_1\eta_n\\
&\quad -2b_4\beta_1\eta_n - c_3\beta_1\eta_n + 4\alpha_0\nu_n + 2b_4\alpha_0\nu_n + c_3\alpha_0\nu_n),
\end{aligned}$$

$$\begin{aligned}
\ell_{n,1} &= \frac{1}{2}\left(2b_2 + c_1 - 2\alpha_0 + 2b_3\alpha_0 + 2\alpha_0^2 + 2b_4\alpha_0^2 + 4\beta_1 + 3b_4\beta_1 + 2\beta_{n+1}\right.\\
&\quad \left.+4(n+1)\beta_{n+1} + 4b_4\beta_{n+1} + 2c_3\beta_{n+1} - 2\eta_n + 2\eta_n^2 - 4\nu_n\right),\\
\theta_{n,1} &= -\beta_{n+1}\left(2b_3 + c_2 - 1 - 2(n+1) + 2\alpha_0 + 2b_4\alpha_0 + \alpha_{n+1}\right.\\
&\quad \left.+2(n+1)\alpha_{n+1} + 2b_4\alpha_{n+1} + c_3\alpha_{n+1} + \eta_n + \eta_{n+1}\right),\\
\theta_{n,0} &= -\beta_{n+1}\left(2b_2 + c_1 - 2\alpha_0 + 2b_3\alpha_0 + 2\alpha_0^2 + 2b_4\alpha_0^2 - \alpha_{n+1}\right.\\
&\quad -2(n+1)\alpha_{n+1} + 2b_3\alpha_{n+1} + c_2\alpha_{n+1} + 2\alpha_0\alpha_{n+1}\\
&\quad +2b_4\alpha_0\alpha_{n+1} + \alpha_{n+1}^2 + 2(n+1)\alpha_{n+1}^2 + 2b_4\alpha_{n+1}^2\\
&\quad +c_3\alpha_{n+1}^2 + 4\beta_1 + 2b_4\beta_1 + \beta_{n+1} + 2(n+1)\beta_{n+1}\\
&\quad +2b_4\beta_{n+1} + c_3\beta_{n+1} + 3\beta_{n+2} + 2(n+1)\beta_{n+2}\\
&\quad +2b_4\beta_{n+2} + c_3\beta_{n+2} - \eta_n + \alpha_{n+1}\eta_n + \eta_n^2\\
&\quad \left.-\eta_{n+1} + \alpha_{n+1}\eta_{n+1} + \eta_{n+1}^2 - 2\nu_n - 2\nu_{n+1}\right).
\end{aligned}$$

PROOF. Substituting (3.1) and (3.2) into (2.10), (2.11), (2.5) and (2.6), we collect the coefficients at the highest powers of x and successively determine (in the order as in the statement) the coefficients of D and of other unknown polynomials. We also use $\eta_n = \alpha_n + \eta_{n-1}$. □

To obtain a system of difference equations for the recurrence coefficients of the orthogonal polynomials related to the weight (1.2) we will use equations (2.5) and (2.6) with the polynomials l_n, Θ_n defined in terms of the data (1.4).

Collecting the coefficients in x we have several equations from which we can successively determine μ_{n+1}, μ_n and μ_{n-1}. The formulas are cumbersome and we omit them. By substituting these relations, we obtain one of the nonlinear difference equations which involves α_n, β_n, η_n and ν_n. The second equation can be obtained by equating the quantities found before, for instance, μ_n and μ_{n-1} with n replaced by $n+1$. By substituting $b_j = 0$, $j = 0, \ldots, 4$, $c_3 = \alpha$, $c_2 = s$, $c_1 = 2t - s$, $c_0 = -2t$, we get the following result.

THEOREM 3.2. *The recurrence coefficients α_n, β_n and their combinations η_n and ν_n given by (3.3) for the weight (1.2) satisfy the following system of nonlinear difference equations:*

$$\begin{aligned}
&\alpha_n^2\left((-\alpha-2n+1)\beta_n+(\alpha+2n+5)\beta_{n+1}\right)-(\alpha+2n+1)\beta_n^2+\alpha_n^4-\alpha_n^3\\
&-\alpha_n\left(\beta_n\left(2\alpha_0+(\alpha+2n-1)\alpha_{n-1}+\eta_{n-2}+\eta_{n-1}-2n+s+1\right)-\right.\\
&\left.\beta_{n+1}\left(2\alpha_0+(\alpha+2n+4)\alpha_{n+1}-\eta_{n-1}+\eta_{n+1}-2n+s-4\right)\right)\\
&+\beta_{n+1}\left((\alpha+2n+5)\beta_{n+1}+(\alpha+2n+5)\beta_{n+2}+(\alpha+2n+3)\alpha_{n+1}^2\right.\\
&+\left(\eta_{n-1}-1\right)\eta_{n-1}+\left(\eta_{n+1}-1\right)\eta_{n+1}-4\nu_{n-1}\\
&+\alpha_{n+1}\left(2\alpha_0+\eta_{n-1}-2\eta_n+\eta_{n+1}-2n+s-3\right)-s\\
&\left.+2\left(\left(\alpha_0-1\right)\alpha_0+2\beta_1+t\right)\right)-\beta_n\left((\alpha+2n-1)\beta_{n-1}+(\alpha+2n-1)\alpha_{n-1}^2-4\beta_{n+1}\right.\\
&+\left(\eta_{n-2}-1\right)\eta_{n-2}+\left(\eta_{n-1}-1\right)\eta_{n-1}-2\nu_{n-2}-2\nu_{n-1}\\
&\left.+\alpha_{n-1}\left(2\alpha_0+\eta_{n-2}+\eta_{n-1}-2n+s+1\right)-s+2\left(\left(\alpha_0-1\right)\alpha_0+2\beta_1+t\right)\right)=0,
\end{aligned}$$

$$\begin{aligned}
&-2\alpha_0\beta_n-2n\alpha_{n-1}\beta_n-\alpha\alpha_{n-1}\beta_n+\alpha_{n-1}\beta_n+2\alpha_0\beta_{n+2}+2n\alpha_{n+2}\beta_{n+2}\\
&+\alpha\alpha_{n+2}\beta_{n+2}+5\alpha_{n+2}\beta_{n+2}-(\alpha+2n-2)\alpha_n^3+(\alpha+2n+6)\alpha_{n+1}^3-\beta_n\eta_{n-2}\\
&+\beta_n\eta_{n-1}-2\beta_n\eta_n-\beta_{n+1}\eta_n+\beta_{n+2}\eta_n+\beta_{n+1}\eta_{n+1}+\beta_{n+2}\eta_{n+2}+2n\beta_n-\beta_n\\
&-2\beta_{n+1}-2n\beta_{n+2}-5\beta_{n+2}+\alpha_n^2\left(-2\alpha_0+4\eta_{n-1}-2\eta_n+2n-s-1\right)\\
&+\alpha_{n+1}^2\left(2\alpha_0+2\eta_n-2n+s-5\right)-s\beta_n+s\beta_{n+2}\\
&+\alpha_n\left(-2\left(\alpha_0-1\right)\alpha_0-4\beta_1-(\alpha+2n)\left(2\beta_n+\beta_{n+1}\right)\right.\\
&\left.-4\alpha_{n+1}\eta_{n-1}-2\left(\eta_n-1\right)\eta_n+4\nu_{n-1}+s-2t\right)\\
&+\alpha_{n+1}\left(2\left(\alpha_0-1\right)\alpha_0+4\beta_1+(\alpha+2n+7)\beta_{n+1}\right.\\
&\left.+(2\alpha+4n+11)\beta_{n+2}+4\beta_n+2\left(\eta_n-1\right)\eta_n-4\nu_{n-1}-s+2t\right)=0.
\end{aligned}$$

REMARK 3.3. For the Laguerre-Hahn class, we first use the following data in (1.6):

$$\begin{aligned}
A(x)&=x^3(x-1)\,,\quad B(x)=b_4x^4+b_3x^3+b_2x^2+b_1x+b_0\,,\\
C(x)&=c_3x^3+c_2x^2+c_1x+c_0\,,\quad D(x)=d_2x^2+d_1x+d_0\,.
\end{aligned}$$

Nevertheless, the calculations can be repeated for a general polynomial $A(x)$ of degree four. For the weight (1.2) we have $B(x) = 0$ and A and C as in (1.4). We gave the final result (the nonlinear difference system) only for the particular case (1.2). The general case is similar with more complicated formulas, which are long enough to present in this paper.

References

[BCE2010] Estelle Basor, Yang Chen, and Torsten Ehrhardt, *Painlevé V and time-dependent Jacobi polynomials*, J. Phys. A **43** (2010), no. 1, 015204, 25, DOI 10.1088/1751-8113/43/1/015204. MR2570057

[BPR2013] A. Branquinho, A. Paiva, and M. N. Rebocho, *Sylvester equations for Laguerre-Hahn orthogonal polynomials on the real line*, Appl. Math. Comput. **219** (2013), no. 17, 9118–9131, DOI 10.1016/j.amc.2013.03.083. MR3047806

[CC2015] Min Chen and Yang Chen, *Singular linear statistics of the Laguerre unitary ensemble and Painlevé III. Double scaling analysis*, J. Math. Phys. **56** (2015), no. 6, 063506, 14, DOI 10.1063/1.4922620. MR3369903

[CD2010] Yang Chen and Dan Dai, *Painlevé V and a Pollaczek-Jacobi type orthogonal polynomials*, J. Approx. Theory **162** (2010), no. 12, 2149–2167, DOI 10.1016/j.jat.2010.07.005. MR2736877

[CFR2019] Yang Chen, Galina Filipuk, and Maria das Neves Rebocho, *Nonlinear difference equations for a modified Laguerre weight: Laguerre-Freud equations and asymptotics*, Jaen J. Approx. **11** (2019), no. 1-2, 47–65. MR4050320

[CG12015] Yang Chen and James Griffin, *Non linear difference equations arising from a deformation of the q-Laguerre weight*, Indag. Math. (N.S.) **26** (2015), no. 1, 266–279, DOI 10.1016/j.indag.2014.10.004. MR3281704

[CG22015] Yang Chen and James Griffin, *Deformed q^{-1}-Laguerre polynomials, recurrence coefficients, and non-linear difference equations*, Acta Phys. Polon. B **46** (2015), no. 9, 1871–1881, DOI 10.5506/APhysPolB.46.1871. MR3403845

[CI1997] Yang Chen and Mourad E. H. Ismail, *Ladder operators and differential equations for orthogonal polynomials*, J. Phys. A **30** (1997), no. 22, 7817–7829, DOI 10.1088/0305-4470/30/22/020. MR1616931

[CI2010] Yang Chen and Alexander Its, *Painlevé III and a singular linear statistics in Hermitian random matrix ensembles. I*, J. Approx. Theory **162** (2010), no. 2, 270–297, DOI 10.1016/j.jat.2009.05.005. MR2581382

[C1978] T. S. Chihara, *An introduction to orthogonal polynomials*, Mathematics and its Applications, Vol. 13, Gordon and Breach Science Publishers, New York-London-Paris, 1978. MR0481884

[Di1988] J. Dini, *Sur les formes linéaires et les polynômes orthogonaux de Laguerre-Hahn*, Thèse de doctorat, Univ. Pierre et Marie Curie, Paris, 1988.

[FR2016] G. Filipuk and M. N. Rebocho, *Discrete Painlevé equations for recurrence coefficients of Laguerre-Hahn orthogonal polynomials of class one*, Integral Transforms Spec. Funct. **27** (2016), no. 7, 548–565, DOI 10.1080/10652469.2016.1160903. MR3507140

[M1995] Alphonse P. Magnus, *Painlevé-type differential equations for the recurrence coefficients of semi-classical orthogonal polynomials*, Proceedings of the Fourth International Symposium on Orthogonal Polynomials and their Applications (Evian-Les-Bains, 1992), J. Comput. Appl. Math. **57** (1995), no. 1-2, 215–237, DOI 10.1016/0377-0427(93)E0247-J. MR1340938

[M1991] P. Maroni, *Une théorie algébrique des polynômes orthogonaux. Application aux polynômes orthogonaux semi-classiques* (French), Orthogonal polynomials and their applications (Erice, 1990), IMACS Ann. Comput. Appl. Math., vol. 9, Baltzer, Basel, 1991, pp. 95–130. MR1270222

[MLC2018] Chao Min, Shulin Lyu, and Yang Chen, *Painlevé III′ and the Hankel determinant generated by a singularly perturbed Gaussian weight*, Nuclear Phys. B **936** (2018), 169–188, DOI 10.1016/j.nuclphysb.2018.09.016. MR3869764

[MC2020] Chao Min and Yang Chen, *Painlevé V and the Hankel determinant for a singularly perturbed Jacobi weight*, Nuclear Phys. B **961** (2020), 115221, 25, DOI 10.1016/j.nuclphysb.2020.115221. MR4167672

[MC2021] Chao Min and Yang Chen, *Differential, difference, and asymptotic relations for Pollaczek-Jacobi type orthogonal polynomials and their Hankel determinants*, Stud. Appl. Math. **147** (2021), no. 1, 390–416, DOI 10.1111/sapm.12392. MR4313159

[S1975] Gábor Szegő, *Orthogonal polynomials*, 4th ed., American Mathematical Society Colloquium Publications, Vol. XXIII, American Mathematical Society, Providence, R.I., 1975. MR0372517

[VA2018] Walter Van Assche, *Orthogonal polynomials and Painlevé equations*, Australian Mathematical Society Lecture Series, vol. 27, Cambridge University Press, Cambridge, 2018. MR3729446

Faculty of Science and Technology, Department of Mathematics, University of Macau, Av. Padre Tomás Pereiro, Taipa, Macau, People's Republic of China
Email address: yayangchen@umac.mo

Institute of Mathematics, University of Warsaw, Banacha 2 Warsaw, 02-097, Poland
Email address: filipuk@mimuw.edu.pl

Departamento de Matemática and CMA-UBI, Universidade da Beira Interior, Rua Marquês d'Ávila e Bolama, 6201-001 Covilhã, Portugal
Email address: mneves@ubi.pt

Contemporary Mathematics
Volume **782**, 2023
https://doi.org/10.1090/conm/782/15725

q-Heun equation and initial-value space of q-Painlevé equation

Shoko Sasaki, Shun Takagi, and Kouichi Takemura

ABSTRACT. We show that the q-Heun equation and its variants appear in the linear q-difference equations associated to some q-Painlevé equations by considering the blow-up associated to their initial-value spaces. We obtain the firstly degenerated Ruijsenaars-van Diejen operator from the linear q-difference equation associated to the q-Painlevé equation of type E_8.

1. Introduction

The Painlevé transcendent is a transcendental solution of one of the six Painlevé equations and it is potentially a member of prospective special functions. The sixth Painlevé equation is a non-linear generalization of the hypergeometric equation, and the hypergeometric equation is a standard form of the second order Fuchsian differential equation with three singularities $\{0, 1, \infty\}$. Heun's differential equation is a standard form of the second order Fuchsian differential equation with four singularities $\{0, 1, t, \infty\}$, and it is written as

$$\frac{d^2y}{dx^2} + \left(\frac{\gamma}{x} + \frac{\delta}{x-1} + \frac{\epsilon}{x-t}\right)\frac{dy}{dx} + \frac{\alpha\beta x - B}{x(x-1)(x-t)}y = 0,$$

under the condition $\gamma + \delta + \epsilon = \alpha + \beta + 1$. The parameter B, which is independent from the local exponents, is called the accessory parameter. It is known that Heun's differential equation is related with the sixth Painlevé equation (see [**2**], [**12**] and references therein).

A q-difference analogue of Heun's differential equation was given by Hahn [**3**] as the generic second-order linear q-difference equation whose coefficients are quadratic polynomials, and we may write it as

$$\begin{aligned}(1.1)\qquad &(x - h_1q^{1/2})(x - h_2q^{1/2})g(x/q) + l_3l_4(x - l_1q^{-1/2})(x - l_2q^{-1/2})g(qx) \\ &- \{(l_3 + l_4)x^2 + Ex + (l_1l_2l_3l_4h_1h_2)^{1/2}(h_3^{1/2} + h_3^{-1/2})\}g(x) = 0\,.\end{aligned}$$

There is one accessory parameter E in the q-Heun equation. Note that the q-Heun equation was used to introduced the variants of the q-hypergeometric equation [**4**]. Several discrete Painlevé equations had been studied in the 1990's. Sakai [**10**] proposed a list of the second order discrete Painlevé equations. In Sakai's list,

2020 *Mathematics Subject Classification.* Primary 39A13; Secondary 33E17.
Key words and phrases. q-Painlevé equation, q-Heun equation, initial-value space, Lax pair.
The third author was supported by JSPS KAKENHI Grant Number JP18K03378.

each member of the q-difference Painlevé equation was labelled by the affine root system from the symmetry. A relationship between the q-Heun equation and the q-Painlevé equation of type $D_5^{(1)}$ (q-$P(D_5^{(1)})$) was pointed out in [**13**], and it was called the q-Painlevé-Heun correspondence.

In this paper, we establish a further relationship between the q-Heun equation and the q-Painlevé equation by using the initial-value space of the q-Painlevé equation. Namely we derive the q-Heun equation by considering the Lax pair and the initial-value space.

Jimbo and Sakai [**5**] introduced a q-analogue of the sixth Painlevé equation by considering the connection preserving deformation of certain q-linear difference equations. It is essentially equivalent to a compatibility condition of a Lax pair, and the Lax pair, which was reformulated by Kajiwara, Noumi and Yamada [**6**], is written as

$$\begin{aligned} L_1 &= \Big\{ \frac{z(g\nu_1-1)(g\nu_2-1)}{qg} - \frac{\nu_1\nu_2\nu_3\nu_4(g-\nu_5/\kappa_2)(g-\nu_6/\kappa_2)}{fg} \Big\} \\ &\quad + \frac{\nu_1\nu_2(z-q\nu_3)(z-q\nu_4)}{q(qf-z)}(g-T_z^{-1}) + \frac{(z-\kappa_1/\nu_7)(z-\kappa_1/\nu_8)}{q(f-z)}\Big(\frac{1}{g}-T_z\Big), \\ L_2 &= \Big(1-\frac{f}{z}\Big)T + T_z - \frac{1}{g}. \end{aligned}$$

Here T_z represents the transformation $z \to qz$ and T represents the time evolution. The parameters are constrained by the relation $\kappa_1^2\kappa_2^2 = q\nu_1\nu_2\dots\nu_8$. We use the notation such as $T(f) = \overline{f}$ and $T^{-1}(g) = \underline{g}$. The time evolution of the parameters is given by $\overline{\kappa}_1 = \kappa_1/q$, $\overline{\kappa}_2 = q\kappa_2$ and $\overline{\nu}_i = \nu_i$ ($i = 1, 2, \dots, 8$). It follows from the compatibility condition for the Lax operators L_1 and L_2 (see [**6**]) that the parameters f and g satisfies

$$f\overline{f} = \nu_3\nu_4\frac{(g-\nu_5/\kappa_2)(g-\nu_6/\kappa_2)}{(g-1/\nu_1)(g-1/\nu_2)}, \quad g\underline{g} = \frac{1}{\nu_1\nu_2}\frac{(f-\kappa_1/\nu_7)(f-\kappa_1/\nu_8)}{(f-\nu_3)(f-\nu_4)}.$$

This is the q-Painlevé equation essentially introduced by Jimbo and Sakai [**5**], which we denote by q-$P(D_5^{(1)})$. The time evolution of the variable f is given by

$$\overline{f} = \nu_3\nu_4\frac{(g-\nu_5/\kappa_2)(g-\nu_6/\kappa_2)}{f(g-1/\nu_1)(g-1/\nu_2)} \tag{1.2}$$

and the value $\overline{f}$ is indefinite in the case $(f, g) = (0, \nu_5/\kappa_2)$ because of the form $0/0$ in the right hand side of Eq. (1.2). To resolve the indefiniteness, we consider the blow-up at $(f, g) = (0, \nu_5/\kappa_2)$. Namely we set $(f, g) = (f_1, f_1g_1 + \nu_5/\kappa_2)$. Then the point $(f, g) = (0, \nu_5/\kappa_2)$ corresponds to the line $f_1 = 0$, and we have

$$\overline{f} = \frac{\nu_3\nu_4 g_1(f_1g_1 + (\nu_5-\nu_6)/\kappa_2)}{(f_1g_1+\nu_5/\kappa_2-1/\nu_1)(f_1g_1+\nu_5/\kappa_2-1/\nu_2)}.$$

Hence the indefiniteness is resolved. The indefiniteness may occur at the following eight points;

$$P_i : (\infty, 1/\nu_i)_{i=1,2}, \quad (\nu_i, \infty)_{i=3,4}, \quad (0, \nu_i/\kappa_2)_{i=5,6}, \quad (\kappa_1/\nu_i, 0)_{i=7,8}.$$

The space of initial values of q-$P(D_5^{(1)})$ is defined by blowing up eight points $P_1, \dots, P_8$ of $\mathbb{P}^1 \times \mathbb{P}^1$.

We now consider the blow-up at P_5 $(0, \nu_5/\kappa_2)$ jointly with the q-difference equation $L_1y(z) = 0$, where L_1 is one of the Lax pair. Note that the equation

$L_1 y(z)=0$ itself (see Eq. (2.1)) is different from the q-Heun equation in Eq. (1.1). We substitute $(f,g)=(f_1, f_1 g_1+\nu_5/\kappa_2)$ into the equation $L_1 y(z)=0$ and set $f_1=0$. Then we have

$$\begin{aligned}&(z-q\nu_3)(z-q\nu_4)y(z/q)+\frac{1}{\nu_1\nu_2}\Big(z-\frac{\kappa_1}{\nu_7}\Big)\Big(z-\frac{\kappa_1}{\nu_8}\Big)y(qz)\\&-\Big[\Big(\frac{1}{\nu_1}+\frac{1}{\nu_2}\Big)z^2+\Big\{qg_1\nu_3\nu_4\Big(1-\frac{\nu_6}{\nu_5}\Big)-\frac{q\nu_5(\nu_3+\nu_4)}{\kappa_2}-\frac{\kappa_1\kappa_2(\nu_7+\nu_8)}{\nu_1\nu_2\nu_5\nu_7\nu_8}\Big\}z\\&\quad+\frac{q\kappa_1{\nu_3}^{1/2}{\nu_4}^{1/2}}{{\nu_1}^{1/2}{\nu_2}^{1/2}{\nu_7}^{1/2}{\nu_8}^{1/2}}\Big\{\Big(\frac{q\nu_6}{\nu_5}\Big)^{1/2}+\Big(\frac{q\nu_6}{\nu_5}\Big)^{-1/2}\Big\}\Big]y(z)=0\end{aligned}$$

(see Eq. (2.4)). Thus we obtain the q-Heun equation with the accessory parameter by considering the equation $L_1 y(z)=0$ on the line $f_1=0$, which is an exceptional curve with respect to the blow-up at the point P_5 on the ininial value space. We can also obtain the q-Heun equation by the blow-up at P_i $(i=1,\dots,8)$.

We establish similar results on the q-Painlevé equation of type $E_6^{(1)}$, $E_7^{(1)}$ and $E_8^{(1)}$ in this paper. The corresponding linear q-difference equations for the cases $E_6^{(1)}$ and $E_7^{(1)}$ are the variants of q-Heun equation which were introduced in [**14**] (see Eqs. (2.5), (2.8)). Note that q-Heun equation and the variants are also obtained by degenerations of Ruijsenaars-van Diejen system [**9, 13, 15**]. The corresponding linear q-difference equations for the case $E_8^{(1)}$ is the firstly degenerated Ruijsenaars-van Diejen operator. Note that Noumi, Ruijsenaars and Yamada [**8**] obtained the non-degenerate Ruijsenaars-van Diejen system from the Lax formalism of the elliptic Painlevé equation e-$P(E_8^{(1)})$.

This paper is organized as follows. In section 2, we recall the q-Painlevé equations q-$P(D_5^{(1)})$, q-$P(E_6^{(1)})$ and q-$P(E_7^{(1)})$, their Lax pairs, and the corresponding initial-value spaces by following the review by Kajiwara, Noumi and Yamada [**6**]. We consider restriction of the variables of initial-value spaces and obtain the q-Heun equation and the variants. In section 3, we recall the q-Painlevé equation of type $E_8^{(1)}$ by following Yamada [**16**], and obtain the firstly degenerated Ruijsenaars-van Diejen operator by considering a suitable limit.

2. q-Painlevé equation, Lax pairs, initial-value space of q-Painlevé equation and q-Heun equation

Lax pairs and the initial-value spaces for discrete Painlevé equations were reviewed by Kajiwara, Noumi and Yamada in [**6**]. We recall them for the q-Painlevé equations of type $D_5^{(1)}$, $E_6^{(1)}$ and $E_7^{(1)}$, and we apply the procedure of the blow-up associated with the the initial-value spaces to the linear q-difference equation related to the Lax pairs.

Throughout this section, we assume that the parameters $\kappa_1,\kappa_2,\nu_1,\nu_2,\dots,\nu_8$ are non-zero and satisfy the relation $\kappa_1^2\kappa_2^2=q\nu_1\nu_2\dots\nu_8$.

2.1. The case $D_5^{(1)}$. The q-Painlevé equation of type $D_5^{(1)}$ (q-$P(D_5^{(1)})$) was given as

$$f\overline{f}=\nu_3\nu_4\frac{(g-\nu_5/\kappa_2)(g-\nu_6/\kappa_2)}{(g-1/\nu_1)(g-1/\nu_2)},\quad g\underline{g}=\frac{1}{\nu_1\nu_2}\frac{(f-\kappa_1/\nu_7)(f-\kappa_1/\nu_8)}{(f-\nu_3)(f-\nu_4)}.$$

The space of initial condition for q-$P(D_5^{(1)})$ was realized by blowing up eight points $P_1, \dots, P_8$ of $\mathbb{P}^1 \times \mathbb{P}^1$, where

$$P_i : (\infty, 1/\nu_i)_{i=1,2}, \quad (\nu_i, \infty)_{i=3,4}, \quad (0, \nu_i/\kappa_2)_{i=5,6}, \quad (\kappa_1/\nu_i, 0)_{i=7,8}.$$

On the other hand, q-$P(D_5^{(1)})$ was realized by the compatibility condition for the Lax pair L_1 and L_2, where

$$\begin{aligned} L_1 = & \Big\{ \frac{z(g\nu_1 - 1)(g\nu_2 - 1)}{qg} - \frac{\nu_1\nu_2\nu_3\nu_4(g - \nu_5/\kappa_2)(g - \nu_6/\kappa_2)}{fg} \Big\} \\ & + \frac{\nu_1\nu_2(z - q\nu_3)(z - q\nu_4)}{q(qf - z)}(g - T_z^{-1}) + \frac{(z - \kappa_1/\nu_7)(z - \kappa_1/\nu_8)}{q(f - z)}\Big(\frac{1}{g} - T_z\Big) \\ L_2 = & \Big(1 - \frac{f}{z}\Big)T + T_z - \frac{1}{g} \end{aligned}$$

We track the linear q-difference equation $L_1 y(z) = 0$, i. e.

$$\begin{aligned} & \Big\{ \frac{z(g\nu_1 - 1)(g\nu_2 - 1)}{qg} - \frac{\nu_1\nu_2\nu_3\nu_4(g - \nu_5/\kappa_2)(g - \nu_6/\kappa_2)}{fg} \Big\} y(z) \\ & + \frac{\nu_1\nu_2(z - q\nu_3)(z - q\nu_4)}{q(qf - z)}(gy(z) - y(z/q)) \\ & + \frac{(z - \kappa_1/\nu_7)(z - \kappa_1/\nu_8)}{q(f - z)}\Big(\frac{y(z)}{g} - y(qz)\Big) = 0, \end{aligned} \tag{2.1}$$

on the process of the blow-up of the point P_i $(i = 1, 2, \dots, 8)$ respectively, and we investigate a relationship with the q-Heun equation

$$\begin{aligned} & (z - h_1 q^{1/2})(z - h_2 q^{1/2}) g(z/q) + l_3 l_4 (z - l_1 q^{-1/2})(z - l_2 q^{-1/2}) g(qz) \\ & - \{(l_3 + l_4)z^2 + Ez + (l_1 l_2 l_3 l_4 h_1 h_2)^{1/2}(h_3^{1/2} + h_3^{-1/2})\} g(z) = 0. \end{aligned} \tag{2.2}$$

The parameter E is called the accessory parameter. Note that details of the calculation in this subsection were performed in [**11**].

<u>The points $P_5 : (0, \nu_5/\kappa_2)$ and $P_6 : (0, \nu_6/\kappa_2)$</u>

We realize the blow-up at $(f, g) = (0, \nu_5/\kappa_2)$ by setting $(f, g) = (f_1, f_1 g_1 + \nu_5/\kappa_2)$. Then the point $P_5 : (0, \nu_5/\kappa_2)$ corresponds to the line $f_1 = 0$. By the transformation $(f, g) = (f_1, f_1 g_1 + \nu_5/\kappa_2)$, the operator L_1 is written as

$$\begin{aligned} L_1 = & \frac{(\nu_1\nu_5 - \kappa_2 + f_1 g_1 \kappa_2 \nu_1)(\nu_2\nu_5 - \kappa_2 + f_1 g_1 \kappa_2 \nu_2) z}{q\kappa_2(f_1 g_1 \kappa_2 + \nu_5)} \\ & - \frac{g_1 \nu_1\nu_2\nu_3\nu_4(f_1 g_1 \kappa_2 + \nu_5 - \nu_6)}{f_1 g_1 \kappa_2 + \nu_5} \\ & + \frac{\nu_1\nu_2(z - q\nu_3)(z - q\nu_4)(f_1 g_1 \kappa_2 + \nu_5 - \kappa_2 T_z^{-1})}{q\kappa_2(qf_1 - z)} \\ & + \frac{(\nu_7 z - \kappa_1)(\nu_8 z - \kappa_1)(\kappa_2 - (f_1 g_1 \kappa_2 + \nu_5)T_z)}{q(f_1 g_1 \kappa_2 + \nu_5)\nu_7\nu_8(f_1 - z)}. \end{aligned}$$

We set $f_1 = 0$ on the equation $L_1 y(z) = 0$. Then we have

(2.3)

$$\nu_1\nu_2(z-q\nu_3)(z-q\nu_4)y(z/q) + \Big(z-\frac{\kappa_1}{\nu_7}\Big)\Big(z-\frac{\kappa_1}{\nu_8}\Big)y(qz) - \Big[(\nu_1+\nu_2)z^2 + \Big\{qg_1\nu_1\nu_2\nu_3\nu_4\Big(1-\frac{\nu_6}{\nu_5}\Big) - \frac{q\nu_1\nu_2\nu_5(\nu_3+\nu_4)}{\kappa_2} - \frac{\kappa_1\kappa_2(\nu_7+\nu_8)}{\nu_5\nu_7\nu_8}\Big\}z + \frac{q^2\nu_1\nu_2\nu_3\nu_4\nu_5}{\kappa_2} + \frac{{\kappa_1}^2\kappa_2}{\nu_5\nu_7\nu_8}\Big\}\Big]y(z) = 0.$$

It follows from the relation $\kappa_1^2\kappa_2^2 = q\nu_1\nu_2\dots\nu_8$ that Eq. (2.3) is written as

$$(2.4)\quad (z-q\nu_3)(z-q\nu_4)y(z/q) + \frac{1}{\nu_1\nu_2}\Big(z-\frac{\kappa_1}{\nu_7}\Big)\Big(z-\frac{\kappa_1}{\nu_8}\Big)y(qz) - \Big[\Big(\frac{1}{\nu_1}+\frac{1}{\nu_2}\Big)z^2 + \Big\{qg_1\nu_3\nu_4\Big(1-\frac{\nu_6}{\nu_5}\Big) - \frac{q\nu_5(\nu_3+\nu_4)}{\kappa_2} - \frac{\kappa_1\kappa_2(\nu_7+\nu_8)}{\nu_1\nu_2\nu_5\nu_7\nu_8}\Big\}z + \frac{q\kappa_1\nu_3^{1/2}{\nu_4}^{1/2}}{{\nu_1}^{1/2}{\nu_2}^{1/2}{\nu_7}^{1/2}{\nu_8}^{1/2}}\Big\{\Big(\frac{q\nu_6}{\nu_5}\Big)^{1/2} + \Big(\frac{q\nu_6}{\nu_5}\Big)^{-1/2}\Big\}\Big]y(z) = 0.$$

We compare it with the q-Heun equation given in Eq. (2.2). Then we obtain the following correspondence

$$h_1 = q^{1/2}\nu_3,\quad h_2 = q^{1/2}\nu_4,\quad h_3 = \frac{q\nu_6}{\nu_5}$$
$$l_1 = \frac{q^{1/2}\kappa_1}{\nu_7},\quad l_2 = \frac{q^{1/2}\kappa_1}{\nu_8},\quad l_3 = \frac{1}{\nu_1},\quad l_4 = \frac{1}{\nu_2}.$$

We may regard the parameter g_1 as an accessory parameter.

On the blow-up at the point $P_6 : (0, \nu_6/\kappa_2)$, we obtain the q-Heun equation where the parameters ν_5 and ν_6 are exchanged.

The points $P_7 : (\kappa_1/\nu_7, 0)$ and $P_8 : (\kappa_1/\nu_8, 0)$

We realize the blow-up at $(f, g) = (\kappa_1/\nu_7, 0)$ by setting $(f, g) = (f_1g_1 + \kappa_1/\nu_7, g_1)$. Then the point $P_7 : (\kappa_1/\nu_7, 0)$ corresponds to the line $g_1 = 0$. By the transformation $(f, g) = (f_1g_1 + \kappa_1/\nu_7, g_1)$, the operator L_1 is written as

$$L_1 = \frac{(g_1\nu_1-1)(g_1\nu_2-1)z}{qg_1} - \frac{\nu_1\nu_2\nu_3\nu_4(g_1\kappa_2-\nu_5)(g_1\kappa_2-\nu_6)\nu_7}{g_1{\kappa_2}^2(\kappa_1+f_1g_1\nu_7)} + \frac{\nu_1\nu_2\nu_7(z-q\nu_3)(z-q\nu_4)(g_1-T_z^{-1})}{q(q\kappa_1+qf_1g_1\nu_7-\nu_7z)} + \frac{(\nu_7z-\kappa_1)(\nu_8z-\kappa_1)(1-g_1T_z)}{qg_1\nu_8(\kappa_1+f_1g_1\nu_7-\nu_7z)}.$$

We consider the equation $L_1y(z) = 0$. It follows from the relation $\kappa_1^2\kappa_2^2 = q\nu_1\nu_2\dots\nu_8$ that divergence as $g_1 \to 0$ in $L_1y(z) = 0$ was avoided and we have

$$-\frac{\nu_1\nu_2\nu_7(z-q\nu_3)(z-q\nu_4)}{q(q\kappa_1-\nu_7z)}y(z/q) + \frac{\nu_8z-\kappa_1}{q\nu_8}y(qz) + \Big\{\frac{\nu_1\nu_2\nu_3\nu_4\nu_7(\nu_5+\nu_6)}{\kappa_2(\kappa_1-\nu_7z)} - \frac{\kappa_1(\nu_1+\nu_2)z}{q(\kappa_1-\nu_7z)} - \frac{\nu_1\nu_2\nu_3\nu_4(\nu_5+\nu_6){\nu_7}^2z}{\kappa_1\kappa_2(\kappa_1-\nu_7z)} + \frac{f_1\nu_7z}{q(\kappa_1-\nu_7z)} - \frac{f_1{\nu_7}^2z}{q\nu_8(\kappa_1-\nu_7z)} + \frac{(\nu_1+\nu_2)\nu_7z^2}{q(\kappa_1-\nu_7z)}\Big\}y(z) = 0$$

by setting $g_1 = 0$. To obtain the q-Heun equation, we transform the dependent variable $y(z)$ by setting $u(z) = y(z)/(\nu_7 z - \kappa_1)$. Then it follows that

$$(z - q\nu_3)(z - q\nu_4)u(z/q) + \frac{q^2}{\nu_1\nu_2}\Big(z - \frac{\kappa_1}{q\nu_7}\Big)\Big(z - \frac{\kappa_1}{\nu_8}\Big)u(qz)$$
$$- \Big[\Big(\frac{q}{\nu_1} + \frac{q}{\nu_2}\Big)z^2 + \Big\{\frac{qf_1}{\nu_1\nu_2}\Big(1 - \frac{\nu_7}{\nu_8}\Big) - \frac{\kappa_1 q}{\nu_7}\Big(\frac{1}{\nu_1} + \frac{1}{\nu_2}\Big) - \frac{q^2\nu_3\nu_4\nu_7(\nu_5+\nu_6)}{\kappa_1\kappa_2}\Big\}z$$
$$+ \frac{q^{3/2}\kappa_1{\nu_3}^{1/2}{\nu_4}^{1/2}}{{\nu_1}^{1/2}{\nu_2}^{1/2}{\nu_7}^{1/2}{\nu_8}^{1/2}}\Big\{\Big(\frac{\nu_6}{\nu_5}\Big)^{1/2} + \Big(\frac{\nu_6}{\nu_5}\Big)^{-1/2}\Big\}\Big]u(z) = 0.$$

We compare it with the q-Heun equation given in Eq. (2.2). Then we obtain the following correspondence

$$h_1 = q^{1/2}\nu_3, \quad h_2 = q^{1/2}\nu_4, \quad h_3 = \frac{\nu_6}{\nu_5}$$
$$l_1 = \frac{\kappa_1}{q^{1/2}\nu_7}, \quad l_2 = \frac{q^{1/2}\kappa_1}{\nu_8}, \quad l_3 = \frac{q}{\nu_1}, \quad l_4 = \frac{q}{\nu_2}.$$

We may regard f_1 as an accessory parameter.

On the blow-up at the point $P_8 : (\kappa_1/\nu_8, 0)$, we obtain the q-Heun equation where the parameters ν_7 and ν_8 are exchanged.

The points $P_1 : (\infty, 1/\nu_1)$ and $P_2 : (\infty, 1/\nu_2)$

We consider the blow-up at the point $P_1 : (\infty, 1/\nu_1)$. Set $(f, g) = (1/f_0, g_0)$. Then the point $(f, g) = (\infty, 1/\nu_1)$ corresponds to the point $(f_0, g_0) = (0, 1/\nu_1)$. We realize the blow-up at $(f_0, g_0) = (0, 1/\nu_1)$ by setting $(f_0, g_0) = (f_1, f_1 g_1 + 1/\nu_1)$. Then the point $P_1 : (\infty, \frac{1}{\nu_1})$ corresponds to the line $f_1 = 0$. We substitute $(f, g) = (1/f_0, g_0)$ and $(f_0, g_0) = (f_1, f_1 g_1 + 1/\nu_1)$ into the equation $L_1 y(z) = 0$. We set $f_1 = 0$. By applying the relation $\kappa_1^2\kappa_2^2 = q\nu_1\nu_2\dots\nu_8$, we obtain

$$(z - q\nu_3)(z - q\nu_4)y(z/q) + \frac{q}{\nu_1\nu_2}\Big(z - \frac{\kappa_1}{\nu_7}\Big)\Big(z - \frac{\kappa_1}{\nu_8}\Big)y(qz)$$
$$- \Big[\Big(\frac{1}{\nu_1} + \frac{q}{\nu_2}\Big)z^2 + \Big\{g_1 q\Big(1 - \frac{\nu_1}{\nu_2}\Big) - \frac{q}{\nu_1}(\nu_3 + \nu_4) - \frac{\kappa_1 q}{\nu_2}\Big(\frac{1}{\nu_7} + \frac{1}{\nu_8}\Big)\Big\}z$$
$$+ \Big(\frac{q^3\nu_3\nu_4{\kappa_1}^2}{\nu_1\nu_2\nu_7\nu_8}\Big)^{1/2}\Big\{\Big(\frac{\nu_6}{\nu_5}\Big)^{1/2} + \Big(\frac{\nu_6}{\nu_5}\Big)^{-1/2}\Big\}\Big]y(z) = 0.$$

We compare it with the q-Heun equation given in Eq. (2.2). Then we obtain the following correspondence

$$h_1 = q^{1/2}\nu_3, \quad h_2 = q^{1/2}\nu_4, \quad h_3 = \frac{\nu_6}{\nu_5}$$
$$l_1 = \frac{q^{1/2}\kappa_1}{\nu_7}, \quad l_2 = \frac{q^{1/2}\kappa_1}{\nu_8}, \quad l_3 = \frac{1}{\nu_1}, \quad l_4 = \frac{q}{\nu_2}.$$

We may regard g_1 as an accessory parameter.

On the blow-up at the point $P_2 : (\infty, 1/\nu_2)$, we obtain the q-Heun equation where the parameters ν_1 and ν_2 are exchanged.

The points $P_3 : (\nu_3, \infty)$ and $P_4 : (\nu_4, \infty)$

We consider the blow-up at the point $P_3 : (\nu_3, \infty)$. Set $(f, g) = (f_0, 1/g_0)$. Then the point $(f, g) = (\nu_3, \infty)$ corresponds to the point $(f_0, g_0) = (\nu_3, 0)$. We realize the blow-up at $(f_0, g_0) = (\nu_3, 0)$ by setting $(f_0, g_0) = (f_1 g_1 + \nu_3, g_1)$. Then the point

$P_3:(\nu_3,\infty)$ corresponds to the line $g_1=0$. We substitute $(f,g)=(f_0,1/g_0)$ and $(f_0,g_0)=(f_1g_1+\nu_3,g_1)$ into the equation $L_1y(z)=0$. Set $g_1=0$. Then we have

$$\begin{aligned}&\frac{\nu_1\nu_2(z-q\nu_4)}{q}y(z/q)-\frac{(\nu_7z-\kappa_1)(\nu_8z-\kappa_1)}{q\nu_7\nu_8(\nu_3-z)}y(qz)\\&+\frac{1}{\kappa_2q(q\nu_3-z)}\Big[q^2\nu_1\nu_2\nu_3\nu_4(\nu_5+\nu_6)+\{qf_1\kappa_2\nu_1\nu_2(1-\nu_4/\nu_3)\\&\qquad-q\kappa_2\nu_3(\nu_1+\nu_2)-q\nu_1\nu_2\nu_4(\nu_5+\nu_6)\}z+\kappa_2(\nu_1+\nu_2)z^2\Big]y(z)=0.\end{aligned}$$

To obtain the q-Heun equation, we set $u(z)=y(z)/(z-q\nu_3)$. Then

$$\begin{aligned}&(z-q^2\nu_3)(z-q\nu_4)u(z/q)+\frac{q^2}{\nu_1\nu_2}\Big(z-\frac{\kappa_1}{\nu_7}\Big)\Big(z-\frac{\kappa_1}{\nu_8}\Big)u(qz)\\&-\Big[q\Big(\frac{1}{\nu_1}+\frac{1}{\nu_2}\Big)z^2+q^2\Big\{f_1\Big(1-\frac{\nu_4}{\nu_3}\Big)-\nu_3\Big(\frac{1}{\nu_1}+\frac{1}{\nu_2}\Big)-\frac{\nu_4}{\kappa_2}(\nu_5+\nu_6)\Big\}z\\&\quad+\frac{q^3\nu_3\nu_4{\nu_5}^{1/2}{\nu_6}^{1/2}}{\kappa_2}\Big\{\Big(\frac{\nu_6}{\nu_5}\Big)^{1/2}+\Big(\frac{\nu_6}{\nu_5}\Big)^{-1/2}\Big\}\Big]u(z)=0.\end{aligned}$$

We compare it with the q-Heun equation given in Eq. (2.2). Then we obtain the following correspondence

$$\begin{aligned}&h_1=q^{3/2}\nu_3,\quad h_2=q^{1/2}\nu_4,\quad h_3=\frac{\nu_6}{\nu_5},\\&l_1=\frac{q^{1/2}\kappa_1}{\nu_7},\quad l_2=\frac{q^{1/2}\kappa_1}{\nu_8},\quad l_3=\frac{q}{\nu_1},\quad l_4=\frac{q}{\nu_2}.\end{aligned}$$

We may regard f_1 as an accessory parameter.

On the blow-up at the point $P_4:(\nu_4,\infty)$, we obtain the q-Heun equation where the parameters ν_3 and ν_4 are exchanged.

The other cases related to the q-Heun equation

We can obtain the q-Heun equation by other restrictions of the parameters. We substitute $f=\kappa_1/\nu_7$ to the equation $L_1y(z)=0$. Then we have

$$\begin{aligned}&(z-q\nu_3)(z-q\nu_4)y(z/q)+\frac{1}{\nu_1\nu_2}\Big(z-\frac{q\kappa_1}{\nu_7}\Big)\Big(z-\frac{\kappa_1}{\nu_8}\Big)y(qz)-\Big[\Big(\frac{1}{\nu_1}+\frac{1}{\nu_2}\Big)z^2\\&+\Big\{\frac{q(\nu_3\nu_7-\kappa_1)(\nu_4\nu_7-\kappa_1)}{\nu_7\kappa_1}g-\frac{\kappa_1q}{\nu_7}\Big(\frac{1}{\nu_1}+\frac{1}{\nu_2}\Big)-\frac{q\nu_3\nu_4\nu_7(\nu_5+\nu_6)}{\kappa_1\kappa_2}\Big\}z\\&\quad+\frac{q^{3/2}\kappa_1{\nu_3}^{1/2}{\nu_4}^{1/2}}{{\nu_1}^{1/2}{\nu_2}^{1/2}{\nu_7}^{1/2}{\nu_8}^{1/2}}\Big\{\Big(\frac{\nu_6}{\nu_5}\Big)^{1/2}+\Big(\frac{\nu_6}{\nu_5}\Big)^{-1/2}\Big\}\Big]y(z)=0.\end{aligned}$$

Hence we obtain the q-Heun equation, and we may regard g as an accessory parameter. We have a similar result by substituting $f=\kappa_1/\nu_8$ to the equation $L_1y(z)=0$.

We substitute $f=\nu_3$ to the equation $L_1y(z)=0$. Then we have

$$\begin{aligned}&(z-\nu_3)(z-q\nu_4)y(z/q)+\frac{1}{\nu_1\nu_2}\Big(z-\frac{\kappa_1}{\nu_7}\Big)\Big(z-\frac{\kappa_1}{\nu_8}\Big)y(qz)\\&-\Big[\Big(\frac{1}{\nu_1}+\frac{1}{\nu_2}\Big)z^2-\Big(\frac{(\nu_3\nu_7-\kappa_1)(\nu_4\nu_7-\kappa_1)}{\nu_1\nu_2\nu_3\nu_7\nu_7}\frac{1}{g}+\frac{\nu_3}{\nu_1}+\frac{\nu_3}{\nu_2}+\frac{q\nu_4\nu_5}{\kappa_2}+\frac{q\nu_4\nu_6}{\kappa_2}\Big)z\\&\quad+\frac{q\nu_3\nu_4{\nu_5}^{1/2}{\nu_6}^{1/2}}{\kappa_2}\Big\{\Big(\frac{\nu_6}{\nu_5}\Big)^{1/2}+\Big(\frac{\nu_6}{\nu_5}\Big)^{-1/2}\Big\}\Big]y(z)=0.\end{aligned}$$

Hence we obtain the q-Heun equation, and we may regard g as an accessory parameter. Note that this procedure was essentially considered in section 3.1 of [**13**]. We have a similar result by substituting $f = \nu_4$ to the equation $L_1 y(z) = 0$.

2.2. The case $E_6^{(1)}$. The q-Painlevé equation of type $E_6^{(1)}$ (q-$P(E_6^{(1)})$) in [**6**] was given as

$$\frac{(fg-1)(\overline{f}g-1)}{f\overline{f}} = \frac{(g-1/\nu_1)(g-1/\nu_2)(g-1/\nu_3)(g-1/\nu_4)}{(g-\nu_5/\kappa_2)(g-\nu_6/\kappa_2)},$$
$$\frac{(fg-1)(f\underline{g}-1)}{g\underline{g}} = \frac{(f-\nu_1)(f-\nu_2)(f-\nu_3)(f-\nu_4)}{(f-\kappa_1/\nu_7)(f-\kappa_1/\nu_8)}.$$

The space of initial condition for q-$P(E_6^{(1)})$ was realized by blowing up eight points $P_1, \dots, P_8$ of $\mathbb{P}^1 \times \mathbb{P}^1$, where

$$P_i : \Big(\nu_i, \frac{1}{\nu_i}\Big)_{i=1,2,3,4}, \quad \Big(0, \frac{\nu_i}{\kappa_2}\Big)_{i=5,6}, \quad \Big(\frac{\kappa_1}{\nu_i}, 0\Big)_{i=7,8}.$$

On the other hand, q-$P(E_6^{(1)})$ was obtained by the compatibility condition for the Lax operators L_1 and L_2, where

$$\begin{aligned} L_1 =& \frac{z(g\nu_1-1)(g\nu_2-1)(g\nu_3-1)(g\nu_4-1)}{g(fg-1)(gz-q)} - \frac{(g\kappa_2/\nu_5-1)(g\kappa_2/\nu_6-1)\kappa_1^2}{qfg\nu_7\nu_8} \\ &+ \frac{(\nu_1-z/q)(\nu_2-z/q)(\nu_3-z/q)(\nu_4-z/q)}{f-z/q}\Big\{\frac{g}{1-gz/q} - T_z^{-1}\Big\} \\ &+ \frac{(\kappa_1/\nu_7-z)(\kappa_1/\nu_8-z)}{q(f-z)}\Big\{\Big(\frac{1}{g}-z\Big) - T_z\Big\}, \\ L_2 =& \Big(1-\frac{f}{z}\Big)T + T_z - \Big(\frac{1}{g}-z\Big). \end{aligned}$$

We track the linear q-difference equation $L_1 y(z) = 0$ on the process of the blow-up of the point P_i ($i = 1, 2, \dots, 8$) respectively, and we investigate a relationship with the variant of the q-Heun equation of degree 3;

$$\begin{aligned} (2.5) \qquad & (z-h_1q^{1/2})(z-h_2q^{1/2})(z-h_3q^{1/2})g(z/q) \\ &+ (z-l_1q^{-1/2})(z-l_2q^{-1/2})(z-l_3q^{-1/2})g(qz) \\ &+ \{-(q^{1/2}+q^{-1/2})z^3 + (h_1+h_2+h_3+l_1+l_2+l_3)z^2 \\ &\quad - Ez + (l_1l_2l_3h_1h_2h_3)^{1/2}(h_4^{-1/2}+h_4^{1/2})\}g(z) = 0. \end{aligned}$$

The variant of the q-Heun equation of degree 3 was introduced in [**14**], and it has one accessory parameter E. The term $h_1 + h_2 + h_3 + l_1 + l_2 + l_3$ in the coefficient of $z^2 g(z)$ is specified by the condition that the singularity $z = 0$ is apparent, which is a fundamental condition of the variant of the q-Heun equation of degree 3. The variant of the q-Heun equation of degree 3 is the eigenfunction of the thirdly degenerated Ruijsenaars-van Diejen operator [**13**] with the eigenvalue E which corresponds to the accessory parameter, while the q-Heun equation is related to the fourthly degenerated Ruijsenaars-van Diejen operator. Note that details of the calculation in this subsection were performed in [**11**].

The points $P_5 : (0, \nu_5/\kappa_2)$ and $P_6 : (0, \nu_6/\kappa_2)$

We realize the blow-up at $(f, g) = (0, \nu_5/\kappa_2)$ by setting $(f, g) = (f_1 g_1, g_1 + \nu_5/\kappa_2)$. Then the point $P_5 : (0, \nu_5/\kappa_2)$ corresponds to the line $g_1 = 0$. We substitute

$(f,g)=(f_1g_1, g_1+\nu_5/\kappa_2)$ into the equation $L_1y(z)=0$, and set $g_1=0$. Then we have

$$
\begin{aligned}
(2.6)\quad &\Big(\frac{\kappa_1}{\nu_7}-z\Big)\Big(\frac{\kappa_1}{\nu_8}-z\Big)y(qz)+\frac{(q\nu_1-z)(q\nu_2-z)(q\nu_3-z)(q\nu_4-z)}{q^2}y(z/q)\\
&+\Big\{-\frac{\kappa_1^2\kappa_2^2(\nu_5-\nu_6)}{f_1\nu_5^2\nu_6\nu_7\nu_8}z+\Big(\frac{\kappa_1}{\nu_7}-z\Big)\Big(\frac{\kappa_1}{\nu_8}-z\Big)z-\frac{\kappa_2(\kappa_1/\nu_7-z)(\kappa_1/\nu_8-z)}{\nu_5}\\
&\quad+\frac{\nu_5(q\nu_1-z)(q\nu_2-z)(q\nu_3-z)(q\nu_4-z)}{q(\nu_5z-q\kappa_2)}\\
&\quad-\frac{(\nu_1\nu_5-\kappa_2)(\nu_2\nu_5-\kappa_2)(\nu_3\nu_5-\kappa_2)(\nu_4\nu_5-\kappa_2)qz^2}{\kappa_2^2\nu_5(\nu_5z-q\kappa_2)}\Big\}y(z)=0\,.
\end{aligned}
$$

To obtain a variant of the q-Heun equation, we apply gauge transformations. Set

$$(\alpha x;q)_\infty=\prod_{i=0}^{\infty}(1-q^i\alpha x)=(1-\alpha x)(1-q\alpha x)(1-q^2\alpha x)\cdots$$

PROPOSITION 2.1.

(i) *If $y(x)$ satisfies $q^{-\lambda}a(x)g(x/q)+b(x)g(x)+q^{\lambda}c(x)g(qx)=0$, then the function $h(x)=x^{\lambda}y(x)$ satisfies $a(x)g(x/q)+b(x)g(x)+c(x)g(qx)=0$.*
(ii) *If $y(x)$ satisfies $(1-\alpha x)a(x)g(x/q)+b(x)g(x)+c(x)g(qx)=0$, then the function $u(x)=(\alpha qx;q)_\infty y(x)$ satisfies $a(x)g(x/q)+b(x)g(x)+(1-\alpha qx)c(x)g(qx)=0$.*

PROOF. If $h(x)=x^{\lambda}y(x)$, then $y(qx)=q^{-\lambda}x^{-\lambda}h(qx)$ and $y(x/q)=q^{\lambda}x^{-\lambda}h(x/q)$. Hence the equation $q^{-\lambda}a(x)h(x/q)+b(x)g(x)+q^{\lambda}c(x)h(qx)=0$ is rewritten as $a(x)y(x/q)+b(x)y(x)+c(x)y(qx)=0$ and we obtain (i).
We can obtain (ii) similarly, as shown in [**7**]. □

We apply a gauge transformation in Proposition 2.1 (ii) to Eq. (2.6). Then we obtain

$$
\begin{aligned}
&\frac{q^{1/2}}{\nu_1}(\nu_1-z)\Big(\frac{\kappa_1}{\nu_7}-z\Big)\Big(\frac{\kappa_1}{\nu_8}-z\Big)\tilde{y}(qz)+\frac{\nu_1}{q^{1/2}}(q\nu_2-z)(q\nu_3-z)(q\nu_4-z)\tilde{y}(z/q)\\
&+q^{1/2}\Big\{-\frac{q^2\nu_1\nu_2\nu_3\nu_4\nu_5}{\kappa_2}-\frac{\kappa_1^2\kappa_2}{\nu_5\nu_7\nu_8}+\Big(\frac{q\nu_1\nu_2\nu_3\nu_5}{\kappa_2}+\frac{q\nu_1\nu_2\nu_4\nu_5}{\kappa_2}+\frac{q\nu_1\nu_3\nu_4\nu_5}{\kappa_2}\\
&+\frac{q\nu_2\nu_3\nu_4\nu_5}{\kappa_2}-\frac{q\nu_1\nu_2\nu_3\nu_4\nu_5^2}{\kappa_2^2}+\frac{\kappa_1\kappa_2}{\nu_5\nu_7}+\frac{\kappa_1\kappa_2}{\nu_5\nu_8}+\frac{\kappa_1^2}{\nu_7\nu_8}+\frac{\kappa_1^2\kappa_2^2}{f_1\nu_5^2\nu_7\nu_8}-\frac{\kappa_1^2\kappa_2^2}{f_1\nu_5\nu_6\nu_7\nu_8}\Big)z\\
&-\Big(\nu_1+\nu_2+\nu_3+\nu_4+\frac{\kappa_1}{\nu_7}+\frac{\kappa_1}{\nu_8}\Big)z^2+\Big(1+\frac{1}{q}\Big)z^3\Big\}\tilde{y}(z)=0\,.
\end{aligned}
$$

By applying the relation $\kappa_1^2\kappa_2^2 = q\nu_1\nu_2\dots\nu_8$ and a gauge transformation in Proposition 2.1 (i), we have

(2.7)
$$(z-q\nu_2)(z-q\nu_3)(z-q\nu_4)u(z/q)+(z-\nu_1)\Big(z-\frac{\kappa_1}{\nu_7}\Big)\Big(z-\frac{\kappa_1}{\nu_8}\Big)u(qz)$$
$$+\Big[-(q^{1/2}+q^{-1/2})z^3+q^{1/2}\Big(\nu_1+\nu_2+\nu_3+\nu_4+\frac{\kappa_1}{\nu_7}+\frac{\kappa_1}{\nu_8}\Big)z^2$$
$$-q^{3/2}\Big\{\nu_1\nu_2\nu_3\nu_4\Big(\frac{\nu_6}{\nu_5}-1\Big)\Big(\frac{1}{f_1}+\frac{\nu_5^2}{\kappa_2^2}\Big)+\nu_1\nu_2\nu_3\nu_4\frac{\nu_5}{\kappa_2}\Big(\frac{1}{\nu_1}+\frac{1}{\nu_2}+\frac{1}{\nu_3}+\frac{1}{\nu_4}\Big)$$
$$+\frac{\kappa_2}{q\nu_5}\Big(\frac{\kappa_1}{\nu_7}+\frac{\kappa_1}{\nu_8}\Big)\Big\}z+\Big(q^3\frac{\nu_1\nu_2\nu_3\nu_4\kappa_1^2}{\nu_7\nu_8}\Big)^{1/2}\Big\{\Big(\frac{q\nu_5}{\nu_6}\Big)^{1/2}+\Big(\frac{q\nu_5}{\nu_6}\Big)^{-1/2}\Big\}\Big]u(z)=0.$$

In fact, if $y(z)$ is a solution of Eq. (2.6) and the parameter λ satisfies $q^\lambda = q^{1/2}/\nu_1$, then the function $u(z) = z^\lambda(z/\nu_1;q)_\infty y(z)$ satisfies Eq. (2.7). We compare Eq. (2.7) with the variant of the q-Heun equation of degree 3 given in Eq. (2.5). Then we obtain the following correspondence

$$h_1 = q^{1/2}\nu_2,\quad h_2 = q^{1/2}\nu_3,\quad h_3 = q^{1/2}\nu_4,\quad h_4 = \frac{q\nu_5}{\nu_6},$$
$$l_1 = q^{1/2}\nu_1,\quad l_2 = \frac{q^{1/2}\kappa_1}{\nu_7},\quad l_3 = \frac{q^{1/2}\kappa_1}{\nu_8}.$$

We may regard f_1 as an accessory parameter.

On the blow-up at the point P_6, we obtain the equation where the parameters ν_5 and ν_6 are exchanged.

The points $P_7 : (\kappa_1/\nu_7, 0)$ and $P_8 : (\kappa_1/\nu_8, 0)$

We realize the blow-up at $(f,g) = (\kappa_1/\nu_7, 0)$ by setting $(f,g) = (f_1g_1 + \kappa_1/\nu_7, g_1)$. Then the point $P_7 : (\kappa_1/\nu_7, 0)$ corresponds to the line $g_1 = 0$. We substitute $(f,g) = (f_1g_1 + \kappa_1/\nu_7, g_1)$ into the equation $L_1y(z) = 0$, and set $g_1 = 0$. Then we have

$$\frac{z-\kappa_1/\nu_8}{q}y(qz)+\frac{(q\nu_1-z)(q\nu_2-z)(q\nu_3-z)(q\nu_4-z)}{q^3(z-q\kappa_1/\nu_7)}y(z/q)$$
$$+\Big[\frac{{\kappa_1}^2q\{-\kappa_2(\nu_5+\nu_6)+\nu_5\nu_6(1-\nu_8/\nu_7)z\}}{q^2\nu_5\nu_6\nu_7\nu_8(z-\kappa_1/\nu_7)}$$
$$+\frac{z\{f_1q(\nu_7-\nu_8)+\nu_8z((q+1)z-q(\nu_1+\nu_2+\nu_3+\nu_4))\}}{q^2\nu_8(z-\kappa_1/\nu_7)}$$
$$+\frac{\kappa_1z\{\kappa_2q(\nu_5+\nu_6)\nu_7+\nu_5\nu_6(-(q\nu_7+\nu_8)z+q(\nu_1+\nu_2+\nu_3+\nu_4)\nu_8)\}}{q^2\nu_5\nu_6\nu_7\nu_8(z-\kappa_1/\nu_7)}\Big]y(z)=0.$$

By setting $\tilde{y}(z)=y(z)/(z-\kappa_1/\nu_7)$, we have

$$\frac{(z-\kappa_1/\nu_8)(qz-\kappa_1/\nu_7)}{q}\tilde{y}(qz)+\frac{(q\nu_1-z)(q\nu_2-z)(q\nu_3-z)(q\nu_4-z)}{q^4}\tilde{y}(z/q)$$
$$+\Big[\frac{{\kappa_1}^2q\{-\kappa_2(\nu_5+\nu_6)+\nu_5\nu_6(1-\nu_8/\nu_7)z\}}{q^2\nu_5\nu_6\nu_7\nu_8}$$
$$+\frac{z\{f_1q(\nu_7-\nu_8)+\nu_8z((q+1)z-q(\nu_1+\nu_2+\nu_3+\nu_4))\}}{q^2\nu_8}$$
$$+\frac{\kappa_1z\{\kappa_2q(\nu_5+\nu_6)\nu_7+\nu_5\nu_6(-(q\nu_7+\nu_8)z+q(\nu_1+\nu_2+\nu_3+\nu_4)\nu_8)\}}{q^2\nu_5\nu_6\nu_7\nu_8}\Big]\tilde{y}(z)=0\,.$$

Let λ be the parameter such that $q^\lambda = q^{3/2}/\nu_1$. Then the function $u(z) = z^\lambda(z/\nu_1;q)_\infty\tilde{y}(z)$ satisfies

$$(z-q\nu_2)(z-q\nu_3)(z-q\nu_4)u(z/q)+(z-\nu_1)\Big(z-\frac{\kappa_1}{q\nu_7}\Big)\Big(z-\frac{\kappa_1}{\nu_8}\Big)u(qz)$$
$$+\Big[-(q^{1/2}+q^{-1/2})z^3+q^{1/2}\Big(\nu_1+\nu_2+\nu_3+\nu_4+\frac{\kappa_1}{q\nu_7}+\frac{\kappa_1}{\nu_8}\Big)z^2$$
$$-q^{1/2}\Big\{\Big(f_1+\frac{\kappa_1^2}{\nu_7^2}\Big)\Big(\frac{\nu_7}{\nu_8}-1\Big)+\frac{\kappa_1}{\nu_7}(\nu_1+\nu_2+\nu_3+\nu_4)+\frac{\kappa_1}{\nu_8}\Big(\frac{\kappa_2}{\nu_5}+\frac{\kappa_2}{\nu_6}\Big)\Big\}z$$
$$+\Big(\frac{q^2\kappa_1^2\nu_1\nu_2\nu_3\nu_4}{\nu_7\nu_8}\Big)^{1/2}\Big\{\Big(\frac{\nu_5}{\nu_6}\Big)^{1/2}+\Big(\frac{\nu_5}{\nu_6}\Big)^{-1/2}\Big\}\Big]u(z)=0\,.$$

We compare it with the variant of the q-Heun equation of degree 3 given in Eq. (2.5). Then we obtain the following correspondence

$$h_1=q^{1/2}\nu_1,\quad h_2=q^{1/2}\nu_3,\quad h_3=q^{1/2}\nu_4,\quad h_4=\frac{\nu_5}{\nu_6},$$
$$l_1=q^{1/2}\nu_1,\quad l_2=\frac{\kappa_1}{q^{1/2}\nu_7},\quad l_3=\frac{q^{1/2}\kappa_1}{\nu_8}.$$

We may regard f_1 as an accessory parameter.

On the blow-up at the point P_8, we obtain the equation where the parameters ν_7 and ν_8 are exchanged.

The points $P_i:(\nu_i,1/\nu_i)$ $(i=1,2,3,4)$

We realize the blow-up at $(f,g)=(\nu_1,1/\nu_1)$ by setting $(f,g)=(f_1g_1+\nu_1,g_1+1/\nu_1)$. Then the point $P_1:(\nu_1,1/\nu_1)$ corresponds to the line $g_1=0$. We substitute $(f,g)=(f_1g_1+\nu_1,g_1+1/\nu_1)$ into the equation $L_1y(z)=0$, and set $g_1=0$. Then we have

$$\frac{(z-\kappa_1/\nu_7)(z-\kappa_1/\nu_8)}{q(z-\nu_1)}y(qz)+\frac{(z-q\nu_2)(z-q\nu_3)(z-q\nu_4)}{q^3}y(z/q)$$
$$+\Big\{-\frac{\kappa_1^2(1-\kappa_2/(\nu_1\nu_5))(1-\kappa_2/(\nu_1\nu_6))}{q\nu_7\nu_8}+\frac{(z-\kappa_1/\nu_7)(z-\kappa_1/\nu_8)}{q}$$
$$+\frac{\nu_1(\nu_2-\nu_1)(\nu_3-\nu_1)(\nu_4-\nu_1)z}{(f_1+\nu_1^2)(z-q\nu_1)}-\frac{q(\nu_2-z/q)(\nu_3-z/q)(\nu_4-z/q)}{z-q\nu_1}\Big\}y(z)=0\,.$$

Let λ be the parameter such that $q^{\lambda} = q^{1/2}/\nu_1$. Then the function $u(z) = z^{\lambda}(z/\nu_1;q)_{\infty}y(z)$ satisfies

$$
\begin{aligned}
&(z-q\nu_2)(z-q\nu_3)(z-q\nu_4)u(z/q) + (z-q\nu_1)\Big(z-\frac{\kappa_1}{\nu_7}\Big)\Big(z-\frac{\kappa_1}{\nu_8}\Big)u(qz) \\
&+ \Big[-(q^{-1/2}+q^{1/2})z^3 + q^{1/2}\Big(q\nu_1+\nu_2+\nu_3+\nu_4+\frac{\kappa_1}{\nu_7}+\frac{\kappa_1}{\nu_8}\Big)z^2 \\
&\quad - q^{3/2}\Big\{\frac{\nu_1(\nu_2-\nu_1)(\nu_3-\nu_1)(\nu_4-\nu_1)}{f_1+\nu_1^2} + \nu_1\Big(\frac{\kappa_1}{\nu_7}+\frac{\kappa_1}{\nu_8}\Big) \\
&\qquad + \nu_2\nu_3\nu_4\Big(-\frac{1}{\nu_1}+\frac{1}{\nu_2}+\frac{1}{\nu_3}+\frac{1}{\nu_4}+\frac{\nu_5}{\kappa_2}+\frac{\nu_6}{\kappa_2}\Big)\Big\}z \\
&\quad + \Big(q^4\frac{\nu_1\nu_2\nu_3\nu_4\kappa_1^2}{\nu_7\nu_8}\Big)^{1/2}\Big\{\Big(\frac{\nu_5}{\nu_6}\Big)^{1/2}+\Big(\frac{\nu_5}{\nu_6}\Big)^{-1/2}\Big\}\Big]u(z) = 0\,.
\end{aligned}
$$

We compare it with the variant of the q-Heun equation of degree 3 given in Eq. (2.5). Then we obtain the following correspondence

$$
\begin{aligned}
&h_1 = q^{1/2}\nu_2, \quad h_2 = q^{1/2}\nu_3 \quad h_3 = q^{1/2}\nu_4, \quad h_4 = \frac{\nu_5}{\nu_6}, \\
&l_1 = q^{3/2}\nu_1, \quad l_2 = \frac{q^{1/2}\kappa_1}{\nu_7}, \quad l_3 = \frac{q^{1/2}\kappa_1}{\nu_8}.
\end{aligned}
$$

We may regard f_1 as an accessory parameter.

On the blow-up at the point P_i ($i = 2, 3, 4$), we obtain the equation where the parameters ν_1 and ν_i are exchanged.

<u>The other cases related to the variant of the q-Heun equation of degree three</u>

We substitute $f = \kappa_1/\nu_7$ to the equation $L_1y(z) = 0$. Then we have

$$
\begin{aligned}
&(z-q\kappa_1/\nu_7)(z-\kappa_1/\nu_8)y(qz) + \frac{(q\nu_1-z)(q\nu_2-z)(q\nu_3-z)(q\nu_4-z)}{q^2}y(z/q) \\
&+ \Big[\Big(1+\frac{1}{q}\Big)z^3 - \Big(\nu_1+\nu_2+\nu_3+\nu_4+\frac{q\kappa_1}{\nu_7}+\frac{\kappa_1}{\nu_8}\Big)z^2 \\
&\quad + q\Big\{\frac{(1-\nu_1\nu_7/\kappa_1)(1-\nu_2\nu_7/\kappa_1)(1-\nu_3\nu_7/\kappa_1)(1-\nu_4\nu_7/\kappa_1)}{(g-\nu_7/\kappa_1)\nu_7/\kappa_1} + \frac{\kappa_1^2}{\nu_7\nu_8} \\
&\quad + \frac{\nu_1\nu_2\nu_3\nu_4\nu_7}{\kappa_1}\Big(-\frac{1}{\nu_1}-\frac{1}{\nu_2}-\frac{1}{\nu_3}-\frac{1}{\nu_4}+\frac{\nu_5}{\kappa_2}+\frac{\nu_6}{\kappa_2}+\frac{\nu_7}{\kappa_1}\Big) + \sum_{1\le i<j\le 4}\nu_i\nu_j\Big\}z \\
&\quad - \frac{q\kappa_1^2\kappa_2(\nu_5+\nu_6)}{\nu_5\nu_6\nu_7\nu_8}\Big\}\Big]y(z) = 0\,.
\end{aligned}
$$

By applying a gauge transformation, we obtain the variant of the q-Heun equation of degree 3, and we may regard g as an accessory parameter. We have a similar result by substituting $f = \kappa_1/\nu_8$ to the equation $L_1y(z) = 0$.

We substitute $f = \nu_1$ to the equation $L_1 y(z) = 0$. Then we have

$$
\begin{aligned}
&q(z-\kappa_1/\nu_7)(z-\kappa_1/\nu_8)y(qz) + \frac{(z-\nu_1)(z-q\nu_2)(z-q\nu_3)(z-q\nu_4)}{q} y(z/q) \\
&+ \Big[(q+1)z^3 - \Big(\nu_1 + q\nu_2 + q\nu_3 + q\nu_4 + \frac{q\kappa_1}{\nu_7} + \frac{q\kappa_1}{\nu_8}\Big)z^2 \\
&\quad + q\Big\{\frac{-(\nu_1-\kappa_1/\nu_7)(\nu_1-\kappa_1/\nu_8)}{g\nu_1} + \nu_1(\nu_2+\nu_3+\nu_4) \\
&\qquad + \frac{\kappa_1^2}{\nu_7\nu_8} + \frac{\kappa_1^2}{\nu_1\nu_7\nu_8}\Big(\frac{\kappa_2}{\nu_5} + \frac{\kappa_2}{\nu_6}\Big)\Big\}z - \frac{q^2\nu_1\nu_2\nu_3\nu_4(\nu_5+\nu_6)}{\kappa_2}\Big\}\Big]y(z) = 0\,.
\end{aligned}
$$

By applying a gauge transformation, we obtain the variant of the q-Heun equation of degree 3, and we may regard g as an accessory parameter. We have similar results by substituting $f = \nu_i$ $(i = 2,3,4)$ to the equation $L_1 y(z) = 0$.

2.3. The case $E_7^{(1)}$. The q-Painlevé equation of type $E_7^{(1)}$ (q-$P(E_7^{(1)})$) in [**6**] was given as

$$
\begin{aligned}
\frac{(fg-\kappa_1/\kappa_2)(\overline{f}g-\kappa_1/(q\kappa_2))}{(fg-1)(\overline{f}g-1)} &= \frac{(g-\nu_5/\kappa_2)(g-\nu_6/\kappa_2)(g-\nu_7/\kappa_2)(g-\nu_8/\kappa_2)}{(g-1/\nu_1)(g-1/\nu_2)(g-1/\nu_3)(g-1/\nu_4)}, \\
\frac{(fg-\kappa_1/\kappa_2)(f\underline{g}-q\kappa_1/\kappa_2)}{(fg-1)(f\underline{g}-1)} &= \frac{(f-\kappa_1/\nu_5)(f-\kappa_1/\nu_6)(f-\kappa_1/\nu_7)(f-\kappa_1/\nu_8)}{(f-\nu_1)(f-\nu_2)(f-\nu_3)(f-\nu_4)}.
\end{aligned}
$$

The space of initial condition for q-$P(E_7^{(1)})$ was realized by blowing up eight points $P_1, \dots, P_8$ of $\mathbb{P}^1 \times \mathbb{P}^1$, where

$$
P_i : \Big(\nu_i, \frac{1}{\nu_i}\Big)_{i=1,2,3,4}, \quad \Big(\frac{\kappa_1}{\nu_i}, \frac{\nu_i}{\kappa_2}\Big)_{i=5,6,7,8}.
$$

On the other hand, q-$P(E_7^{(1)})$ was obtained by the compatibility condition for the Lax operators L_1 and L_2, where

$$
\begin{aligned}
L_1 = {} & \frac{q(\kappa_1-\kappa_2)(g\kappa_2-\nu_5)(g\kappa_2-\nu_6)(g\kappa_2-\nu_7)(g\kappa_2-\nu_8)}{g\kappa_1\kappa_2^2(fg\kappa_2-\kappa_1)(g\kappa_2 z-\kappa_1)} \\
& - \frac{q(\kappa_1-\kappa_2)(g\nu_1-1)(g\nu_2-1)(g\nu_3-1)(g\nu_4-1)}{g(fg-1)\kappa_1\nu_1\nu_2\nu_3\nu_4(gz-q)} \\
& + \frac{(q\nu_1-z)(q\nu_2-z)(q\nu_3-z)(q\nu_4-z)\{(g\kappa_2 z-\kappa_1 q)-\kappa_1(gz-q)T_z^{-1}\}}{q\kappa_1\nu_1\nu_2\nu_3\nu_4(fq-z)z^2(q-gz)} \\
& - \frac{q(\kappa_1-\nu_5 z)(\kappa_1-\nu_6 z)(\kappa_1-\nu_7 z)(\kappa_1-\nu_8 z)\{\kappa_1(gz-1)-(g\kappa_2 z-\kappa_1)T_z\}}{\kappa_1^4(f-z)z^2(g\kappa_2 z-\kappa_1)}, \\
L_2 = {} & (1-zg\kappa_2/\kappa_1)T_z - (1-zg) + z(z-f)gT\,.
\end{aligned}
$$

We track the linear q-difference equation $L_1 y(z) = 0$ on the process of the blow-up of the point P_i $(i = 1,2,\dots,8)$ respectively, and we investigate a relationship with the variant of the q-Heun equation of degree 4;

(2.8)
$$
\begin{aligned}
&(z-h_1q^{1/2})(z-h_2q^{1/2})(z-h_3q^{1/2})(z-h_4q^{1/2})g(z/q)\\
&+(z-l_1q^{-1/2})(z-l_2q^{-1/2})(z-l_3q^{-1/2})(z-l_4q^{-1/2})g(qz)\\
&+\Big[-(q^{1/2}+q^{-1/2})z^4+(h_1+h_2+h_3+h_4+l_1+l_2+l_3+l_4)z^3+Ez^2\\
&\quad+(h_1h_2h_3h_4l_1l_2l_3l_4)^{1/2}\{(h_1^{-1}+h_2^{-1}+h_3^{-1}+h_4^{-1}+l_1^{-1}+l_2^{-1}+l_3^{-1}+l_4^{-1})z\\
&\quad-(q^{1/2}+q^{-1/2})\}\Big]g(z)=0\,.
\end{aligned}
$$

The variant of the q-Heun equation of degree 4 was introduced in [**14**], and it has one accessory parameter E. The terms in the coefficients of $z^3g(z)$ and $zg(z)$ are specified by the condition that the singularities $z=0$ and $z=\infty$ are apparent, which is a fundamental condition of the variant of the q-Heun equation of degree 4. The variant of the q-Heun equation of degree 4 is the eigenfunction of the secondly degenerated Ruijsenaars-van Diejen operator [**13**] with the eigenvalue E which corresponds to the accessory parameter. Note that details of the calculation in this subsection were performed in [**11**].

The points $P_i:(\nu_i,1/\nu_i)$ $(i=1,2,3,4)$

We realize the blow-up at $(f,g)=(\nu_1,1/\nu_1)$ by setting $(f,g)=(f_1g_1+\nu_1,g_1+1/\nu_1)$. Then the point $P_1:(\nu_1,1/\nu_1)$ corresponds to the line $g_1=0$. We substitute $(f,g)=(f_1g_1+\nu_1,g_1+1/\nu_1)$ into the equation $L_1y(z)=0$, and set $g_1=0$. Then we have

$$
\begin{aligned}
&\frac{q(\kappa_1-\nu_5z)(\kappa_1-\nu_6z)(\kappa_1-\nu_7z)(\kappa_1-\nu_8z)}{\kappa_1^4(\nu_1-z)z^2}y(qz)\\
&+\frac{(q\nu_2-z)(q\nu_3-z)(q\nu_4-z)}{q\nu_1\nu_2\nu_3\nu_4z^2}y(z/q)\\
&+\Big\{-\frac{q(\kappa_1-\kappa_2)(\nu_1-\nu_2)(\nu_1-\nu_3)(\nu_1-\nu_4)}{\kappa_1(f_1+\nu_1^2)\nu_2\nu_3\nu_4(q\nu_1-z)}\\
&\quad+\frac{q(\kappa_2-\nu_1\nu_5)(\kappa_2-\nu_1\nu_6)(\kappa_2-\nu_1\nu_7)(\kappa_2-\nu_1\nu_8)}{\kappa_1\kappa_2^2\nu_1^2(\kappa_1\nu_1-\kappa_2z)}\\
&\quad-\frac{(q\nu_2-z)(q\nu_3-z)(q\nu_4-z)(q\kappa_1\nu_1-\kappa_2z)}{\kappa_1q\nu_1\nu_2\nu_3\nu_4(q\nu_1-z)z^2}\\
&\quad-\frac{q(\kappa_1-\nu_5z)(\kappa_1-\nu_6z)(\kappa_1-\nu_7z)(\kappa_1-\nu_8z)}{\kappa_1^3z^2(\kappa_1\nu_1-\kappa_2z)}\Big\}y(z)=0\,.
\end{aligned}
$$

Let λ be the parameter such that $q^{\lambda} = q^{3/2}\kappa_2/\kappa_1$. Then the function $u(z) = z^{\lambda}y(z)/(q\nu_1 - z)$ satisfies

$$\begin{aligned}
&\Big(z - \frac{\kappa_1}{\nu_5}\Big)\Big(z - \frac{\kappa_1}{\nu_6}\Big)\Big(z - \frac{\kappa_1}{\nu_7}\Big)\Big(z - \frac{\kappa_1}{\nu_8}\Big)u(qz)\\
&+ (z - q^2\nu_1)(z - q\nu_2)(z - q\nu_3)(z - q\nu_4)u(z/q)\\
&+ \Big[-(q^{1/2} + q^{-1/2})z^4 + q^{1/2}\Big(q\nu_1 + \nu_2 + \nu_3 + \nu_4 + \frac{\kappa_1}{\nu_5} + \frac{\kappa_1}{\nu_6} + \frac{\kappa_1}{\nu_7} + \frac{\kappa_1}{\nu_8}\Big)z^3\\
&\quad - q^{3/2}\Big(\frac{K_1}{\kappa_2(f_1 + \nu_1^2)} + K_2\Big)z^2 + \frac{q^3\kappa_1\nu_1\nu_2\nu_3\nu_4}{\kappa_2}\Big\{q^{-1/2}\Big(\frac{1}{q\nu_1} + \frac{1}{\nu_2} + \frac{1}{\nu_3} + \frac{1}{\nu_4}\\
&\quad + \frac{\nu_5}{\kappa_1} + \frac{\nu_6}{\kappa_1} + \frac{\nu_7}{\kappa_1} + \frac{\nu_8}{\kappa_1}\Big)z - (q^{1/2} + q^{-1/2})\Big\}\Big]u(z) = 0,
\end{aligned}$$

where

$$\begin{aligned}
K_1 &= \nu_1(\kappa_1 - \kappa_2)(\nu_1 - \nu_2)(\nu_1 - \nu_3)(\nu_1 - \nu_4),\\
K_2 &= \kappa_1\nu_1\Big(\frac{-\nu_1 + \nu_2 + \nu_3 + \nu_4}{\kappa_2} + \frac{1}{\nu_5} + \frac{1}{\nu_6} + \frac{1}{\nu_7} + \frac{1}{\nu_8}\Big)\\
&\quad + \nu_2\nu_3\nu_4\Big(-\frac{1}{\nu_1} + \frac{1}{\nu_2} + \frac{1}{\nu_3} + \frac{1}{\nu_4} + \frac{\nu_5 + \nu_6 + \nu_7 + \nu_8}{\kappa_2}\Big).
\end{aligned}$$

Note that we used the relation $\kappa_1^2\kappa_2^2 = q\nu_1\nu_2 \dots \nu_8$.

We compare it with the variant of the q-Heun equation of degree 4 given in Eq. (2.8). Then we obtain the following correspondence

$$\begin{aligned}
&h_1 = q^{3/2}\nu_1, \quad h_2 = q^{1/2}\nu_2, \quad h_3 = q^{1/2}\nu_3, \quad h_4 = q^{1/2}\nu_4,\\
&l_1 = \frac{q^{1/2}\kappa_1}{\nu_5}, \quad l_2 = \frac{q^{1/2}\kappa_1}{\nu_6}, \quad l_3 = \frac{q^{1/2}\kappa_1}{\nu_7}, \quad l_4 = \frac{q^{1/2}\kappa_1}{\nu_8}.
\end{aligned}$$

We may regard f_1 as an accessory parameter.

On the blow-up at the point P_i ($i = 2, 3, 4$), we obtain the equation where the parameters ν_1 and ν_i are exchanged.

The points $P_i : (\kappa_1/\nu_i, \nu_i/\kappa_2)$ $(i = 5, 6, 7, 8)$

We realize the blow-up at $(f, g) = (\kappa_1/\nu_5, \nu_5/\kappa_2)$ by setting $(f, g) = (f_1g_1 + \kappa_1/\nu_5, g_1 + \nu_5/\kappa_2)$. Then the point $P_5 : (\kappa_1/\nu_5, \nu_5/\kappa_2)$ corresponds to the line $g_1 = 0$. We substitute $(f, g) = (f_1g_1 + \kappa_1/\nu_5, g_1 + \nu_5/\kappa_2)$ into the equation $L_1y(z) = 0$, and set $g_1 = 0$. Then we have

$$\begin{aligned}
&\frac{q\nu_5(\kappa_1 - \nu_6z)(\kappa_1 - \nu_7z)(\kappa_1 - \nu_8z)}{\kappa_1^4z^2}y(qz)\\
&+ \frac{\nu_5(q\nu_1 - z)(q\nu_2 - z)(q\nu_3 - z)(q\nu_4 - z)}{q\nu_1\nu_2\nu_3\nu_4z^2(q\kappa_1 - \nu_5z)}y(z/q)\\
&+ \Big\{-\frac{q(\kappa_1 - \kappa_2)(\nu_5 - \nu_6)(\nu_5 - \nu_7)(\nu_5 - \nu_8)}{\kappa_1(\kappa_1\kappa_2 + f_1\nu_5^2)(\kappa_1 - \nu_5z)}\\
&\quad - \frac{\kappa_2\nu_5(q\nu_1 - z)(q\nu_2 - z)(q\nu_3 - z)(q\nu_4 - z)}{q\kappa_1\nu_1\nu_2\nu_3\nu_4z^2(q\kappa_2 - \nu_5z)}\\
&\quad + \frac{q(\kappa_2 - \nu_1\nu_5)(\kappa_2 - \nu_2\nu_5)(\kappa_2 - \nu_3\nu_5)(\kappa_2 - \nu_4\nu_5)}{\kappa_1\kappa_2\nu_1\nu_2\nu_3\nu_4\nu_5(q\kappa_2 - \nu_5z)}\\
&\quad - \frac{q\nu_5(\kappa_2 - \nu_5z)(\kappa_1 - \nu_6z)(\kappa_1 - \nu_7z)(\kappa_1 - \nu_8z)}{\kappa_1^3\kappa_2z^2(\kappa_1 - \nu_5z)}\Big\}y(z) = 0.
\end{aligned}$$

Let λ be the parameter such that $q^\lambda = q^{3/2}\kappa_2/\kappa_1$. Then the function $u(z) = z^\lambda y(z)/(\kappa_1 - \nu_5 z)$ satisfies

$$\begin{aligned}
&(z-q\nu_1)(z-q\nu_2)(z-q\nu_3)(z-q\nu_4)u(z/q)\\
&+\Big(z-\frac{\kappa_1}{q\nu_5}\Big)\Big(z-\frac{\kappa_1}{\nu_6}\Big)\Big(z-\frac{\kappa_1}{\nu_7}\Big)\Big(z-\frac{\kappa_1}{\nu_8}\Big)u(qz)\\
&+\Big[-(q^{1/2}+q^{-1/2})z^4+q^{1/2}\Big(\nu_1+\nu_2+\nu_3+\nu_4+\frac{\kappa_1}{q\nu_5}+\frac{\kappa_1}{\nu_6}+\frac{\kappa_1}{\nu_7}+\frac{\kappa_1}{\nu_8}\Big)z^3\\
&-q^{1/2}\Big(\frac{K_1}{\kappa_2\nu_5(f_1\nu_5^2+\kappa_1\kappa_2)}+K_2\Big)z^2+\frac{q^2\kappa_1\nu_1\nu_2\nu_3\nu_4}{\kappa_2}\Big\{q^{-1/2}\Big(\frac{1}{\nu_1}+\frac{1}{\nu_2}+\frac{1}{\nu_3}+\frac{1}{\nu_4}\\
&+\frac{q\nu_5}{\kappa_1}+\frac{\nu_6}{\kappa_1}+\frac{\nu_7}{\kappa_1}+\frac{\nu_8}{\kappa_1}\Big)z-(q^{1/2}+q^{-1/2})\Big\}\Big]u(z)=0,
\end{aligned}$$

where

$$\begin{aligned}
K_1 &= q\nu_1\nu_2\nu_3\nu_4(\kappa_1-\kappa_2)(\nu_5-\nu_6)(\nu_5-\nu_7)(\nu_5-\nu_8),\\
K_2 &= \frac{\kappa_1\kappa_2}{\nu_5}\Big(\frac{\nu_1+\nu_2+\nu_3+\nu_4}{\kappa_2}-\frac{1}{\nu_5}+\frac{1}{\nu_6}+\frac{1}{\nu_7}+\frac{1}{\nu_8}\Big)\\
&\quad+q\nu_1\nu_2\nu_3\nu_4\frac{\nu_5}{\kappa_2}\Big(\frac{1}{\nu_1}+\frac{1}{\nu_2}+\frac{1}{\nu_3}+\frac{1}{\nu_4}+\frac{-\nu_5+\nu_6+\nu_7+\nu_8}{\kappa_2}\Big).
\end{aligned}$$

Note that we used the relation $\kappa_1^2\kappa_2^2 = q\nu_1\nu_2\dots\nu_8$.

We compare it with the variant of the q-Heun equation of degree 4 given in Eq. (2.8). Then we obtain the following correspondence

$$\begin{aligned}
&h_1 = q^{1/2}\nu_1, \quad h_2 = q^{1/2}\nu_2, \quad h_3 = q^{1/2}\nu_3, \quad h_4 = q^{1/2}\nu_4,\\
&l_1 = \frac{\kappa_1}{q^{1/2}\nu_5}, \quad l_2 = \frac{q^{1/2}\kappa_1}{\nu_6}, \quad l_3 = \frac{q^{1/2}\kappa_1}{\nu_7}, \quad l_4 = \frac{q^{1/2}\kappa_1}{\nu_8}.
\end{aligned}$$

We may regard f_1 as an accessory parameter.

On the blow-up at the point P_i ($i = 6, 7, 8$), we obtain the equation where the parameters ν_5 and ν_i are exchanged.

The other cases related to the variant of the q-Heun equation of degree four

We substitute $f = \nu_1$ to the equation $L_1 y(z) = 0$. Let λ be the parameter such that $q^\lambda = q^{1/2}\kappa_2/\kappa_1$. Then the function $u(z) = z^\lambda y(z)$ satisfies

$$\begin{aligned}
&\Big(z-\frac{\kappa_1}{\nu_5}\Big)\Big(z-\frac{\kappa_1}{\nu_6}\Big)\Big(z-\frac{\kappa_1}{\nu_7}\Big)\Big(z-\frac{\kappa_1}{\nu_8}\Big)u(qz)\\
&+(z-\nu_1)(z-q\nu_2)(z-q\nu_3)(z-q\nu_4)u(z/q)\\
&+\Big[-(q^{1/2}+q^{-1/2})z^4+q^{1/2}\Big(\frac{\nu_1}{q}+\nu_2+\nu_3+\nu_4+\frac{\kappa_1}{\nu_5}+\frac{\kappa_1}{\nu_6}+\frac{\kappa_1}{\nu_7}+\frac{\kappa_1}{\nu_8}\Big)z^3\\
&\quad-q^{1/2}\Big(\frac{(\nu_1-\kappa_1/\nu_5)(\nu_1-\kappa_1/\nu_6)(\nu_1-\kappa_1/\nu_7)(\nu_1-\kappa_1/\nu_8)(\kappa_1-\kappa_2)}{\nu_1^2(g\nu_1\kappa_2-\kappa_1)}+K_2\Big)z^2\\
&\quad+\frac{q^2\kappa_1\nu_1\nu_2\nu_3\nu_4}{\kappa_2}\Big\{q^{-1/2}\Big(\frac{q}{\nu_1}+\frac{1}{\nu_2}+\frac{1}{\nu_3}+\frac{1}{\nu_4}+\frac{\nu_5}{\kappa_1}+\frac{\nu_6}{\kappa_1}+\frac{\nu_7}{\kappa_1}+\frac{\nu_8}{\kappa_1}\Big)z\\
&\qquad-(q^{1/2}+q^{-1/2})\Big\}\Big]u(z)=0,
\end{aligned}$$

where K_2 is a value which does not depend on z nor g. Hence we obtain the variant of the q-Heun equation of degree 4, and we may regard g as an accessory parameter. We have similar results by substituting $f = \nu_i$ ($i = 2, 3, 4$) to the equation $L_1 y(z) = 0$.

We substitute $f = \kappa_1/\nu_5$ to the equation $L_1 y(z) = 0$. Let λ be the parameter such that $q^\lambda = q^{1/2}\kappa_2/\kappa_1$. Then we have

$$
\begin{aligned}
&(z-q\nu_1)(z-q\nu_2)(z-q\nu_3)(z-q\nu_4)u(z/q)\\
&+\Big(z-\frac{q\kappa_1}{\nu_5}\Big)\Big(z-\frac{\kappa_1}{\nu_6}\Big)\Big(z-\frac{\kappa_1}{\nu_7}\Big)\Big(z-\frac{\kappa_1}{\nu_8}\Big)u(qz)\\
&+\Big[-(q^{1/2}+q^{-1/2})z^4+q^{1/2}\Big(\nu_1+\nu_2+\nu_3+\nu_4+\frac{q\kappa_1}{\nu_5}+\frac{\kappa_1}{\nu_6}+\frac{\kappa_1}{\nu_7}+\frac{\kappa_1}{\nu_8}\Big)z^3\\
&-q^{3/2}\Big(\frac{(\kappa_1-\kappa_2)(\kappa_1-\nu_1\nu_5)(\kappa_1-\nu_1\nu_6)(\kappa_1-\nu_1\nu_7)(\kappa_1-\nu_1\nu_8)}{\kappa_1^2\kappa_2\nu_5(\nu_5-\kappa_1 g)}+K_2\Big)z^2\\
&+\frac{q^3\kappa_1\nu_1\nu_2\nu_3\nu_4}{\kappa_2}\Big\{q^{-1/2}\Big(\frac{1}{\nu_1}+\frac{1}{\nu_2}+\frac{1}{\nu_3}+\frac{1}{\nu_4}+\frac{\nu_5}{q\kappa_1}+\frac{\nu_6}{\kappa_1}+\frac{\nu_7}{\kappa_1}+\frac{\nu_8}{\kappa_1}\Big)z\\
&\qquad\qquad -(q^{1/2}+q^{-1/2})\Big\}\Big]u(z)=0\,.
\end{aligned}
$$

where K_2 is a value which does not depend on z nor g. Hence we obtain the variant of the q-Heun equation of degree 4, and we may regard g as an accessory parameter. We have similar results by substituting $f = \kappa_1/\nu_i$ $(i = 6, 7, 8)$ to the equation $L_1 y(z) = 0$.

3. On associated linear q-difference equation related with q-$P(E_8^{(1)})$

We review a Lax pair of the q-Painlevé equation of type $E_8^{(1)}$ (q-$P(E_8^{(1)})$) by following Yamada [**16**]. Let $u_1, u_2, \dots, u_8$, h_1 and h_2 be the non-zero parameters which satisfy the condition $h_1^2 h_2^2 = q u_1 u_2 \dots u_8$. Set

$$
U(z)=\prod_{i=1}^{8}(z-u_i),
$$

and define the quartic polynomials $P_n(x)$ and $P_d(x)$ by the condition

$$
\begin{aligned}
&\Big(z-\frac{h_2}{z}\Big)P_n\Big(z+\frac{h_2}{z}\Big)=\frac{U(z)}{z^3}-\Big(\frac{z}{h_2}\Big)^3U\Big(\frac{h_2}{z}\Big),\\
&\Big(z-\frac{h_2}{z}\Big)P_d\Big(z+\frac{h_2}{z}\Big)=z^5U\Big(\frac{h_2}{z}\Big)-\Big(\frac{h_2}{z}\Big)^5U(z).
\end{aligned}
\tag{3.1}
$$

Set

$$
\begin{aligned}
&\psi_n(f_1,f_2;g)=(f_1-g)(f_2-g)-\Big(\frac{h_1}{q}-h_2\Big)(h_1-h_2)\frac{1}{h_2},\\
&\psi_d(f_1,f_2;g)=\Big(\frac{f_1}{h_1}q-\frac{g}{h_2}\Big)\Big(\frac{f_2}{h_1}-\frac{g}{h_2}\Big)-\Big(\frac{q}{h_1}-\frac{1}{h_2}\Big)\Big(\frac{1}{h_1}-\frac{1}{h_2}\Big)h_2,\\
&V(f_1,f_2;g)=q\psi_n(f_1,f_2;g)P_d(g)-h_1^2h_2^4\psi_d(f_1,f_2;g)P_n(g),\\
&f(u)=u+\frac{h_1}{u},\ g(u)=u+\frac{h_2}{u},\ \overline{f}(u)=u+\frac{h_1}{qu},\ \overline{g}(u)=u+\frac{qh_2}{u},\\
&\varphi(f,g)=(f-g)\Big(\frac{f}{h_1}-\frac{g}{h_2}\Big)-(h_1-h_2)\Big(\frac{1}{h_1}-\frac{1}{h_2}\Big).
\end{aligned}
$$

A Lax pair for the q-Painlevé equation of type $E_8^{(1)}$ introduced in [**16**] is written as

$$\begin{aligned} L_1 &= \frac{q^5 U(z/q)}{(z^2-h_1q^2)(f-f(z/q))}\Big[T_z^{-1} - \frac{g-g(qh_1/z)}{g-g(z/q)}\Big] \\ &+ \frac{z^8 U(h_1/z)}{h_1^4(z^2-h_1)(f-f(z))}\Big[T_z - \frac{g-g(z)}{g-g(h_1/z)}\Big] \\ &+ \frac{(h_1-h_2)z^2(z^2-qh_1)V(\overline{f}(z/q),f;g)}{qh_1^3h_2^3g\varphi(f,g)(g-g(h_1/z))(g-g(z/q))}, \\ L_2 &= (g-g(z/q))T_z^{-1} - (g-g(qh_1/z)) + (f-f(z/q))(h_1/z - z/q^2)T\cdot T_z^{-1}. \end{aligned}$$

Here T_z represents the transformation $z \to qz$ and T represents the time evolution. The q-Painlevé equation of type $E_8^{(1)}$ (q-$P(E_8^{(1)})$) for the variables f and g was obtained by the compatibility condition for the operators L_1 and L_2 in [**16**], and the expression of q-$P(E_8^{(1)})$ was also described there. It seems that the definitions of the equation q-$P(E_8^{(1)})$ and the Lax pair in [**6**] is less explicit than those in [**6**], and we use the definition in [**16**].

We pick up the q-difference equation $L_1 y(z) = 0$, i. e.

$$\begin{aligned} &\frac{q^5 U(z/q)}{(z^2-h_1q^2)(f-f(z/q))}\Big[y(z/q) - \frac{g-g(qh_1/z)}{g-g(z/q)}y(z)\Big] \\ &+ \frac{z^8 U(h_1/z)}{h_1^4(z^2-h_1)(f-f(z))}\Big[y(qz) - \frac{g-g(z)}{g-g(h_1/z)}y(z)\Big] \\ &+ \frac{(h_1-h_2)z^2(z^2-qh_1)V(\overline{f}(z/q),f;g)}{qh_1^3h_2^3g\varphi(f,g)\{g-g(h_1/z)\}\{g-g(z/q)\}}y(z) = 0, \end{aligned} \tag{3.2}$$

and consider a limit with respect to the parameters f and g. Namely, set

$$f = f(u_1+\varepsilon c_1) = u_1 + \varepsilon c_1 + \frac{h_1}{u_1+\varepsilon c_1}, \quad g = g(u_1+\varepsilon c_2) = u_1 + \varepsilon c_2 + \frac{h_2}{u_1+\varepsilon c_2} \tag{3.3}$$

and consider the limit $\varepsilon \to 0$ in Eq. (3.2). Then

$$f - f(z) = -\frac{(z-u_1)(u_1z-h_1)}{u_1z} + O(\varepsilon), \; g - g(z) = -\frac{(z-u_1)(u_1z-h_2)}{u_1z} + O(\varepsilon)$$

as $\varepsilon \to 0$. We set

$$U_7(z) = \prod_{i=2}^{8}(z-u_i).$$

Then we have $U(z) = (z-u_1)U_7(z)$. On the coefficients of $y(z/q)$ and $y(qz)$ in Eq. (3.2), we have

$$\begin{aligned} \frac{q^5U(z/q)}{(z^2-h_1q^2)(f-f(z/q))} &= \frac{-q^5u_1zU_7(z/q)}{(z^2-h_1q^2)(u_1z-qh_1)} + O(\varepsilon), \\ \frac{z^8U(h_1/z)}{h_1^4(z^2-h_1)(f-f(z))} &= \frac{z^8u_1U_7(h_1/z)}{h_1^4(z^2-h_1)(z-u_1)} + O(\varepsilon). \end{aligned}$$

The limit of the coefficient of $y(z)$ is more complicated. By applying a gauge transformation for $y(z)$ and calculating the limit of Eq. (3.2) as $\varepsilon \to 0$, we obtain the following lemma.

LEMMA 3.1. *Set*

$$y(z) = \frac{(z-qu_1)(u_1z-h_1)}{z}\tilde{y}(z). \tag{3.4}$$

Write f and g as Eq. (3.3). Then Eq. (3.2) tends to

$$B^-(z)\tilde{y}(z/q) + B^+(z)\tilde{y}(qz) + B^0(z)\tilde{y}(z) = \Big(C_0\frac{c_2}{c_1-c_2} + C_0'\Big)\tilde{y}(z) \tag{3.5}$$

as $\varepsilon \to 0$, where

$$B^-(z) = \frac{(z-q^2u_1)}{q^2h_1z^2(z^2-qh_1)(z^2-q^2h_1)}\prod_{j=2}^{8}(z-qu_j),$$

$$B^+(z) = \frac{q(qu_1z-h_1)}{h_1^5z^2(z^2-qh_1)(z^2-h_1)}\prod_{j=2}^{8}(u_jz-h_1)$$

$$B^0(z) = \frac{-qz(h_1^{1/2}-qu_1)}{2h_1^{7/2}(z-h_1^{1/2})(z-qh_1^{1/2})}\prod_{j=2}^{8}(h_1^{1/2}-u_j)$$
$$+\frac{qz(h_1^{1/2}+qu_1)}{2h_1^{1/2}(z+h_1^{1/2})(z+qh_1^{1/2})}\prod_{j=2}^{8}(h_1^{1/2}+u_j) - \frac{q^{3/2}}{h_1}(u_1u_2u_3u_4u_5u_6u_7u_8)^{1/2}$$
$$\cdot\Big[(q+1)\Big(\frac{z^2}{(qh_1)^2}+\frac{1}{z^2}\Big) - \Big(\frac{z}{qh_1}+\frac{1}{z}\Big)\Big\{q\frac{u_1}{h_1}+\frac{1}{qu_1}+\sum_{j=2}^{8}\Big(\frac{u_j}{h_1}+\frac{1}{u_j}\Big)\Big\}\Big],$$

$$C_0 = \frac{q(h_2-h_1)}{h_1^2u_1(h_1-u_1^2)(h_2-u_1^2)}\prod_{j=2}^{8}(u_1-u_j)$$

and C_0' is the constant which does not depend on z, c_1 and c_2.

PROOF. As $\varepsilon \to 0$, we have

$$\varphi(f,g) = -\frac{(h_1-h_2)(h_1-u_1^2)(h_2-u_1^2)(c_1-c_2)}{u_1^3h_1h_2}\varepsilon + O(\varepsilon^2),$$
$$\psi_n(\overline{f}(z/q),f;g) = \frac{(h_1-h_2)(u_1z-qh_2)(h_2z-h_1u_1)}{qh_2u_1^2z} + O(\varepsilon),$$
$$\psi_d(\overline{f}(z/q),f;g) = -\frac{(h_1-h_2)(u_1z-qh_2)(h_2z-h_1u_1)}{h_1^2h_2^2z} + O(\varepsilon)$$

and

$$\frac{\psi_n(\overline{f}(z/q),f;g)}{\psi_d(\overline{f}(z/q),f;g)} = -\frac{h_1^2h_2}{qu_1^2} + \frac{h_1^2h_2}{qu_1^3}\Big\{\frac{h_2(h_1-u_1^2)(h_2+u_1^2)(u_1z-h_1)(z-qu_1)}{u_1^2(h_1-h_2)(u_1z-h_2q)(h_2z-h_1u_1)}(c_1-c_2)+2c_2\Big\}\varepsilon + O(\varepsilon^2). \tag{3.6}$$

Set $u = u_1 + c_2\varepsilon$. Then $g = u + h_2/u$ and it follows from Eq. (3.1) that

$$\begin{aligned}
&V(\overline{f}(z/q), f; g)\\
&= q\psi_n(\overline{f}(z/q), f; g)P_d(u + h_2/u) - h_1^2h_2^4\psi_d(\overline{f}(z/q), f; g)P_n(h_2, u + h_2/u)\\
&= q\psi_n(\overline{f}(z/q), f; g)\frac{u^3(u^2 - u_1^2)U(h_2/u)}{u - h_2/u}\\
&\quad + q\psi_n(\overline{f}(z/q), f; g)\frac{u^3u_1^2U(h_2/u)}{u - h_2/u} + h_1^2h_2\psi_d(\overline{f}(z/q), f; g)\frac{u^3U(h_2/u)}{u - h_2/u}\\
&\quad - q\psi_n(\overline{f}(z/q), f; g)\frac{(h_2/u)^5U(u)}{u - h_2/u} - h_1^2h_2^4\psi_d(\overline{f}(z/q), f; g)\frac{U(u)/u^3}{u - h_2/u}.
\end{aligned}$$

We define the equivalence $A(\varepsilon) \simeq B(\varepsilon)$ by

$$\lim_{\varepsilon\to 0}\frac{A(\varepsilon)}{B(\varepsilon)} = 1.$$

Then $U(u) = U(u_1 + \varepsilon c_2) \simeq \varepsilon c_2U_7(u_1)$ and $U(h_2/u) \simeq -(u_1 - h_2/u_1)U_7(h_2/u_1)$. Hence

$$\begin{aligned}
&q\psi_n(\overline{f}(z/q), f; g)\frac{u^3(u^2 - u_1^2)U(h_2/u)}{u - h_2/u}\\
&\simeq -2qu_1^4U_7(h_2/u_1)c_2\psi_n(\overline{f}(z/q), f; g)\varepsilon \simeq 2u_1^2h_1^2h_2c_2U_7(h_2/u_1)\psi_d(\overline{f}(z/q), f; g)\varepsilon.
\end{aligned}$$

It follows from Eq. (3.6) that

$$\begin{aligned}
&\psi_n(\overline{f}(z/q), f; g)q\frac{u^3u_1^2U(h_2/u)}{u - h_2/u} + h_1^2h_2\psi_d(\overline{f}(z/q), f; g)\frac{u^3U(h_2/u)}{u - h_2/u}\\
&\simeq -\psi_d(\overline{f}(z/q), f; g)U_7(h_2/u_1)\\
&\quad\cdot\left\{\frac{h_1^2h_2^2(h_1 - u_1^2)(h_2 + u_1^2)(u_1z - h_1)(z - qu_1)}{(h_1 - h_2)(u_1z - h_2q)(h_2z - h_1u_1)}(c_1 - c_2) + 2h_1^2h_2u_1^2c_2\right\}\varepsilon.
\end{aligned}$$

Similarly, we have

$$\begin{aligned}
&- q\psi_n(\overline{f}(z/q), f; g)\frac{(h_2/u)^5U(u)}{u - h_2/u} - h_1^2h_2^4\psi_d(\overline{f}(z/q), f; g)\frac{U(u)/u^3}{u - h_2/u}\\
&\simeq q\frac{h_1^2h_2}{qu_1^2}(h_2/u_1)^5\varepsilon c_2\frac{U_7(u_1)}{u_1 - h_2/u_1}\psi_d(\overline{f}(z/q), f; g)\\
&\quad - h_1^2h_2^4\frac{\varepsilon c_2}{u_1^3}\frac{U_7(u_1)}{u_1 - h_2/u_1}\psi_d(\overline{f}(z/q), f; g)\\
&= -\frac{h_1^2h_2^4(u_1 + h_2/u_1)}{u_1^5}U_7(u_1)\psi_d(\overline{f}(z/q), f; g)c_2\varepsilon.
\end{aligned}$$

Thus

$$V(\overline{f}(z/q),f;g)\simeq-\frac{h_1^2h_2^4}{u_1^6}(h_2+u_1^2)U_7(u_1)\psi_d(\overline{f}(z/q),f;g)c_2\varepsilon$$
$$-\psi_d(\overline{f}(z/q),f;g)U_7(h_2/u_1)\frac{h_1^2h_2^2(h_1-u_1^2)(h_2+u_1^2)(u_1z-h_1)(z-qu_1)}{(h_1-h_2)(u_1z-h_2q)(h_2z-h_1u_1)}(c_1-c_2)\varepsilon$$
$$\simeq\frac{h_2^2(h_1-h_2)(h_2+u_1^2)(u_1z-qh_2)(h_2z-h_1u_1)}{u_1^6z}U_7(u_1)c_2\varepsilon$$
$$+\frac{(h_1-u_1^2)(h_2+u_1^2)(u_1z-h_1)(z-qu_1)}{z}U_7(h_2/u_1)(c_1-c_2)\varepsilon.$$

On the other hand, we have

$$\frac{(h_1-h_2)z^2(z^2-qh_1)}{qh_1^3h_2^3g\varphi(f,g)(g-g(h_1/z))(g-g(z/q))}\simeq\frac{-u_1^6}{\varepsilon(c_1-c_2)h_1h_2^2}$$
$$\cdot\frac{z^4(z^2-qh_1)}{(h_2+u_1^2)(h_1-u_1^2)(h_2-u_1^2)(h_2z-h_1u_1)(z-qu_1)(u_1z-h_1)(u_1z-h_2q)}.$$

Therefore

$$\frac{(h_1-h_2)z^2(z^2-qh_1)V(\overline{f}(z/q),f;g)}{qh_1^3h_2^3g\varphi(f,g)(g-g(h_1/z))(g-g(z/q))}$$
$$\simeq-\frac{(h_1-h_2)U_7(u_1)z^3(z^2-qh_1)c_2}{h_1(-u_1^2+h_2)(-u_1^2+h_1)(z-qu_1)(u_1z-h_1)(c_1-c_2)}$$
$$-\frac{u_1^6U_7(h_2/u_1)z^3(z^2-qh_1)}{h_1h_2^2(-u_1^2+h_2)(h_2z-h_1u_1)(u_1z-h_2q)}.$$

By substituting it into Eq. (3.2), we have

$$\frac{-q^5u_1zU_7(z/q)}{(z^2-h_1q^2)(u_1z-qh_1)}y(z/q)+\frac{q^5u_1zU_7(z/q)}{(z^2-h_1q^2)}\frac{(h_2z-qh_1u_1)}{h_1(z-qu_1)(u_1z-h_2q)}y(z)$$
$$+\frac{z^8u_1U_7(h_1/z)}{h_1^4(z^2-h_1)(z-u_1)}y(qz)-\frac{z^8u_1U_7(h_1/z)}{h_1^4(z^2-h_1)}\frac{h_1(u_1z-h_2)}{(h_2z-h_1u_1)(u_1z-h_1)}y(z)$$
$$-\frac{(h_1-h_2)U_7(u_1)z^3(z^2-qh_1)c_2}{h_1(h_1-u_1^2)(h_2-u_1^2)(z-qu_1)(u_1z-h_1)(c_1-c_2)}y(z)$$
$$-\frac{u_1^6U_7(h_2/u_1)z^3(z^2-qh_1)}{h_1h_2^2(h_2-u_1^2)(h_2z-h_1u_1)(u_1z-h_2q)}y(z)=0.$$

Recall that the function $\tilde{y}(z)$ was defined in Eq. (3.4) and it follows that $y(z/q)=z^{-1}q(z/q-qu_1)(u_1z/q-h_1)\tilde{y}(z/q)$ and $y(qz)=z^{-1}q^{-1}(qz-qu_1)(qu_1z-h_1)\tilde{y}(qz)$. Then we obtain

$$\frac{-q^{-3}(z-q^2u_1)\prod_{j=2}^8(z-qu_j)}{z^2(z^2-qh_1)(z^2-h_1q^2)}\tilde{y}(z/q)-\frac{(qu_1z-h_1)\prod_{j=2}^8(u_jz-h_1)}{h_1^4z^2(z^2-qh_1)(z^2-h_1)}\tilde{y}(qz)$$
$$+H(z)\tilde{y}(z)-\frac{(h_1-h_2)c_2\prod_{j=2}^8(u_1-u_j)}{u_1h_1(h_1-u_1^2)(h_2-u_1^2)(c_1-c_2)}\tilde{y}(z)=0, \tag{3.7}$$

where

$$H(z) = \frac{q^5(h_2 z - q h_1 u_1)(u_1 z - h_1) U_7(z/q)}{h_1 z^2 (z^2 - q h_1)(z^2 - q^2 h_1)(u_1 z - h_2 q)} - \frac{z^5(u_1 z - h_2)(z - q u_1) U_7(h_1/z)}{h_1^3 (z^2 - q h_1)(z^2 - h_1)(h_2 z - h_1 u_1)} - \frac{u_1^5 U_7(h_2/u_1)(z - q u_1)(u_1 z - h_1)}{h_1 h_2^2 (h_2 - u_1^2)(h_2 z - h_1 u_1)(u_1 z - q h_2)}.$$

Since it is shown that the poles $z = \pm(qh_1)^{1/2}, h_1u_1/h_2, qh_2/u_1$ of the function $H(z)$ are cancelled, we have

$$\begin{aligned} H(z) = {} & \frac{z(h_1^{1/2} - qu_1)(h_1^{1/2} - u_2)(h_1^{1/2} - u_3)\cdots(h_1^{1/2} - u_8)}{2h_1^{5/2}(z - h_1^{1/2})(z - qh_1^{1/2})} \\ & - \frac{z(h_1^{1/2} + qu_1)(h_1^{1/2} + u_2)(h_1^{1/2} + u_3)\cdots(h_1^{1/2} + u_8)}{2h_1^{5/2}(z + h_1^{1/2})(z + qh_1^{1/2})} \\ & + h_1 h_2 (q+1)(z^2/(qh_1)^2 + z^{-2}) - (z/(qh_1) + z^{-1}) h_2 (qu_1 + u_2 \\ & + u_3 + \cdots + u_8 + h_1/(qu_1) + h_1/u_2 + h_1/u_3 + \cdots + h_1/u_8) + C, \end{aligned} \tag{3.8}$$

where C is a constant which does not depend on the variable z and the parameter c_1 and c_2. Recall that there is a relation $h_1^2 h_2^2 = q u_1 u_2 u_3 u_4 u_5 u_6 u_7 u_8$. By substituting Eq. (3.8) and $h_2 = (q u_1 u_2 u_3 u_4 u_5 u_6 u_7 u_8)^{1/2}/h_1$ into Eq. (3.7), we obtain the lemma. □

Set $z = (qh_1)^{1/2} x$, $\tilde{y}(z) = u(x)$, $u_1 = q^{-1} h_1^{1/2} v_1$, $u_2 = h_1^{1/2} v_2, \dots, u_8 = h_1^{1/2} v_8$ and $E = C_0 c_2/(c_1 - c_2) + C_0'$ in Eq. (3.5). Then we have

$$\begin{aligned} & \frac{\prod_{j=1}^8 (x - q^{1/2} v_j)}{q x^2 (x^2 - 1)(x^2 - q)} u(x/q) + \frac{\prod_{j=1}^8 (q^{1/2} v_j x - 1)}{q x^2 (x^2 - 1)(q x^2 - 1)} u(qx) \\ & \quad + \tilde{B}^0(x) u(x) = E u(x), \end{aligned} \tag{3.9}$$

where

$$\begin{aligned} \tilde{B}^0(x) = {} & \frac{-q^{1/2} x}{2(x - q^{-1/2})(x - q^{1/2})} \prod_{j=1}^8 (1 - v_j) + \frac{q^{1/2} x}{2(x + q^{-1/2})(x + q^{1/2})} \prod_{j=1}^8 (1 + v_j) \\ & - (v_1 v_2 v_3 v_4 v_5 v_6 v_7 v_8)^{1/2} \Big[(q+1)\Big(x^2 + \frac{1}{x^2}\Big) - q^{1/2}\Big(x + \frac{1}{x}\Big) \sum_{j=1}^8 \Big(v_j + \frac{1}{v_j}\Big) \Big]. \end{aligned}$$

We compare Eq. (3.9) with the firstly degenerated Ruijsenaars-van Diejen operator $\tilde{A}^{\langle 1 \rangle}(h; z)$ discussed in [**13, 15**]. We explain $\tilde{A}^{\langle 1 \rangle}(h; z)$. Let $\exp(\pm i a_- \partial_z)$ be the shift operator such that $\exp(\pm i a_- \partial_z) f(z) = f(z \pm i a_-)$. The firstly degenerated Ruijsenaars-van Diejen operator $\tilde{A}^{\langle 1 \rangle}(h; z)$ is given by

$$\tilde{A}^{\langle 1 \rangle}(h; z) = \tilde{V}^{\langle 1 \rangle}(h; z) \exp(-i a_- \partial_z) + \tilde{W}^{\langle 1 \rangle}(h; z) \exp(i a_- \partial_z) + U^{\langle 1 \rangle}(h; z), \tag{3.10}$$

where

$$\tilde{V}^{\langle 1\rangle}(h;z) = \frac{\prod_{n=1}^{8}(1 - e^{-2\pi i z}e^{2\pi i \tilde{h}_n}e^{-\pi a_-})}{e^{-2\pi a_-}e^{-4\pi i z}(1 - e^{-4\pi i z})(1 - e^{-4\pi i z}e^{-2\pi a_-})},$$

$$\tilde{W}^{\langle 1\rangle}(h;z) = \frac{\prod_{n=1}^{8}(1 - e^{2\pi i z}e^{2\pi i \tilde{h}_n}e^{-\pi a_-})}{e^{-2\pi a_-}e^{4\pi i z}(1 - e^{4\pi i z})(1 - e^{4\pi i z}e^{-2\pi a_-})},$$

$$U^{\langle 1\rangle}(h;z) = \frac{\prod_{n=1}^{8}(e^{2\pi i \tilde{h}_n} - 1)}{2(1 - e^{2\pi i z}e^{\pi a_-})(1 - e^{-2\pi i z}e^{\pi a_-})} + \frac{\prod_{n=1}^{8}(e^{2\pi i \tilde{h}_n} + 1)}{2(1 + e^{2\pi i z}e^{\pi a_-})(1 + e^{-2\pi i z}e^{\pi a_-})}$$
$$+ e^{-\pi a_-}\prod_{n=1}^{8} e^{\pi i \tilde{h}_n} \cdot \Big[(e^{2\pi i z} + e^{-2\pi i z})\sum_{n=1}^{8}(e^{2\pi i \tilde{h}_n} + e^{-2\pi i \tilde{h}_n})$$
$$- (e^{\pi a_-} + e^{-\pi a_-})(e^{4\pi i z} + e^{-4\pi i z})\Big].$$

Set $x = e^{2\pi i z}$, $q = e^{-2\pi a_-}$ and $v_n = e^{2\pi i \tilde{h}_n}$. Then we have $\exp(\pm i a_- \partial_z) f(e^{2\pi i z}) = f(e^{2\pi i(z \pm i a_-)}) = f(q^{\pm 1}x)$. It follows from a straightforward calculation that the equation $\tilde{A}^{\langle 1\rangle}(h;z)f(z) = Ef(z)$ is equivalent to Eq. (3.9). Therefore we obtain the following theorem.

THEOREM 3.2. *The q-difference equation in Eq.* (3.9), *which was obtained from the associated linear equation of the q-Painlevé equation of type $E_8^{(1)}$, is equivalent to the equation $\tilde{A}^{\langle 1\rangle}(h;z)f(z) = Ef(z)$, where $\tilde{A}^{\langle 1\rangle}(h;z)$ is the firstly degenerated Ruijsenaars-van Diejen operator in Eq.* (3.10) *and E is an arbitrary complex number.*

Acknowledgment

The authors thank the referee for valuable comments.

References

[1] Pascal Baseilhac, Satoshi Tsujimoto, Luc Vinet, and Alexei Zhedanov, *The Heun-Askey-Wilson algebra and the Heun operator of Askey-Wilson type*, Ann. Henri Poincaré **20** (2019), no. 9, 3091–3112, DOI 10.1007/s00023-019-00821-3. MR3995898

[2] Boris Dubrovin and Andrei Kapaev, *A Riemann-Hilbert approach to the Heun equation*, SIGMA Symmetry Integrability Geom. Methods Appl. **14** (2018), Paper No. 093, 24, DOI 10.3842/SIGMA.2018.093. MR3851769

[3] Wolfgang Hahn, *On linear geometric difference equations with accessory parameters*, Funkcial. Ekvac. **14** (1971), 73–78. MR326213

[4] Mitsuo Kato, *Algebraic transformations of ${}_3F_2$*, Funkcial. Ekvac. **51** (2008), no. 2, 221–243, DOI 10.1619/fesi.51.221. MR2440940

[5] Michio Jimbo and Hidetaka Sakai, *A q-analog of the sixth Painlevé equation*, Lett. Math. Phys. **38** (1996), no. 2, 145–154, DOI 10.1007/BF00398316. MR1403067

[6] Kenji Kajiwara, Masatoshi Noumi, and Yasuhiko Yamada, *Geometric aspects of Painlevé equations*, J. Phys. A **50** (2017), no. 7, 073001, 164, DOI 10.1088/1751-8121/50/7/073001. MR3609039

[7] Ryuya Matsunawa, Tomoki Sato, and Kouichi Takemura, *Variants of confluent q-hypergeometric equations*, Geometric and harmonic analysis on homogeneous spaces and applications, Springer Proc. Math. Stat., vol. 366, Springer, Cham, [2021] ©2021, pp. 161–180, DOI 10.1007/978-3-030-78346-4_10. MR4378951

[8] Masatoshi Noumi, Simon Ruijsenaars, and Yasuhiko Yamada, *The elliptic Painlevé Lax equation vs. van Diejen's 8-coupling elliptic Hamiltonian*, SIGMA Symmetry Integrability Geom. Methods Appl. **16** (2020), Paper No. 063, 16, DOI 10.3842/SIGMA.2020.063. MR4120359

[9] S. N. M. Ruijsenaars, *Integrable BC_N analytic difference operators: hidden parameter symmetries and eigenfunctions*, New trends in integrability and partial solvability, NATO Sci. Ser. II Math. Phys. Chem., vol. 132, Kluwer Acad. Publ., Dordrecht, 2004, pp. 217–261, DOI 10.1007/978-94-007-1023-8_9. MR2153340

[10] Hidetaka Sakai, *Rational surfaces associated with affine root systems and geometry of the Painlevé equations*, Comm. Math. Phys. **220** (2001), no. 1, 165–229, DOI 10.1007/s002200100446. MR1882403

[11] Shoko Sasaki, *Lax pair of q-Painlevé equation and q-Heun equation* (in Japanese), Master's thesis, 2021.

[12] Kouichi Takemura, *Middle convolution and Heun's equation*, SIGMA Symmetry Integrability Geom. Methods Appl. **5** (2009), Paper 040, 22, DOI 10.3842/SIGMA.2009.040. MR2506172

[13] Kouichi Takemura, *Degenerations of Ruijsenaars–van Diejen operator and q-Painlevé equations*, J. Integrable Syst. **2** (2017), no. 1, xyx008, 27, DOI 10.1093/integr/xyx008. MR3710451

[14] Kouichi Takemura, *On q-deformations of the Heun equation*, SIGMA Symmetry Integrability Geom. Methods Appl. **14** (2018), Paper No. 061, 16, DOI 10.3842/SIGMA.2018.061. MR3815312

[15] J. F. van Diejen, *Difference Calogero-Moser systems and finite Toda chains*, J. Math. Phys. **36** (1995), no. 3, 1299–1323, DOI 10.1063/1.531122. MR1317442

[16] Yasuhiko Yamada, *Lax formalism for q-Painlevé equations with affine Weyl group symmetry of type $E_n^{(1)}$*, Int. Math. Res. Not. IMRN **17** (2011), 3823–3838, DOI 10.1093/imrn/rnq232. MR2836394

Department of Mathematics, Faculty of Science and Engineering, Chuo University, 1-13-27 Kasuga, Bunkyo-ku, Tokyo 112-8551, Japan

Department of Mathematics, Faculty of Science and Engineering, Chuo University, 1-13-27 Kasuga, Bunkyo-ku, Tokyo 112-8551, Japan

Department of Mathematics, Ochanomizu University, 2-1-1 Otsuka, Bunkyo-ku, Tokyo 112-8610, Japan

Email address: takemura.kouichi@ocha.ac.jp

Contemporary Mathematics
Volume **782**, 2023
https://doi.org/10.1090/conm/782/15726

Differential transcendence of solutions for q-difference equation of Ramanujan function

Hiroshi Ogawara

ABSTRACT. No non-trivial solution for the second-order linear q-difference equation of the Ramanujan function, one of q-analogues of the Airy function, satisfies any algebraic differential equation over the rational function field. Our proof is based on a criterion for differential transcendence of solutions for difference Riccati equations, which is constructed by S. Nishioka.

1. Introduction

In this paper, we consider the following linear q-difference equation

$$qxy(q^2x) - y(qx) + y(x) = 0. \tag{1.1}$$

The equation (1.1) has a solution $A_q(x)$ called the Ramanujan function,

$$A_q(x) = \sum_{n=0}^{\infty} \frac{q^{n^2}}{(q;q)_n}(-x)^n. \tag{1.2}$$

The Ramanujan function $A_q(x)$ is introduced by S. Ramanujan and appeared in the lost notebook [**Ram88**]. When $x = -1, -q$, we have the q-series

$$A_q(-1) = G(q) = \sum_{n=0}^{\infty} \frac{q^{n^2}}{(q;q)_n}, \quad A_q(-q) = H(q) = \sum_{n=0}^{\infty} \frac{q^{n^2+n}}{(q;q)_n},$$

which appear in the Rogers-Ramanujan identities.

M. E. H. Ismail and C. Zhang found $A_q(x)$ is one of q-analogues of the Airy function in a study of asymptotic behavior of the q-Hermite polynomial [**Ism05, IZ07**]. Another q-analogue of the Airy function is the q-Airy function, which was found by K. Kajiwara, T. Masuda, M. Noumi, Y. Ohta and Y. Yamada as a special solution of the q-Painlevé II equation of type $A_6^{(1)}$ or $(A_1 + A_1')^{(1)}$ [**KMN$^+$04, HKW06**]. In [**Mor14**], T. Morita constructed the connection formula of (1.1) between the fundamental system of solutions around $x = 0$ and that of solutions around $x = \infty$ by using the q-Borel-Laplace resummation method. Morita also

2020 *Mathematics Subject Classification.* Primary 39A13; Secondary 12H10, 33D15.
Key words and phrases. Differential transcendence, Transcendental transcendence, Hyper-transcendence, Ramanujan function, q-difference equation, Difference Riccati equation.

gave the connection formula between the Ramanujan function $\mathrm{A}_q(x)$ and the q-Airy function $\mathrm{Ai}_q(x)$,

$$(1.3) \qquad \mathrm{A}_{q^2}\left(-\frac{q^3}{x^2}\right) = \frac{1}{(q,-1;q)_\infty}\left\{\theta_q\left(\frac{x}{q}\right)\mathrm{Ai}_q(-x) + \theta_q\left(-\frac{x}{q}\right)\mathrm{Ai}_q(x)\right\}.$$

In [**Nis18**], S. Nishioka gave a criterion for differential transcendence of solutions for difference Riccati equations and showed that no non-trivial solution for the q-Airy equation satisfies any algebraic differential equation over the rational function field. Although the Ramanujan function and the q-Airy equation have the relation (1.3), we cannot obtain directly differential transcendence of $A_q(x)$, for it is not clear whether the sum of differentially transcendental functions is also differentially transcendental. Thus we shall employ the criterion of Nishioka in order to show differential transcendence of solutions for (1.1). First, we show transcendence of solutions for iterated Riccati equations associated with (1.1) in a similar method to the proof of transcendence of solutions for the q-Airy equation in [**Nis17**]. Then, according to the criterion of Nishioka, differential transcendence of solutions for (1.1) is reduced to non-existence of rational solutions for the third-order linear q-difference equation,

$$(1.4) \qquad -q^6xh(q^3x) + q^2(1-q^2x)h(q^2x) - q(1-q^2x)h(qx) + qxh(x) + \frac{1}{x} = 0.$$

We show (1.4) has no rational solution and hence obtain the following theorem.

THEOREM 1.1. *Let $q \in \mathbb{C}^\times$ be not a root of unity. Then no non-trivial solution for* (1.1) *satisfies any algebraic differential equation over $\mathbb{C}(x)$.*

In Section 2, we prepare notions of difference and differential fields. In order to apply the criterion for differential transcendence of Nishioka, we shall transform (1.1) into the difference Riccati equation of Tietze's normal form in Section 3. In Section 4, we show there exists no algebraic solution for iterated difference Riccati equations associated with (1.1). Finally, we prove Theorem 1.1 in Section 5. Furthermore, we show that the equation (1.1) is unsolvable by using the transcendence of solutions.

2. Preliminaries

Throughout this paper, any field is of characteristic 0 and $q \in \mathbb{C}^\times$ not a root of unity. A pair $\mathcal{K} = (K, \tau)$ is called a *difference field* if K is a field and τ is an isomorphism of K into itself, namely an injective endomorphism. The endomorphism τ is called the *transforming operator* and K the *underlying field.* When difference fields $\mathcal{K} = (K, \tau_K)$ and $\mathcal{L} = (L, \tau_L)$ satisfy $K \subset L$ and $\tau_L|_K = \tau_K$, we call $\mathcal{L}/\mathcal{K}$ a *difference field extension.* In this case, we say $\mathcal{L}$ is a *difference overfield* of $\mathcal{K}$ and $\mathcal{K}$ is a *difference subfield* of $\mathcal{L}$. For brevity, we often omit the subscripts L and K from τ. We define a difference intermediate field in the proper way. Let $\mathcal{K} = (K, \tau)$ be a difference field and $\mathcal{L} = (L, \tau)$ a difference overfield of $\mathcal{K}$. For a subset $B \subset L$, the difference subfield $\left(K\left(\bigcup_{i\geq 0}\tau^i(B)\right), \tau\right)$ of $\mathcal{L}$ is denoted by $\mathcal{K}\langle B\rangle_{\mathcal{L}}$ or by $\mathcal{K}\langle B\rangle$ for brevity. For a difference field $\mathcal{K} = (K, \tau)$, we define $\mathcal{K}^{(k)} = (K, \tau^k)$.

A triplet $\mathcal{K} = (K, D, \tau)$ is called a *DTC field* if the pair (K, τ) is a difference field and D is a derivation from K to itself, namely $D(a + b) = D(a) + D(b)$ and $D(ab) = D(a)b + aD(b)$ for $a, b \in K$. A DTC field $\mathcal{L} = (L, D_L, \tau_L)$ is called a *DTC overfield* of a DTC field $\mathcal{K} = (K, D_K, \tau_K)$ and $\mathcal{K}$ a *DTC subfield* of $\mathcal{L}$ if $K \subset L$,

$\tau_L|_K = \tau_K$ and $D_L|_K = D_K$. In this case, $\mathcal{L}/\mathcal{K}$ denotes the DTC field extension. Let $\mathcal{L}/\mathcal{K}$ be a DTC field extension and B a subset of L. The minimum DTC subfield of $\mathcal{L}$ including K and B is denoted by $\mathcal{K}\langle B\rangle_{\mathcal{L}}$. For a DTC field extension $\mathcal{L}/\mathcal{K}$, we say an element $y \in L$ is *differentially algebraic* over K if the transcendence degree of $K(y, Dy, D^2y, \ldots)/K$ is finite.

For the rational function field $\mathbb{C}(x)/\mathbb{C}$, the place containing the irreducible linear polynomial $x - c$ is denoted by P_c. The discrete valuation associated with P_c is denoted by v_c.

We prepare the following proposition (see [**Hen97**, Lemma 10] or [**Nis10**, Lemma 8]) for use in a proof in Section 4.

PROPOSITION 2.1. *Let C be an algebraically closed field and x a transcendental element over C. Let $q \in C^\times$ be not a root of unity. Put the difference field $\mathcal{K} = (C(x), \tau : f(x) \mapsto f(qx))$. Suppose that $\mathcal{L} = (L, \tau)$ is a difference overfield of $\mathcal{K}$ and $L/C(x)$ is finite. Then there exists $z \in L$ such that $z^n = x, L = C(z)$ and $\tau(z) = q^{1/n}z$, where n is the degree of the extension $L/C(x)$ and $q^{1/n} \in C^\times$ is some n-th root of q.*

3. Transformation into difference Riccati equation of Tietze's normal form

Let $\mathcal{K} = (K, \tau)$ be a difference field. We consider a second-order linear difference equation

$$\tau^2(y) + a\tau(y) + by = 0, \qquad (a, b \in K). \tag{3.1}$$

Putting $z = \tau(y)/y$, we obtain the following difference Riccati equation

$$\tau(z) = \frac{-az - b}{z}. \tag{3.2}$$

For a non-singular matrix $A \in \mathrm{GL}_2(K)$, we define a linear fractional transformation as

$$A * z = \frac{az + b}{cz + d}, \qquad A = \begin{pmatrix} a & b \\ c & d \end{pmatrix}.$$

We also define

$$A_i = \begin{pmatrix} a^{(i)} & b^{(i)} \\ c^{(i)} & d^{(i)} \end{pmatrix} = (\tau^{i-1}A)(\tau^{i-2}A)\cdots(\tau A)A, \quad (i = 1, 2, \ldots).$$

In this paper, $\mathrm{Eq}(A, i)/\mathcal{K}$ denotes the equation over $\mathcal{K}$,

$$\tau^i(y) = A_i * y.$$

Then (3.2) is expressed as

$$\tau(z) = \begin{pmatrix} -a & -b \\ 1 & 0 \end{pmatrix} * z.$$

If a difference Riccati equation is expressed as

$$\tau(y) = \begin{pmatrix} 1 & r \\ 1 & 0 \end{pmatrix} * y,$$

we say the equation is of *Tietze's normal form*. Any difference Riccati equation associated with (3.1) is transformed into either one of Tietze's normal form or the difference equation $\tau^2(y) = y$ by linear fractional transformations (see [**Nis18**, Proposition 4]).

For the linear q-difference equation of the Ramanujan function (1.1), we set $g = \tau(y)/y$, where τ is the q-shift operator $\tau : f(x) \mapsto f(qx)$. Then we have the following q-difference Riccati equation,

$$\tau(g) = A * g, \qquad A = \begin{pmatrix} 1/qx & -1/qx \\ 1 & 0 \end{pmatrix}. \tag{3.3}$$

We put

$$P = \begin{pmatrix} 1 & -1 \\ 1 & 0 \end{pmatrix}, \quad f = P * g. \tag{3.4}$$

Then we have

$$\tau(f) = \tau(P)AP^{-1} * f, \qquad \tau(P)AP^{-1} = \frac{1}{qx}\begin{pmatrix} 1 & -qx \\ 1 & 0 \end{pmatrix}. \tag{3.5}$$

Therefore the equation (1.1) is transformed into the q-difference Riccati equation of Tietze's normal form,

$$\tau(f) = B * f, \qquad B = \begin{pmatrix} 1 & -qx \\ 1 & 0 \end{pmatrix}. \tag{3.6}$$

For the matrix B, we define

$$B_i = \begin{pmatrix} \alpha^{(i)} & \beta^{(i)} \\ \gamma^{(i)} & \delta^{(i)} \end{pmatrix} = (\tau^{i-1}B)(\tau^{i-2}B)\cdots(\tau B)B, \quad (i = 1, 2, \ldots,). \tag{3.7}$$

When $i = 1, 2$, we have

$$B_1 = \begin{pmatrix} \alpha^{(1)} & \beta^{(1)} \\ \gamma^{(1)} & \delta^{(1)} \end{pmatrix} = \begin{pmatrix} 1 & -qx \\ 1 & 0 \end{pmatrix},$$
$$B_2 = \begin{pmatrix} \alpha^{(2)} & \beta^{(2)} \\ \gamma^{(2)} & \delta^{(2)} \end{pmatrix} = \begin{pmatrix} 1 - q^2x & -qx \\ 1 & -qx \end{pmatrix}.$$

The definition of B_i implies $B_i = (\tau B_{i-1})B = (\tau^{i-1}B)B_{i-1}$. Then we obtain

$$(\tau B_{i-1})B = \begin{pmatrix} \tau(\alpha^{(i-1)}) + \tau(\beta^{(i-1)}) & -qx\tau(\alpha^{(i-1)}) \\ \tau(\gamma^{(i-1)}) + \tau(\delta^{(i-1)}) & -qx\tau(\gamma^{(i-1)}) \end{pmatrix},$$
$$(\tau^{i-1}B)B_{i-1} = \begin{pmatrix} \alpha^{(i-1)} - q^ix\gamma^{(i-1)} & \beta^{(i-1)} - q^ix\delta^{(i-1)} \\ \alpha^{(i-1)} & \beta^{(i-1)} \end{pmatrix}.$$

Hence we find the recurrence relations

$$\begin{aligned}
\alpha^{(i)} &= \alpha^{(i-1)} - q^ix\gamma^{(i-1)} = \alpha^{(i-1)} - q^ix\alpha^{(i-2)} \quad (i \geq 3), \\
\beta^{(i)} &= -qx\tau(\alpha^{(i-1)}), \\
\gamma^{(i)} &= \alpha^{(i-1)}, \\
\delta^{(i)} &= -qx\tau(\gamma^{(i-1)}) = \beta^{(i-1)}.
\end{aligned}$$

For brevity, we employ an equivalent relation of power series for their orders. For $a, b \in \mathbb{C}[\![x]\!]$ and $n \in \mathbb{Z}_{\geq 0}$, we define $a \equiv b \mod x^n$ as $a - b \in x^n\mathbb{C}[\![x]\!]$. When $a_{i,j} \equiv b_{i,j} \mod x^{n_{i,j}}$ for each entry $a_{i,j}$ of a matrix $(a_{i,j})_{i,j} \in M_n(\mathbb{C}[\![x]\!])$, we define $(a_{i,j})_{i,j} \equiv (b_{i,j} \mod x^{n_{i,j}})_{i,j}$.

Proposition 3.1. *For each $i \in \mathbb{Z}_{\geq 2}$,*

$$\begin{pmatrix} \alpha^{(i)} & \beta^{(i)} \\ \gamma^{(i)} & \delta^{(i)} \end{pmatrix} \equiv \begin{pmatrix} 1 \mod x & -qx \mod x^2 \\ 1 \mod x & -qx \mod x^2 \end{pmatrix}. \tag{3.8}$$

PROOF. The assertion holds true when $i = 2$. By induction on i, we have

$$\begin{pmatrix}\alpha^{(i)} & \beta^{(i)}\\ \gamma^{(i)} & \delta^{(i)}\end{pmatrix} = \begin{pmatrix}\alpha^{(i-1)} - q^i x\gamma^{(i-1)} & -qx\tau(\alpha^{(i-1)})\\ \alpha^{(i-1)} & \beta^{(i-1)}\end{pmatrix}$$
$$\equiv \begin{pmatrix}1 \mod x & -qx \mod x^2\\ 1 \mod x & -qx \mod x^2\end{pmatrix}.$$

□

4. Transcendence of solutions

In this section, let τ be the q-shift operator $f(x) \mapsto f(qx)$ and $\mathcal{K} = (\mathbb{C}(x), \tau)$. We consider the k-th iterated q-difference Riccati equations,

$$\text{(4.1)} \qquad \mathrm{Eq}(B,k)/\mathcal{K}: \quad \tau^k(f) = B_k * f, \qquad (k = 1, 2, \ldots).$$

First, we shall show there exists no rational solution for $\mathrm{Eq}(B,1)/\mathcal{K}$.

PROPOSITION 4.1. *$\mathrm{Eq}(B,1)/\mathcal{K}$ has no solution in $\mathbb{C}(x)$.*

PROOF. Assume that $f \in \mathbb{C}(x)$ is a solution for $\mathrm{Eq}(B,1)/\mathcal{K}$, that is,

$$f\tau(f) = f - qx.$$

We take relatively prime polynomials $P, Q \in \mathbb{C}[x]$ such that $f = P/Q$. Then we have

$$\text{(4.2)} \qquad P\tau(P) = P\tau(Q) - qxQ\tau(Q).$$

Since $\tau(P)$ and $\tau(Q)$ are relatively prime, we find $P \mid qx\tau(Q)$ and $\tau(Q) \mid P$. This means $\deg_x Q \le \deg_x P \le \deg_x Q + 1$. Putting $n = \deg_x Q$, we have $\deg_x P = n, n+1$. It follows from (4.2) that

- if $\deg_x P = n$, then $2n = \max\{2n, 2n+1\} = 2n+1$,
- if $\deg_x P = n+1$, then $2n+2 \le \max\{2n+1, 2n+1\} = 2n+1$.

This is a contradiction. □

We extend the q-shift operator τ to the field of formal Laurent series $\mathbb{C}((x))$ defined as $\tau\left(\sum_i c_i x^i\right) = \sum_i c_i q^i x^i$.

PROPOSITION 4.2. *Let $f^{(n)} \in \mathbb{C}((x))$ be a solution for $\mathrm{Eq}(B,n)/\mathcal{K}$. Then $v_0(f^{(n)})$ is equal to either 0 or 1. Moreover, for each $n \in \mathbb{Z}_{\ge 1}$,*

- *$\mathrm{Eq}(B,n)/\mathcal{K}$ has a unique solution $f^{(n)}$ in $\mathbb{C}((x))$ satisfying $v_0(f^{(n)}) = 0$,*
- *$\mathrm{Eq}(B,n)/\mathcal{K}$ has a unique solution $f^{(n)}$ in $\mathbb{C}((x))$ satisfying $v_0(f^{(n)}) = 1$.*

If $v_0(f^{(1)}) = v_0(f^{(2)}) = \cdots = v_0(f^{(n)}) \cdots$, then $f^{(1)} = f^{(2)} = \cdots = f^{(n)} = \cdots$ holds.

PROOF. Suppose that $f = \sum_{i=r}^{\infty} f_i x^i$ is a solution for $\mathrm{Eq}(B,n)/\mathcal{K}$ with $f_r \ne 0$. First, we show r is equal to either 0 or 1. The equation $\mathrm{Eq}(B,n)/\mathcal{K}$ is written as

$$\text{(4.3)} \qquad \tau^n(f)(\gamma^{(n)} f + \delta^{(n)}) = \alpha^{(n)} f + \beta^{(n)}.$$

By Proposition 3.1, we have

$$\begin{aligned}&\tau^n(f)(\gamma^{(n)}f+\delta^{(n)})\\&=\left(\sum_{i=r}^{\infty}f_iq^{ni}x^i\right)\left((1+\cdots)\sum_{i=r}^{\infty}f_ix^i+\begin{Bmatrix}0 & (n=1)\\ -qx+\cdots & (n\geq 2)\end{Bmatrix}\right),\\&\alpha^{(n)}f+\beta^{(n)}=(1+\cdots)\left(\sum_{i=r}^{\infty}f_ix^i\right)-qx+\cdots.\end{aligned}$$

In the case $r<0$, by comparing the coefficients of x^{2r}, we have $q^{nr}f_r^2=0$. Hence f_r must vanish, a contradiction. In the case $r>1$, by comparing the coefficients of x, we have $0=-q$, a contradiction. Hence r must be equal to either 0 or 1.

(Uniqueness and existence in the case $r=1$) Suppose that $v_0(f)=1$. Comparing the coefficients of x in (4.3), we obtain $f_1=q$. For $i\in\mathbb{Z}_{\geq 2}$, each coefficient f_i is uniquely determined by f_1, ..., f_{i-1}. Therefore we conclude f is a unique solution satisfying $v_0(f)=1$.

(Uniqueness and existence in the case $r=0$) Suppose that $v_0(f)=0$. Comparing the coefficients of the constant terms in (4.3), we obtain $f_0^2=f_0$, and hence $f_0=1$. For $i\in\mathbb{Z}_{\geq 1}$, each coefficient f_i is uniquely determined by f_0, ..., f_{i-1}. Therefore we conclude f is a unique solution satisfying $v_0(f)=0$.

(Identity) Suppose that the values $v_0(f^{(n)})$ are same. Let $f^{(n)}$ be a solution for $\mathrm{Eq}(B,n)/\mathcal{K}$. Then $f^{(1)}$ is also a solution to $\mathrm{Eq}(B,n)/\mathcal{K}$ for each $n\in\mathbb{Z}_{\geq 1}$. From the uniqueness, we obtain $f^{(n)}=f^{(1)}$. □

THEOREM 4.3. *For each $k\in\mathbb{Z}_{\geq 1}$, $\mathrm{Eq}(B_k,1)/\mathcal{K}^{(k)}$ has no algebraic solution over $\mathbb{C}(x)$.*

PROOF. Assume that there exists $k\in\mathbb{Z}_{\geq 1}$ such that $\mathrm{Eq}(B_k,1)/\mathcal{K}^{(k)}$ has an algebraic solution f over $\mathbb{C}(x)$. Let $\mathcal{L}=(L,\tau_k)$ be the difference overfield generated by f and $\mathcal{K}^{(k)}$, that is, $\mathcal{L}=\mathcal{K}^{(k)}\langle f\rangle$. Then f satisfies

$$\tau_k(f)=\frac{\alpha^{(k)}f+\beta^{(k)}}{\gamma^{(k)}f+\delta^{(k)}}\in\mathbb{C}(x,f),$$

which implies the underlying field L is generated by only f, that is, $L=\mathbb{C}(x,f)$. Since L/K is finite, it follows from Proposition 2.1 that there exists $z\in L$ such that $L=\mathbb{C}(z)$, $z^r=x$, $\tau_k(z)=pz$, where $r=[L:\mathbb{C}(x)]$ and $p\in\mathbb{C}^{\times}$ satisfies $p^r=q^k$. Regarding f as a rational function of z, we can take relatively prime polynomials $P,Q\in\mathbb{C}[z]^{\times}$ such that $f=P/Q$. Then we have

$$\frac{\tau_k(P)}{\tau_k(Q)}=\frac{\alpha^{(k)}P+\beta^{(k)}Q}{\gamma^{(k)}P+\delta^{(k)}Q}.$$

Note that each entry of B_k is a polynomial in $\mathbb{C}[z^r]$. Since $\tau_k(P)$ and $\tau_k(Q)$ are relatively prime, there exists $R\in\mathbb{C}[z]^{\times}$ such that

$$R\begin{pmatrix}\tau_k(P)\\ \tau_k(Q)\end{pmatrix}=\begin{pmatrix}\alpha^{(k)} & \beta^{(k)}\\ \gamma^{(k)} & \delta^{(k)}\end{pmatrix}\begin{pmatrix}P\\ Q\end{pmatrix}.$$

Here, we compute the determinant of B_k,

$$\begin{aligned}\det B_k&=\left(\tau^{k-1}\det B\right)\left(\tau^{k-2}\det B\right)\cdots(\tau\det B)\det B\\&=q^{k(k+1)/2}x^k\neq 0.\end{aligned}$$

Then we obtain

$$\begin{pmatrix} P \\ Q \end{pmatrix} = q^{-k(k+1)/2} \frac{R}{z^{rk}} \begin{pmatrix} \delta^{(k)} & -\beta^{(k)} \\ -\gamma^{(k)} & \alpha^{(k)} \end{pmatrix} \begin{pmatrix} \tau_k(P) \\ \tau_k(Q) \end{pmatrix}. \tag{4.4}$$

Let w_c be the discrete valuation of $\mathbb{C}(z)/\mathbb{C}$ associated with the place P_c containing $z-c$. We find R has no zeros except $z=0$. Indeed, if there exists $c \in \mathbb{C}^\times$ such that $w_c(R) > 0$, then it follows

$$\begin{aligned} w_c(P) &= w_c(R) + w_c(\delta^{(k)}\tau_k(P) - \beta^{(k)}\tau_k(Q)) > 0, \\ w_c(Q) &= w_c(R) + w_c(-\gamma^{(k)}\tau_k(P) + \alpha^{(k)}\tau_k(Q)) > 0, \end{aligned}$$

a contradiction. Hence R is written as the form $R = cz^l$ for $c \in \mathbb{C}^\times, l \geq 0$. Then the exponent l is not greater than rk. Indeed, if $l > rk$, then we have

$$\begin{aligned} w_0(P) &= w_0\left(\frac{R}{z^{rk}}\right) + w_0(\delta^{(k)}\tau_k(P) - \beta^{(k)}\tau_k(Q)) > 0, \\ w_0(Q) &= w_0\left(\frac{R}{z^{rk}}\right) + w_c(-\gamma^{(k)}\tau_k(P) + \alpha^{(k)}\tau_k(Q)) > 0, \end{aligned}$$

a contradiction. We rewrite (4.4) as

$$z^{rk-l}\begin{pmatrix} P \\ Q \end{pmatrix} = cq^{-k(k+1)/2} \begin{pmatrix} \delta^{(k)} & -\beta^{(k)} \\ -\gamma^{(k)} & \alpha^{(k)} \end{pmatrix} \begin{pmatrix} \tau_k(P) \\ \tau_k(Q) \end{pmatrix}. \tag{4.5}$$

Since P and Q are relatively prime, $w_0(P) = 0$ or $w_0(Q) = 0$.

Expanding $f \in \mathbb{C}(z)$ as a Laurent series, we put $f = \sum_{i=m}^{\infty} f_i z^i \in \mathbb{C}((z))$ with $f_m \neq 0$. We shall show $f \in \mathbb{C}((x))$ by contradiction. It is obvious in the case $r = 1$. Assume that $r \geq 2$ and $f \notin \mathbb{C}((x))$. Then there exists an integer i such that $f_i \neq 0$ and $r \nmid i$. For $\xi = \sum_{i=m}^{\infty} \xi_i z^i \in \mathbb{C}((z)) \backslash \mathbb{C}((x))$, we define $\operatorname{div}_r(\xi)$ as

$$\operatorname{div}_r(\xi) = \min\{i \in \mathbb{Z}_{\geq m} \mid \xi_i \neq 0,\ r \nmid i\}.$$

Let us consider the two cases, $l = rk$ and $l < rk$.

[Case $l = rk$] From Proposition 3.1 and (4.5), we obtain $w_0(P) > 0$ and $w_0(Q) = 0$. Then we have

$$w_0(P) = w_0(\delta^{(k)}\tau_k(P) - \beta^{(k)}\tau_k(Q)) = w_0(\beta^{(k)}\tau_k(Q)) = r.$$

Hence $w_0(f) = w_0(P/Q) = r$ holds. The Taylor expansion of f is written as $f = \sum_{i \geq r} f_i z^i$ with $f_r \neq 0$. By Proposition 3.1, we put

$$\begin{aligned} \alpha^{(k)} &= 1 + z^r \hat{\alpha}^{(k)}(z^r), & \beta^{(k)} &= -qz^r + z^{2r}\hat{\beta}^{(k)}(z^r), \\ \gamma^{(k)} &= 1 + z^r \hat{\gamma}^{(k)}(z^r), & \delta^{(k)} &= \begin{Bmatrix} 0 & (k=1) \\ -qz^r + z^{2r}\hat{\delta}^{(k)}(z^r) & (k \geq 2) \end{Bmatrix}, \end{aligned}$$

where $\hat{\alpha}^{(k)}(z^r), \hat{\beta}^{(k)}(z^r), \hat{\gamma}^{(k)}(z^r), \hat{\delta}^{(k)}(z^r) \in \mathbb{C}[z^r]$. By assumption, we put

$$\operatorname{div}_r(f) = \lambda r + \mu \tag{4.6}$$

for $0 < \mu < r$ and $\lambda \geq 1$. Then we have

$$\alpha^{(k)} f + \beta^{(k)} = (1 + z^r \hat{\alpha}^{(k)}(z^r))(f_r z^r + \cdots + f_{\lambda r} z^{\lambda r} + f_{\lambda r+\mu} z^{\lambda r+\mu} + \cdots) + (-qz^r + z^{2r}\hat{\beta}^{(k)}(z^r)), \tag{4.7}$$

$$\begin{aligned}\tau_k(f)(\gamma^{(k)} f + \delta^{(k)}) &= (f_r q^k z^r + \cdots + f_{\lambda r} q^{\lambda k} z^{\lambda r} + f_{\lambda r+\mu} p^{\lambda r+\mu} z^{\lambda r+\mu} + \cdots) \\ &\times \left((1 + z^r \hat{\gamma}^{(k)}(z^r))(f_r z^r + \cdots + f_{\lambda r} z^{\lambda r} + f_{\lambda r+\mu} z^{\lambda r+\mu} + \cdots) \right. \\ &\left. + \begin{cases} 0 & (k=1) \\ -qz^r + z^{2r}\hat{\delta}^{(k)}(z^r) & (k \geq 2) \end{cases} \right).\end{aligned} \tag{4.8}$$

The equation (4.7) implies $\mathrm{div}_r(\alpha^{(k)} f + \beta^{(k)}) = \lambda r + \mu$. On the other hand, the equation (4.8) implies $\mathrm{div}_r(\tau_k(f)(\gamma^{(k)} f + \delta^{(k)})) \geq \lambda r + \mu + r$. Hence, $\lambda r + \mu \geq \lambda r + \mu + r$, namely $0 \geq r$, a contradiction.

[Case $l < rk$] It follows from (4.5) that

$$z^{rk-l} Q = cq^{-k(k+1)/2}(-\gamma^{(k)} \tau_k(P) + \alpha^{(k)} \tau_k(Q)).$$

By Proposition 3.1, we have

$$0 \equiv -\tau_k(P) + \tau_k(Q) \mod z.$$

Hence we find the constant terms of both P and Q are the same value. Since P and Q are relatively prime, the constant terms are non-zero, that is, $w_0(P) = w_0(Q) = 0$. Then we write f for the form $f = \sum_{i=0}^{\infty} f_i z^i$ with $f_0 = 1$. Here, we shall show $f_i = 0$ for $i = 1, 2, \ldots, r-1$ by induction. We have

$$\alpha^{(k)} f + \beta^{(k)} \equiv 1 + f_1 z \mod z^2,$$

$$\tau_k(f)(\gamma^{(k)} f + \delta^{(k)}) \equiv (1 + f_1 pz)(1 + f_1 z) \mod z^2.$$

Comparing the coefficients of z, we obtain $f_1 = f_1 + f_1 p$, which implies $f_1 = 0$. Let h be an integer with $0 < h < r$ and suppose $f_1 = f_2 = \cdots = f_{h-1} = 0$. Then we have

$$\alpha^{(k)} f + \beta^{(k)} \equiv 1 + f_h z^h \mod z^{h+1},$$

$$\tau_k(f)(\gamma^{(k)} f + \delta^{(k)}) \equiv (1 + f_h p^h z^h)(1 + f_h z^h) \mod z^{h+1}.$$

Comparing the coefficients of z^h, we obtain $f_h = f_h + f_h p^h$, which implies $f_h = 0$. By induction, we find $f_1 = f_2 = \cdots = f_{r-1} = 0$.

By assumption, we put $\mathrm{div}_r(f) = \lambda r + \mu$ for $0 < \mu < r$ and $\lambda \geq 1$. In particular, it follows $f_{\lambda r+\mu} \neq 0$. Then we have

$$\begin{aligned}\alpha^{(k)} f + \beta^{(k)} &= (1 + z^r \hat{\alpha}^{(k)}(z^r))(1 + f_r z^r + \cdots + f_{\lambda r} z^{\lambda r} + f_{\lambda r+\mu} z^{\lambda r+\mu} + \cdots) \\ &+ (-qz^r + z^{2r}\hat{\beta}^{(k)}(z^r)),\end{aligned}$$

$$\begin{aligned}\tau_k(f)(\gamma^{(k)} f + \delta^{(k)}) &= (1 + f_r q^k z^r + \cdots + f_{\lambda r} q^{k\lambda} z^{\lambda r} + f_{\lambda r+\mu} p^{\lambda r+\mu} z^{\lambda r+\mu} + \cdots) \\ &\times \left((1 + z^r \hat{\gamma}^{(k)}(z^r))(1 + f_r z^r + \cdots + f_{\lambda r} z^{\lambda r} + f_{\lambda r+\mu} z^{\lambda r+\mu} + \cdots) \right. \\ &\left. + \begin{cases} 0 & (k=1) \\ -qz^r + z^{2r}\hat{\delta}^{(k)}(z^r) & (k \geq 2) \end{cases} \right).\end{aligned}$$

Comparing the coefficients of $z^{\lambda r+\mu}$, we obtain $f_{\lambda r+\mu} = f_{\lambda r+\mu} + f_{\lambda r+\mu}p^{\lambda r+\mu}$, which implies $f_{\lambda r+\mu} = 0$, a contradiction.

Therefore $f \in \mathbb{C}((x))$ holds true. It follows that $\mathbb{C}((z^r)) \cap \mathbb{C}(z) = \mathbb{C}(z^r) = \mathbb{C}(x)$, so we have $f \in \mathbb{C}(x)$. Since the order of f at $x = 0$ is equal to either 1 or 0, it follows from Proposition 4.2 that f is a rational solution for $\mathrm{Eq}(B,1)/\mathcal{K}$. This contradicts Proposition 4.1. □

5. Differential transcendence of solutions

In order to show differential transcendence of solutions for (1.1), we apply the following criterion of Nishioka to $\mathrm{Eq}(B,1)/\mathcal{K}$.

THEOREM 5.1 ([**Nis18**, Theorem 9]). *Let $\mathcal{F} = (F, D_0, \tau_0)$ be a DTC field with $D_0\tau_0 = s\tau_0 D_0$ for a certain $s \in F^\times$ and F/K an algebraic function field of one variable. Let*

$$A = \begin{pmatrix} 1 & r \\ 1 & 0 \end{pmatrix} \in M_2(F), \quad (r \neq 0, D_0 r \neq 0). \tag{5.1}$$

Suppose that there exists a place P of F/K such that $v_P(\tau_0^i r) > 0$ for all $i \in \mathbb{Z}_{\geq 0}$. Additionally, suppose that for any $i \in \mathbb{Z}_{\geq 1}$, $\mathrm{Eq}(A,i)/(F,\tau_0) : \tau_0^i(y) = A_i * y$ has no algebraic solution over F. Let $\mathcal{U} = (U, D, \tau)$ be a DTC overfield of $\mathcal{F}$ with $D\tau = s\tau D$. If there exists a solution $f \in \mathcal{U}$ of $\mathrm{Eq}(A,1)/(F,\tau_0) : \tau_0(y) = A * y$ which is differentially algebraic over F, then there exists $g \in F$ such that

(5.2)

$$\tau^2(sr)\tau(s)s\tau^3(g) + (\tau(r)+1)\tau(s)s\tau^2(g) - (\tau(r)+1)s\tau(g) - rg + s\tau\left(\frac{D(r)}{r}\right) = 0.$$

Putting $F = \mathbb{C}(x), D_0 = d/\,dx\,, \tau_0 : f(x) \mapsto f(qx), s = q, K = \mathbb{C}$ in Theorem 5.1, we obtain immediately the following corollary of q-difference Riccati equations of Tietze's normal form.

COROLLARY 5.2. *Let*

$$A = \begin{pmatrix} 1 & r \\ 1 & 0 \end{pmatrix} \in M_2(\mathbb{C}(x)), \quad (r \notin \mathbb{C}). \tag{5.3}$$

Suppose that there exists a place P of the rational function field $\mathbb{C}(x)/\mathbb{C}$ such that $v_P(r(q^i x)) > 0$ for all $i \in \mathbb{Z}_{\geq 0}$. Additionally, suppose that for any $i \in \mathbb{Z}_{\geq 1}$, $\mathrm{Eq}(A,i)/(\mathbb{C}(x),\tau_0) : y(q^i x) = A_i * y(x)$ has no algebraic solution over $\mathbb{C}(x)$. Let $\mathcal{U} = (U, D, \tau)$ be a DTC overfield of $\mathcal{F} = (\mathbb{C}(x), d/\,dx\,, \tau_0)$ with $D\tau = q\tau D$. If there exists a solution $f \in \mathcal{U}$ of $\mathrm{Eq}(A,1)/(\mathbb{C}(x),\tau_0) : y(qx) = A * y(x)$ which is differentially algebraic over $\mathbb{C}(x)$, then there exists $g \in \mathbb{C}(x)$ such that

(5.4)

$$q^3 r(q^2x)g(q^3x) + q^2(r(qx)+1)g(q^2x) - q(r(qx)+1)g(qx) - r(x)g(x) + q\frac{r'(qx)}{r(qx)} = 0.$$

THEOREM 5.3. *Let τ be the q-shift operator and $\mathcal{F} = (\mathbb{C}(x), d/\,dx\,, \tau)$. Consider the q-difference Riccati equation,*

$$\tau(f) = B * f, \qquad B = \begin{pmatrix} 1 & -qx \\ 1 & 0 \end{pmatrix}. \tag{5.5}$$

Let $\mathcal{U} = (U, D, \tau)$ be a DTC overfield of $\mathcal{F}$ with $D\tau = q\tau D$. Then (5.5) has no non-trivial differentially algebraic solution in $\mathcal{U}$ over $\mathbb{C}(x)$.

PROOF. Let v_0 be the discrete valuation of $\mathbb{C}(x)/\mathbb{C}$ at $x = 0$. Obviously, we have $v_0(\tau^i(-qx)) = v_0(-q^{i+1}x) > 0$ for each $i \in \mathbb{Z}_{\geq 0}$. Recall that $\mathrm{Eq}(B,i)/(\mathbb{C}(x), \tau)$ has no algebraic solution over $\mathbb{C}(x)$ for each $i \in \mathbb{Z}_{\geq 1}$ by Theorem 4.3. Assume that there exists a solution in $\mathcal{U}$ to (5.5) which is differentially algebraic over $\mathbb{C}(x)$. Corollary 5.2 implies that there exists $h \in \mathbb{C}(x)$ such that

$$-q^6 xh(q^3x) + q^2\bigl(1 - q^2x\bigr)h(q^2x) - q\bigl(1 - q^2x\bigr)h(qx) + qxh(x) + \frac{1}{x} = 0. \tag{5.6}$$

We write the rational function h as a Laurent series at $x = 0$,

$$h = \sum_{i=m}^{\infty} h_i x^i, \quad (h_i \in \mathbb{C}, h_m \neq 0).$$

If $m \neq -1$, comparing the coefficients of x^m yields

$$q^{2+2m}h_m - q^{1+m}h_m = 0.$$

Since q is not a root of unity, h_m must vanish, a contradiction. Hence $m = -1$ holds. Comparing the coefficients of x^{-1} yields

$$h_{-1} - h_{-1} + 1 = 0,$$

which implies a contradiction. □

PROOF OF THEOREM 1.1. Assume that there exists a non-trivial solution y for (1.1) satisfying an algebraic differential equation over $\mathbb{C}(x)$. Let A and P be the matrices defined by (3.3) and (3.4), respectively. Putting $g(x) = y(qx)/y(x)$, we see that g satisfies $g(qx) = A * g$ and g is differentially algebraic over $\mathbb{C}(x)$. Putting $f = P * g$, we find f is a solution for (5.5). Since f is a linear fractional transformation of g, we find f is differentially algebraic over $\mathbb{C}(x)$. This is a contradiction by Theorem 5.3. □

In addition, Theorem 4.3 suggests unsolvability of (1.1) with respect to q-difference equations. In order to explain unsolvability of difference equations, we quote generalized Liouvillian extensions from [**Nis16**], which are originated with C. H. Franke [**Fra63, Fra66**].

DEFINITION 5.4 ([**Nis16**, Definition 10]). Let $\mathcal{N}/\mathcal{K}$ be a difference field extension of inversive difference fields and τ a transforming operator of $\mathcal{N}/\mathcal{K}$. We call $\mathcal{N}/\mathcal{K}$ a *generalized Liouvillian extension* if there exists a chain of inversive difference fields,

$$\mathcal{K} = \mathcal{K}_0 \subset \mathcal{K}_1 \subset \cdots \subset \mathcal{K}_n = \mathcal{N},$$

such that for each $i = 1, 2, \ldots, n$, $\mathcal{K}_i$ is the inversive closure of $\mathcal{K}_{i-1}\langle e^{(i)}\rangle$, where $e^{(i)}$ satisfies one of the following,

(i) $e^{(i)}$ is algebraic over K_{i-1},
(ii) $\tau(e^{(i)}) = e^{(i)} + \beta$ for some $\beta \in K_{i-1}$,
(iii) $\tau(e^{(i)}) = \alpha e^{(i)}$ for some $\alpha \in K_{i-1}$.

For $p \in \mathbb{Z}_{\geq 1}$, we say $\mathcal{N}/\mathcal{K}$ is a pLE if $\mathcal{N}^{(p)}/\mathcal{K}^{(p)}$ is a generalized Liouvillian extension. We say a difference equation is *unsolvable* over $\mathcal{K}$ if no non-trivial solution is contained in any pLE $\mathcal{N}/\mathcal{K}$.

THEOREM 5.5. *Let $q \in \mathbb{C}^\times$ be not a root of unity and $\mathcal{K} = (\mathbb{C}(x), \tau : f(x) \mapsto f(qx))$. Then (1.1) is unsolvable over $\mathcal{K}$. Moreover, for each non-trivial solution $y(x)$ to (1.1), $y(x)$ and $y(qx)$ are algebraic independent over any pLE $\mathcal{N}/\mathcal{K}$.*

PROOF. The assertion is straightforwardly proved by Theorem 4.3 and [**Nis16**, Corollary 15]. □

References

[Fra63] Charles H. Franke, *Picard-Vessiot theory of linear homogeneous difference equations*, Trans. Amer. Math. Soc. **108** (1963), 491–515, DOI 10.2307/1993595. MR155819

[Fra66] Charles H. Franke, *Solvability of linear homogeneous difference equations by elementary operations*, Proc. Amer. Math. Soc. **17** (1966), 240–246, DOI 10.2307/2035098. MR190133

[Hen97] Peter A. Hendriks, *An algorithm for computing a standard form for second-order linear q-difference equations*, J. Pure Appl. Algebra **117/118** (1997), 331–352, DOI 10.1016/S0022-4049(97)00017-0. Algorithms for algebra (Eindhoven, 1996). MR1457845

[HKW06] Taro Hamamoto, Kenji Kajiwara, and Nicholas S. Witte, *Hypergeometric solutions to the q-Painlevé equation of type* $(A_1 + A_1')^{(1)}$, Int. Math. Res. Not., posted on 2006, Art. ID 84619, 26, DOI 10.1155/IMRN/2006/84619. MR2264726

[Ism05] Mourad E. H. Ismail, *Asymptotics of q-orthogonal polynomials and a q-Airy function*, Int. Math. Res. Not. **18** (2005), 1063–1088, DOI 10.1155/IMRN.2005.1063. MR2149641

[IZ07] Mourad E. H. Ismail and Changgui Zhang, *Zeros of entire functions and a problem of Ramanujan*, Adv. Math. **209** (2007), no. 1, 363–380, DOI 10.1016/j.aim.2006.05.007. MR2294226

[KMN+04] K. Kajiwara, T. Masuda, M. Noumi, Y. Ohta, and Y. Yamada, *Hypergeometric solutions to the q-Painlevé equations*, Int. Math. Res. Not. **47** (2004), 2497–2521, DOI 10.1155/S1073792804140919. MR2077840

[Mor14] Takeshi Morita, *The Stokes phenomenon for the q-difference equation satisfied by the Ramanujan entire function*, Ramanujan J. **34** (2014), no. 3, 329–346, DOI 10.1007/s11139-013-9519-6. MR3231315

[Nis10] Seiji Nishioka, *Solvability of difference Riccati equations by elementary operations*, J. Math. Sci. Univ. Tokyo **17** (2010), no. 2, 159–178. MR2759757

[Nis16] Seiji Nishioka, *Proof of unsolvability of q-Bessel equation using valuations*, J. Math. Sci. Univ. Tokyo **23** (2016), no. 4, 763–789. MR3588262

[Nis17] Seiji Nishioka, *Transcendence of solutions of q-Airy equation*, Josai Mathematical Monographs **10** (2017), 129–137.

[Nis18] Seiji Nishioka, *Differential transcendence of solutions of difference Riccati equations and Tietze's treatment*, J. Algebra **511** (2018), 16–40, DOI 10.1016/j.jalgebra.2018.06.008. MR3834761

[Ram88] Srinivasa Ramanujan, *The lost notebook and other unpublished papers*, Springer-Verlag, Berlin; Narosa Publishing House, New Delhi, 1988. With an introduction by George E. Andrews. MR947735

MATHEMATICAL SCIENCE EDUCATION CENTER, KUMAMOTO UNIVERSITY, 2-40-1, KUROKAMI, CHUO-KU, KUMAMOTO, 860-8555, JAPAN

Email address: hogawara@kumamoto-u.ac.jp

Contemporary Mathematics
Volume **782**, 2023
https://doi.org/10.1090/conm/782/15727

On the positive powers of q-analogs of Euler series

Changgui Zhang

ABSTRACT. The most simple and famous divergent power series coming from ODE may be the so-called Euler series $\sum_{n\geq 0}(-1)^n\, n!\, x^{n+1}$, that, as well as all its positive powers, is Borel-summable in any direction excepted the negative real half-axis (see [**2**] or [**7**]). By considering a family of linear q-difference operators associated with a given first order non-homogenous q-difference equation, it will be shown that the summability order of q-analoguous counterparties of Euler series depends upon of the degree of power under consideration.

1. Introduction

It is well-known that, given any direction starting from the origin of the complex plane, the set of all Borel-summable series in this direction forms a differential algebra; see [**7**, Proposition 1.3.4.2] or [**2**, Theorem 36]. This is not the same for the summability with respect to the power series solutions of q-difference equations. Indeed, if $q > 1$ and

$$\hat{E}_q(x) = \sum_{n\geq 0}(-1)^n q^{n(n-1)/2} x^n\,, \tag{1.1}$$

the divergent power series $\hat{E}_q(x)$ is Gq-summable of only one level but its square $\hat{E}_q(x)^2$ is Gq-summable of two levels; see [**11**, §4.3.8] and [**8**, Théorème 2.2.1]. The first purpose of this paper is to obtain a linear q-difference equation for every positive power of the power series $\hat{E}_q(x)$. This will allow one to see the different level of summability for each of these corresponding power series.

In what follows, let q be a given nonzero complex number, and let σ_q be the associated q-difference operator defined by $\sigma_q f(x) = f(qx)$. By direct computation, one finds the series defined by (1.1) satisfies term by term the following linear q-difference equation:

$$(x\sigma_q + 1)y = 1\,. \tag{1.2}$$

Now, write (1.2) into the form $x\sigma_q y = 1 - y$ and then square both sides. As $(\sigma_q y)^2 = \sigma_q(y^2)$, one obtains that $x^2\sigma_q(y^2) - y^2 = 1 - 2y$. By using again (1.2), one gets the following equation:

$$(x\sigma_q + 1)(x^2\sigma_q - 1)(y^2) = x + 1 - 2(x\sigma_q + 1)y\,,$$

2020 *Mathematics Subject Classification.* Primary 39B32, 40G10, 30E15.

This work was partially supported by Labex CEMPI (Centre Européen pour les Mathématiques, la Physique et leurs Interaction).

what implies that

$$(1.3) \qquad (x\sigma_q + 1)(x^2\sigma_q - 1)\hat{E}_q(x)^2 = x - 1\,.$$

We shall explain how to obtain a linear q-difference equation for $\hat{E}_q(x)^n$. For doing that, we shall start by recalling some non-commutative rings of q-difference operators and then arrive at a n-th order such operator associated with any given n-th power of a power series solution of some first order non-homogenous q-difference equation. See Theorem 2.1 in § 2 and its generalisation stated in Theorem 4.3 in § 4. Section 3 is reserved to the proofs of Theorem 2.1 and two related Lemmas. In the last section, we shall apply, respectively, both Theorems 2.1 and 4.3 to the above-mentioned power series $\hat{E}_q(x)$ and another q-analog $\hat{E}(x;q)$ of the classic Euler series. At the end of the paper will be outlined some results about the summability of both $\hat{E}_q(x)$ and $\hat{E}(x;q)$ when $q > 1$; see Theorem 5.3.

Throughout the present paper, we shall limit ourself to the q-summation method studied in our previous works [**11**] and [**8**], by means of a q-Laplace integral involving the multivalued function $e^{\log^2 x/2\ln q}$. There exists another summation, using a Jacobi theta function; see [**10**]. Comparing both summations can be seen as part of Stokes analysis, and this would be interesting to find identities on the associated special functions.

2. Notation and statements

As usual, we will denote by $\mathbb{C}\{x\}$ or $\mathbb{C}[[x]]$ the integral domain constituting of the germs of holomorphic functions at $x = 0$ in the complex plane or that of the power series of indeterminate x, respectively. Their respective corresponding field of fractions will be denoted by $\mathbb{C}[x^{-1}]\{x\}$ or $\mathbb{C}[x^{-1}][[x]]$. One has $\mathbb{C}\{x\} \subset \mathbb{C}[[x]]$, as well as $\mathbb{C}[x^{-1}]\{x\} \subset \mathbb{C}[x^{-1}][[x]]$, in a natural way via the classic Taylor or Laurent series.

Let R be one of the above-mentioned $\mathbb{C}$-vector spaces $\mathbb{C}\{x\}$, $\mathbb{C}[[x]]$, $\mathbb{C}[x^{-1}]\{x\}$ and $\mathbb{C}[x^{-1}][[x]]$. By replacing x with qx, the q-difference operator σ_q acts as an automorphism over R. Let $\sigma_q^k = \sigma_{q^k}$ for any integer k; in particular, σ_q^0 is simply the identity map on R. We will denote by $R[\sigma_q]$ the set of the (linear) q-difference operators whose coefficients belong to R, *i.e.* $L \in R[\sigma_q]$ if $L = \sum_{k=0}^n a_k\sigma_q^k$, where $a_k \in R$. By identifying each element of R with the corresponding multiplication map in R, the set $R[\sigma_q]$ constitutes itself a non-commutative ring for the composition of operators. Given L_1, $L_2 \in R[\sigma_q]$, we will write L_1L_2 in stead of $L_1 \circ L_2$.

Furthermore, we will denote by $\nu_0(f) \in \mathbb{Z} \cup \{\infty\}$ the valuation of any given element f of the field $\mathbb{C}[x^{-1}][[x]]$ "at $x = 0$". This means that (i) $\nu_0(f) = \infty$ iff f is the identically vanishing series, and (ii) $\nu_0(f) = \nu \in \mathbb{Z}$ iff $f \equiv c\,x^\nu \mod x^{\nu+1}\,\mathbb{C}[[x]]$ for some nonzero complex number c.

Throughout the whole section, P will be some given nonzero power series belonging to the field $\mathbb{C}[x^{-1}][[x]]$. Define the associated family of power series $\{P_n\}_{n\geq 0}$ as follows: $P_0 = 0$, $P_1 = P$, and

$$(2.1) \qquad P_n = \sum_{k=0}^{n-1}(-x\sigma_q)^k P \quad \text{for} \quad n \geq 2.$$

By noticing that $(-x\sigma_q)^k P = (-x)^k\, q^{k(k-1)/2}\, \sigma_q^k P$, it follows from (2.1) that, for any pair $(n,\ell) \in \mathbb{Z}_{\geq 0} \times \mathbb{Z}_{>0}$,

$$\begin{aligned} P_{n+\ell} - P_n &= \sum_{k=n}^{n+\ell-1} (-x)^k\, q^{k(k-1)/2}\, \sigma_q^k P \\ &\equiv (-1)^n\, c\, q^{n\nu + n(n-1)/2} x^{\nu+n} \mod x^{\nu+n+1}\,\mathbb{C}[[x]]\,, \end{aligned}$$

where $c \neq 0$ and $\nu = \nu_0(P) \in \mathbb{Z}$. This implies that

$$P_m \neq P_n \quad \text{if} \quad m \neq n. \tag{2.2}$$

Given any positive integer $n \geq 1$, let $\{L_{n,k}^P\}_{1\leq k\leq n} \subset \left(\mathbb{C}[x^{-1}][[x]]\right)[\sigma_q]$ be the associated family of q-difference operators defined in the following manner:

$$L_{n,1}^P = \frac{1}{P_1}\left(x^n\sigma_q - (-1)^n\right) = \frac{1}{P}\left(x^n\sigma_q - (-1)^n\right); \tag{2.3}$$

$$L_{n,k+1}^P = \frac{1}{P_{k+1}}\left(x^{n-k}\sigma_q - (-1)^{n-k}\right) L_{n,k}^P, \quad 1 \leq k < n\,. \tag{2.4}$$

Specifically, letting $k = n$, one gets the following form of the n-th order q-difference operator $L_{n,n}^P \in \left(\mathbb{C}[x^{-1}][[x]]\right)[\sigma_q]$:

$$L_{n,n}^P = \frac{1}{P_n}\left(x\sigma_q + 1\right)\frac{1}{P_{n-1}}\left(x^2\sigma_q - 1\right)\cdots\frac{1}{P_1}\left(x^n\sigma_q - (-1)^n\right)\,. \tag{2.5}$$

THEOREM 2.1. *Let n be an integer ≥ 1, let P, $f \in \mathbb{C}[x^{-1}][[x]]$ with $P \neq 0$, and suppose that the following q-difference equation is satisfied:*

$$x\sigma_q f + f = P\,. \tag{2.6}$$

Then:

$$L_{n,n}^P(f^n) = (-1)^{n(n-1)/2}\,, \tag{2.7}$$

where $L_{n,n}^P$ is the q-difference operator defined by (2.5).

Theorem 2.1 will be proved with the help of the following lemmas.

LEMMA 2.2. *Let $(n,k) \in \mathbb{Z}^2$ such that $1 \leq k \leq n$, and let P, f be as in Theorem* 2.1. *For $0 \leq j \leq k$, define $A_{k;j} \in \mathbb{C}[x^{-1}][[x]]$ by*

$$A_{k;j} = \prod_{\substack{\ell\neq j\\ 0\leq \ell\leq k}} (P_j - P_\ell)\,. \tag{2.8}$$

One has:

$$L_{n,k}^P(f^n) = (-1)^{k(2n-k+1)/2}\sum_{j=0}^{k}\frac{1}{A_{k;j}}(f - P_j)^n\,. \tag{2.9}$$

LEMMA 2.3. *Let n be a nonnegative integer, $\mathfrak{C}$ an extension field of $\mathbb{C}$, and let $(\alpha_0, \alpha_1, \ldots, \alpha_n) \in \mathfrak{C}^{n+1}$ such that $\alpha_j \neq \alpha_\ell$ for $j \neq \ell$. Define the $(n+1)$-tuple $(a_0, a_1, \ldots, a_n) \in \mathfrak{C}^{n+1}$ as follows:*

$$a_j = \prod_{\substack{\ell\neq j\\ 0\leq \ell\leq n}} (\alpha_j - \alpha_\ell)$$

($a_0 = 1$ if $n = 0$). The following identity holds in the ring $\mathfrak{C}[T]$ of the polynomial functions over $\mathfrak{C}$:

$$\sum_{j=0}^{n} \frac{1}{a_j} (T - \alpha_j)^n = (-1)^n . \tag{2.10}$$

3. Proofs

PROOF OF LEMMA 2.2. To simplify the presentation, we will write $L_{n,k}$ instead of $L_{n,k}^P$, and set $\epsilon_{n,k} = (-1)^{k(2n-k+1)/2}$. One has:

$$\epsilon_{n;k+1} = (-1)^{n-k} \epsilon_{n,k} . \tag{3.1}$$

Let us proceed by induction on k. For $k = 1$, since $P_1 = P$, one has:

$$\epsilon_{n,1} = (-1)^n; \quad A_{1;0} = -P , \quad A_{1;1} = P .$$

Thus, (2.3) implies that

$$L_{n,1}(f^n) = \frac{1}{P} (x^n \sigma_q(f^n) - (-1)^n f^n) .$$

At the same time, by considering (2.6), one gets that

$$x^n \sigma_q(f^n) = (x\sigma_q f)^n = (P - f)^n ;$$

In this way, it follows that

$$L_{n,1}(f^n) = \frac{\epsilon_{n,1}}{P} ((f - P)^n - f^n) ,$$

that is exactly what wanted in (2.9) for $k = 1$.

Now, suppose that equality (2.9) holds for some integer k between 1 and $n-1$. By using relation (2.4), one can express $L_{n,k+1}(f^n)$ as follows:

$$L_{n,k+1}(f^n) = \frac{\epsilon_{n,k}}{P_{k+1}} \sum_{j=0}^{k} (B_{n,k;j} - C_{n,k;j}) , \tag{3.2}$$

where

$$B_{n,k;j} = x^{n-k} \sigma_q \left(\frac{1}{A_{k;j}} (f - P_j)^n \right) \tag{3.3}$$

and

$$C_{n,k;j} = \frac{(-1)^{n-k}}{A_{k;j}} (f - P_j)^n . \tag{3.4}$$

In view of (2.1), one finds that $x\sigma_q P_j = P - P_{j+1}$, hence:

$$x\sigma_q(f - P_j) = x\sigma_q f - x\sigma_q P_j = (P - f) - (P - P_{j+1}) = -(f - P_{j+1}).$$

Thus, it follows that

$$B_{n,k;j} = \frac{(x\sigma_q(f - P_j))^n}{x^k \sigma_q A_{k;j}} = \frac{(-1)^n}{x^k \sigma_q A_{k;j}} (f - P_{j+1})^n .$$

By using the expression of $A_{k;j}$ given in (2.8), one has:

$$x^k \sigma_q A_{k;j} = \prod_{\substack{\ell \neq j \\ 0 \le \ell \le k}} (x\sigma_q(P_j - P_\ell)) = (-1)^k \prod_{\substack{\ell \neq j \\ 0 \le \ell \le k}} (P_{j+1} - P_{\ell+1}) .$$

This is to say that

$$x^k \sigma_q A_{k;j} = \frac{(-1)^k}{P_{j+1}} A_{k+1,j+1} \,.$$

So, one can write (3.3) as follows:

$$B_{n,k;j} = \frac{(-1)^{n-k} P_{j+1}}{A_{k+1;j+1}} (f - P_{j+1})^n \,. \tag{3.5}$$

Rewrite (3.2) into the following form:

$$L_{n,k+1}(f^n) = \frac{\epsilon_{n,k}}{P_{k+1}} \left(\sum_{j=0}^{k-1} (B_{n,k;j} - C_{n,k;j+1}) + B_{n,k;k} - C_{n,k;0} \right) . \tag{3.6}$$

Let $j < k$; one has:

$$A_{k+1,j+1} = (P_{j+1} - P_{k+1})\, A_{k,j+1}.$$

Replace j by $j+1$ in (3.4), and make use of (3.5). One gets that

$$B_{n,k;j} - C_{n,k;j+1} = \frac{(-1)^{n-k} P_{k+1}}{A_{k+1;j+1}} (f - P_{j+1})^n \,.$$

Furthermore, one sees that $A_{k+1;0} = -P_{k+1}\, A_{k;0}$. By letting $j = k$ and $j = 0$ in (3.5) and (3.4) respectively, one has:

$$B_{n,k;k} = \frac{(-1)^{n-k} P_{k+1}}{A_{k+1;k+1}} (f - P_{k+1})^n$$

and

$$C_{n,k;0} = -\frac{(-1)^{n-k} P_{k+1}}{A_{k+1;0}} (f - P_0)^n \,.$$

Thus, in view of (3.1), equality (3.6) implies that

$$L_{n,k+1}(f^n) = \epsilon_{n,k+1} \sum_{j=0}^{k+1} \frac{1}{A_{k+1;j}} (f - P_j)^n \,,$$

which corresponds to (2.9) in which k was replaced by $k+1$. ☐

PROOF OF LEMMA 2.3. We give here two proofs.

PROOF BY INDUCTION. If $n = 0$, as $a_0 = 1$, equality (2.10) becomes evident. Suppose that (2.10) holds for some index $n \geq 0$, and consider an $(n+2)$-uplet $(\alpha_0, \ldots, \alpha_{n+1})$ where, as before, $\alpha_j \neq \alpha_\ell$ if $j \neq \ell$. Set

$$\mathcal{P}(T) = \sum_{j=0}^{n+1} \frac{1}{\tilde{a}_j} (T - \alpha_j)^{n+1}, \quad \tilde{a}_j = \prod_{\substack{\ell \neq j \\ 0 \leq \ell \leq n+1}} (\alpha_j - \alpha_\ell).$$

Let k be an integer such that $0 \leq k \leq n+1$. If one writes

$$a_{k;j} = \prod_{\substack{\ell \neq k, \ell \neq j \\ 0 \leq \ell \leq n+1}} (\alpha_j - \alpha_\ell)\,,$$

one has $a_{k;j} = \tilde{a}_j / (\alpha_j - \alpha_k)$ for $j \neq k$, what implies that

$$\mathcal{P}(\alpha_k) = - \sum_{\substack{j \neq k \\ 0 \leq j \leq n+1}} \frac{1}{a_{k;j}} (\alpha_k - \alpha_j)^n. \tag{3.7}$$

By applying the induction hypothesis to the $(n+1)$-uplet $(\alpha_0,\ldots,\alpha_{k-1},\alpha_{k+1},\ldots$ $alpha_{n+1})$ appeared in the right hand side of (3.7), one obtains that $\mathcal{P}(\alpha_k) = (-1)^{n+1}$. As $\mathcal{P}$ is a polynomial with $\deg \mathcal{P} \le n+1$, one finds that $\mathcal{P}(T) = (-1)^{n+1}$ identically.

PROOF BY LAGRANGE POLYNOMIALS. [1] Let $\mathbb{K} = \mathfrak{C}(T)$ be the field of the rational functions over $\mathfrak{C}$. Let X be a new indeterminate, and set

$$F(X) = (T-X)^n, \quad \Lambda_j(X) = \prod_{\substack{\ell \neq j \\ 0 \le \ell \le n}} (X - \alpha_\ell)\,.$$

By noticing that $F(X) \in \mathbb{K}[X]$ be such that $\deg F = n < n+1$, the Lagrange interpolation formula [**3**, Chap. IV, §2, p. 26] implies that

$$F(X) = \sum_{j=0}^{n} \frac{F(\alpha_j)}{a_j} \Lambda_j(X)\,.$$

And now one compares the coefficients of X^n for both sides of the above equality. As that of Λ_j is equal to 1, one finds that

$$(-1)^n = \sum_{j=0}^{n} \frac{F(\alpha_j)}{a_j} = \sum_{j=0}^{n} \frac{(T-\alpha_j)^n}{a_j}\,,$$

which is exactly the expected identity (2.10). □

END OF THE PROOF OF THEOREM 2.1. In view of Lemma 2.2, letting $k = n$ in (2.9) yields that

$$L_{n,n}^{P}(f^n) = (-1)^{n(n+1)/2} \sum_{j=0}^{n} \frac{1}{A_{n;j}} (f - P_j)^n\,,$$

where

$$A_{n;j} = \prod_{\substack{\ell \neq j \\ 0 \le \ell \le n}} (P_j - P_\ell)\,.$$

By (2.2), one knows that $P_j \neq P_\ell$ for $j \neq \ell$. Thus, by applying Lemma 2.3 with $\mathfrak{C} = \mathbb{C}[x^{-1}][[x]]$, $X = f$ and $\alpha_j = P_j$, one gets that

$$\sum_{j=0}^{n} \frac{1}{A_{n;j}} (f - P_j)^n = (-1)^n.$$

This permits to finish the proof of Theorem 2.1. □

4. One generalisation of Theorem 2.1

Instead of (2.6), let us consider the following slightly more general q-difference equation:

$$\alpha\,\sigma_q y + y = \beta\,, \tag{4.1}$$

[1] We would like to thank our friend and collaborator J. Sauloy for this elegant proof.

where both α and β are nonzero power series belonging to the field $\mathbb{C}[x^{-1}][[x]]$. In the same spirit as in (2.1), define the sequence $\{\beta_n\}_{n\geq 0}$ in $\mathbb{C}[x^{-1}][[x]]$ as follows:

$$\beta_0 = 0; \qquad \beta_n = \sum_{k=0}^{n-1} (-\alpha\,\sigma_q)^k\,\beta\,, \quad n \geq 1\,. \tag{4.2}$$

PROPOSITION 4.1. *For any given positive integer n, the following conditions are equivalent.*

(1) *$\beta_j \neq 0$ for any positive integer j such that $0 < j \leq n$.*
(2) *$\beta_j \neq \beta_\ell$ for any couple of integers (j,ℓ) such that $0 \leq \ell < j \leq n$.*

PROOF. Indeed, by (4.2), it follows that

$$\beta_j - \beta_\ell = \sum_{k=\ell}^{j-1} (-\alpha\,\sigma_q)^k\,\beta = (-\alpha\,\sigma_q)^\ell \beta_{j-\ell}\,. \tag{4.3}$$

As $\alpha \neq 0$, l'operator $(-\alpha\,\sigma_q)$ is an automorphism on the $\mathbb{C}$-vector space $\mathbb{C}[x^{-1}][[x]]$. So, this is the same for its ℓ-th power or iteration $(-\alpha\,\sigma_q)^\ell$. This implies the equivalence between the conditions stated in Proposition 4.1. □

Given a nonzero $f \in \mathbb{C}[x^{-1}][[x]]$, one remembers that $\nu_0(f)$ denotes the valuation of f at $x = 0$, that is the lowest degree of the terms of f. By using (4.2), one obtains that, for $n > 0$:

$$\nu_0(\beta_n) = \begin{cases} \nu_0(\beta) & \text{if } \nu_0(\alpha) > 0; \\ (n-1)\nu_0(\alpha) + \nu_0(\beta) & \text{if } \nu_0(\alpha) < 0. \end{cases}$$

This gives the following statement for α, $\beta \in \mathbb{C}[x^{-1}][[x]] \setminus \{0\}$.

REMARK 4.2. Both conditions considered in Proposition 4.1 are necessarily fulfilled if $\nu_0(\alpha) \neq 0$.

THEOREM 4.3. *Let $n \in \mathbb{Z}_{>0}$, and let α, β and $f \in \mathbb{C}[x^{-1}][[x]]$. Suppose that f satisfies the q-difference equation in (4.1) and that $\beta_j \neq 0$ for any positive integer j such that $0 < j \leq n$. One has:*

$$L_n^{\alpha,\beta}(f^n) = (-1)^{n(n-1)/2}\,, \tag{4.4}$$

where $L_n^{\alpha,\beta}$ is the n-th order q-difference operator defined by

$$L_n^{\alpha,\beta} = \frac{1}{\beta_n}\,(\alpha\,\sigma_q + 1)\,\frac{1}{\beta_{n-1}}\,(\alpha^2\,\sigma_q - 1)\cdots\frac{1}{\beta_1}\,(\alpha^n\,\sigma_q - (-1)^n)\,. \tag{4.5}$$

PROOF. Replace (k,P) with (n,β) in (2.8), and define:

$$A_{n;j} = \prod_{\substack{\ell\neq j\\ 0\leq\ell\leq n}} (\beta_j - \beta_\ell)\,, \quad 0 \leq j \leq n.$$

By taking into account Proposition 4.1, it follows that $A_{n;j} \neq 0$. Thus, one might proceed in the same way as for the proof of Theorem 2.1. We omit the details. □

5. About the summability of the powers of q-Euler series

Let us come back to the power series $\hat{E}_q(x)$ defined by (1.1), which satisfies the q-difference equation stated in (1.2). Letting $P = 1$, Theorem 2.1 implies immediately the following result.

REMARK 5.1. Given any integer $n \geq 2$, the n-th power $\hat{E}_q(x)^n$ satisifies the following identity:

$$(5.1)\quad (x\sigma_q + 1)\,\frac{1}{P_{n-1}}\,(x^2\sigma_q - 1)\ldots(x^n\sigma_q - (-1)^n)\hat{E}_q(x)^n = (-1)^{n(n-1)/2}\,P_n\,,$$

where $P_k = 1 - x + \cdots + q^{(k-1)(k-2)}(-x)^{k-1}$ for $1 \leq k \leq n$.

In particular, when $n = 2$, as $P_2 = 1 - x$, (5.1) takes the form of (1.3).

Furthermore, it might be interesting to notice that, if $q = 1$, one has $\hat{E}_1(x) = \frac{1}{1+x}$ and, in this case, the above identity in (5.1) is simply equivalent to the following elementary relation:

$$\prod_{k=1}^{n} \frac{x^k - (-1)^k}{\sum_{j=0}^{k-1}(-x)^j} = (-1)^{n(n-1)/2}\,(x+1)^n\,.$$

This observation related with the specific case of $q = 1$ may be also made for Theorem 4.3, by assuming, for example, both α and β to belong to the field $\mathbb{C}[x^{-1}]\{x\}$ of the germs of meromorphic functions at $x = 0$ in $\mathbb{C}$ or to the sub-field $\mathbb{C}(x)$ of the rational functions over $\mathbb{C}$.

Our next remark goes to another q-analog of the following so-called Euler series:

$$(5.2)\qquad \hat{E}(x) = \sum_{n\geq 0}(-1)^n\, n!\, x^{n+1}\,.$$

This power series is divergent for all $x \in \mathbb{C}\setminus\{0\}$ but Borel-summable in every direction excepted $\mathbb{R}^-$. It satisfies the following first order ODE:

$$(5.3)\qquad (x\,\delta + 1)y = x, \quad \delta = x\partial_x = x\frac{d}{dx}.$$

Letting $Y = \hat{E}(x)^2$, one can check that

$$(5.4)\qquad (x\,\delta + 1 - x)\,(x\,\delta + 2)Y = 2x^2.$$

Indeed, if $y = \hat{E}(x)$, one deduces from (5.3) that $x\delta(y^2) = 2x\,y\,\delta y = 2y(x-y) = 2xy - 2y^2$. This means that $(x\delta + 2)(y^2) = 2x\,y$. Thus, applying again (5.3) yields that $(x\,\delta + 1)\,\frac{1}{2x}\,(x\,\delta + 2)\,y^2 = x$. In this way, one gets (5.4), using the identity $(x\,\delta + 1)\,\frac{1}{2x} = \frac{1}{2x}\,(x\,\delta + 1 - x)$ in the non-commutative ring $\mathbb{C}(x)[\delta]$ of the differential operators over $\mathbb{C}(x)$.

Besides, the correspondances $n \leftrightarrow \frac{1-q^n}{1-q}$ and $n! \leftrightarrow \frac{(1-q)\cdots(1-q^n)}{(1-q)^n}$ suggest one to consider the following q-analog of the Euler series:

$$(5.5)\qquad \hat{E}(x;q) = x + \sum_{n\geq 1}(-1)^n \frac{(1-q)\cdots(1-q^n)}{(1-q)^n} x^{n+1}.$$

This can be written in term of a basic hypergeometric series as follows:

$$\hat{E}(x;q) = x\,{}_2\phi_1(q, q; 0; q, -\frac{x}{1-q}) = x\,{}_2\phi_0(q^{-1}, q^{-1}; -; q^{-1}, \frac{x}{1-q^{-1}}).$$

See [**6**, (1.2.22), p. 4] for the general definition of ${}_r\phi_s(\ldots;\ldots;q,z)$.

Let $\delta_q = \frac{\sigma_q - 1}{q-1} = x\,\Delta_q$, where Δ_q is defined in [**1**, p. 488, (10.2.3)]. By observing that $n = \frac{\delta x^n}{x^n}$ and $\frac{1-q^n}{1-q} = \frac{\delta_q x^n}{x^n}$, a direct computation shows that $\hat{E}(x;q)$ satisfies the q-analog of (5.3) as follows: $(x\,\delta_q + 1)y = x$. If one writes

$$\alpha = \frac{x}{q-1-x}\,, \qquad \beta = (q-1)\,\alpha\,, \tag{5.6}$$

it follows from the above that

$$(\alpha\,\sigma_q + 1)\,\hat{E}(x;q) = \beta\,. \tag{5.7}$$

REMARK 5.2. Let α be as in (5.6). The following identity holds in the field $\mathbb{C}[x^{-1}][[x]]$:

$$\left(\alpha\,\delta_q + \frac{1}{q-1-x}\right)\left(\alpha\,\delta_q - \frac{1}{x} + \frac{1}{q-1-x}\right)\hat{E}(x;q)^2 = \alpha\,(\sigma_q\alpha - 1)\,. \tag{5.8}$$

Moreover, when $q \to 1$, (5.8) is reduced into the following equivalent form of (5.4):

$$\left(\delta + \frac{1}{x}\right)\left(\delta + \frac{2}{x}\right)\hat{E}(x)^2 = 2\,. \tag{5.9}$$

To obtain (5.8), one can apply (4.2) for $n = 1$ and $n = 2$, where α and β are defined by (5.6). This gives that

$$\beta_1 = (q-1)\,\alpha, \quad \beta_2 = \beta - \alpha\,\sigma_q\beta = (q-1)\,\alpha\,(1 - \sigma_q\alpha)\,.$$

Thus, it follows from applying Theorem 4.3 to (5.7) that $\hat{E}(x;q)^2$ satisfies the following q-difference equation:

$$L\left(\hat{E}(x;q)^2\right) = \alpha\,(\sigma_q\alpha - 1)\,, \tag{5.10}$$

where

$$L = \frac{1}{(q-1)^2}\,(\alpha\,\sigma_q + 1)\left(\alpha\,\sigma_q - \frac{1}{\alpha}\right). \tag{5.11}$$

By replacing σ_q with $(q-1)\delta_q + 1$ in (5.11), one gets that

$$L = \left(\alpha\,\delta_q + \frac{1+\alpha}{q-1}\right)\left(\alpha\,\delta_q + \frac{\alpha^2-1}{(q-1)\,\alpha}\right).$$

As $\alpha + 1 = \frac{q-1}{q-1-x}$ and $\alpha^2 - 1 = \frac{(q-1)(2x-q+1)}{(q-1-x)^2}$, one deduces immediately (5.8) from (5.10). The limit equation form given in (5.9) can be obtained from (5.8) by noticing both limits $\alpha \to -1$ and $\delta_q \to \delta$ for $q \to 1$.

In the rest of this paper, we will suppose that $q > 1$ even if, in most cases, the hypothesis $|q| > 1$ may be really enough; see [**9**] and [**10**]. Let us recall some results about the summability of power series solutions of q-difference equations. Given any linear q-difference operator L of the following form:

$$L = \sum_{j=0}^{n} a_j\sigma_q^j \in \mathbb{C}\{x\}[\sigma_q]\,, \quad a_0a_n \neq 0\,,$$

one definies its associated Newton polygon $\mathcal{NP}(L)$ as being the convex hull of the set $\{(j,m) : m \geq \nu_0(a_j)\}$ in the strip $[0,n] \times [0,+\infty)$. By following [**8**, (3.1.2)] and [**11**, Proposition 5.14], one knows that, if $\mathcal{NP}(L)$ admits only integer slopes, saying

$\kappa_1 \le \kappa_2 \le \cdots \le \kappa_n$, then there exist $\nu \in \mathbb{Z}$, $(h_1, h_2, \ldots, h_n) \in (1 + x\mathbb{C}\{x\})^n$ and $(c_0, c_1, \ldots, c_n) \in (\mathbb{C} \setminus \{0\})^{n+1}$ such that

$$L = c_0\, x^\nu\, h_1\, (x^{\kappa_1}\sigma_q + c_1)\, h_2\, (x^{\kappa_2}\sigma_q + c_2) \ldots h_n\, (x^{\kappa_n}\sigma_q + c_n)\,. \tag{5.12}$$

Furthermore, let $K^+ = \mathbb{Z}_{>0} \cap \{\kappa_j : 1 \le j \le n\}$, and define $\vec{k}$ as follows:

$$\vec{k} = \begin{cases} \emptyset & \text{if} \quad K^+ = \emptyset \\ (k_1, \ldots, k_m) & \text{if} \quad K^+ = \{k_\ell : 1 \le \ell \le m\} \text{ and } k_1 < \cdots < k_m\,. \end{cases} \tag{5.13}$$

By considering [**8**, § 3.3.5], it follows that any power series $\hat{f} \in \mathbb{C}[[x]]$ such that $L(\hat{f}) \in \mathbb{C}\{x\}$ remains convergent or is Gq-summable of order $\vec{k} = (k_1, \ldots, k_m) \in \mathbb{Z}_{>0}^m$ depending on whether $\vec{k} = \emptyset$ or not.

THEOREM 5.3. *Given any integer $n \ge 2$, the n-th power $\hat{E}_q(x)^n$ is Gq-summable of order $(1, 2, \ldots, n)$. And this is the same for $\hat{E}(x; q)^n$.*

PROOF. This follows from taking into account (5.1) or applying Theorem 4.3 to (5.7) with the help of Remark 4.2. □

Applying [**2**, Theorem 36] or [**7**, Proposition 1.3.4.2] implies that every power $\hat{E}(x)^n$ of the Euler series is Borel-summable at the same level, contrarily to what happens in the case of their respective q-analog $\hat{E}(x; q)^n$; see Theorem 5.3. Furthermore, thanks to [**4**, Theorem 3.15], one knows that the Gq-sum of $\hat{E}(x; q)$ tends toward the Borel-sum of $\hat{E}(x)$ when $q \to 1^+$. In our coming work [**5**], it would be shown that, given one generic direction d and two power series $\hat{f}_1$ and $\hat{f}_2$ such that $L_j(\hat{f}_j) \in \mathbb{C}\{x\}$ for some $L_j \in \mathbb{C}\{x\}[\sigma_q]$, where $j = 1$ or 2, the Gq-sum of $\hat{f}_1 \hat{f}_2$ along d is equal to the product of the Gq-sum of $\hat{f}_1$ along d with that of $\hat{f}_2$. In this way, one would obtain that the Gq-sum of $\hat{E}_q(x)^n$ or $\hat{E}(x; q)^n$ can be really expressed as the n-th power of that of $\hat{E}_q(x)$ or $\hat{E}(x; q)$, respectively.

References

[1] George E. Andrews, Richard Askey, and Ranjan Roy, *Special functions*, Encyclopedia of Mathematics and its Applications, vol. 71, Cambridge University Press, Cambridge, 1999, DOI 10.1017/CBO9781107325937. MR1688958

[2] Werner Balser, *Formal power series and linear systems of meromorphic ordinary differential equations*, Universitext, Springer-Verlag, New York, 2000. MR1722871

[3] N. Bourbaki, *Éléments de mathématique. XI. Première partie: Les structures fondamentales de l'analyse. Livre II: Algèbre. Chapitre IV: Polynomes et fractions rationnelles. Chapitre V: Corps commutatifs* (French), Actualités Scientifiques et Industrielles [Current Scientific and Industrial Topics], No. 1102, Hermann & Cie, Paris, 1950. MR0035759

[4] Lucia Di Vizio and Changgui Zhang, *On q-summation and confluence* (English, with English and French summaries), Ann. Inst. Fourier (Grenoble) **59** (2009), no. 1, 347–392. MR2514868

[5] T. Dreyfus and C. Zhang, *A q-summation process that is a morphism of difference ring*, work in progress, 2021.

[6] George Gasper and Mizan Rahman, *Basic hypergeometric series*, 2nd ed., Encyclopedia of Mathematics and its Applications, vol. 96, Cambridge University Press, Cambridge, 2004. With a foreword by Richard Askey, DOI 10.1017/CBO9780511526251. MR2128719

[7] Bernard Malgrange, *Sommation des séries divergentes* (French, with English summary), Exposition. Math. **13** (1995), no. 2-3, 163–222. MR1346201

[8] F. Marotte and C. Zhang, *Multisommabilité des séries entières solutions formelles d'une équation aux q-différences linéaire analytique* (French), Ann. Inst. Fourier (Grenoble) **50** (2000), no. 6, 1859–1890 (2001). MR1817385

[9] Julien Roques, *On classical irregular q-difference equations*, Compos. Math. **148** (2012), no. 5, 1624–1644, DOI 10.1112/S0010437X12000292. MR2982442

[10] Jean-Pierre Ramis, Jacques Sauloy, and Changgui Zhang, *Local analytic classification of q-difference equations* (English, with French summary), Astérisque **355** (2013), vi+151. MR3185985

[11] Changgui Zhang, *Développements asymptotiques q-Gevrey et séries Gq-sommables* (French, with English and French summaries), Ann. Inst. Fourier (Grenoble) **49** (1999), no. 1, vi–vii, x, 227–261. MR1688144

Laboratoire P. Painlevé (UMR – CNRS 8524), Département Math., FST, Université de Lille, Cité scientifique, 59655 Villeneuve d'Ascq cedex, France

Email address: changgui.zhang@univ-lille.fr

Contemporary Mathematics
Volume **782**, 2023
https://doi.org/10.1090/conm/782/15728

Summability of formal solutions for a family of linear moment integro-differential equations

Maria Suwińska

ABSTRACT. A general family of linear moment integro-differential equations with coefficients depending only on one variable is taken into consideration. The results presented in this paper concern the summability of formal solutions of such equations. Usual tools of the summability theory are utilized. The main result is a direct generalization of existing conclusions in regards to standard linear integro-differential equations.

1. Introduction

The main focus of this paper is to analyze the summability properties of a family of generalized moment integro-differential equations given by

$$\left(1 - \sum_{i\in\mathcal{K}} \sum_{q=0}^{p_i} a_{iq}(z)\partial_{m_1,t}^{-i}\partial_{m_2,z}^{q}\right) u(t,z) = \hat{f}(t,z),$$

where $\mathcal{K} \subset \{1, 2, \ldots, \kappa\}$ is a nonempty, finite subset of indices, $p_i \geq 0$ for all $i \in \mathcal{K}$, the inhomogeneity $\hat{f}$ is a formal power series in the variable (t, z) and all coefficients $a_{iq}(z)$ are holomorphic on a certain neighborhood of the origin. The differentiation and integration are given by moment operators.

Moment differentiation given by the formula

$$\partial_{m,t}\left(\sum_{n=0}^{\infty} \frac{u_n}{m(n)} t^n\right) = \sum_{n=0}^{\infty} \frac{u_{n+1}}{m(n)} t^n,$$

with $(m(n))_{n\geq 0}$ being a sequence of positive real numbers corresponding to a pair of kernel functions, was first considered in [**3**] by W. Balser and M. Yoshino. The family of operators created in this manner is very broad and includes, but is by no means limited to, the standard differentiation of convergent power series when $(m(n))_{n\geq 0} = (n!)_{n\geq 0}$ and Caputo fractional derivative $\partial_t^{1/k}$ for the sequence $(m(n))_{n\geq 0} = (\Gamma(1 + \frac{n}{k}))_{n\geq 0}$.

This more general approach to differentiation caused an emergence of studies concerning the so called moment differential equations. In [**8**–**10**] S. Michalik studied solutions of general Cauchy problems with constant coefficients and their

2020 *Mathematics Subject Classification.* Primary 35C10, 35G10.
Key words and phrases. Summability, formal solutions, moment derivatives, moment partial differential equations, moment integro-differential equations.

summability. Additionally, results on the Gevrey order of formal solutions of moment partial differential equations, an issue connected to the problem of summability, have been published in [**11, 16**].

An even more general approach on moment differentiation and the problems mentioned above deals with the framework of strongly regular sequences, at the same time further broadening the family of considered operators and the definition of summability. For more information on this approach see [**5–7**].

Please note that the results presented in Section 5 are in fact a particular case of the ones presented in [**7**]. In this paper we depart from the notion of the strongly regular sequences for the sake of simplicity. It is nevertheless worth mentioning that working in the framework of strongly regular sequences requires additional assumptions regarding behavior of the anti-derivative. Very similar results concerning standard partial differential equations also exist. The k-summability of the formal solution of the heat equation has been studied by W. Balser and M. Loday-Richaud in [**2**]. More general families of linear integro-differential equations have been considered by P. Remy in [**12–14**].

2. Notation

By $\mathbb{N}$ we denote the set of all positive integers. Then $\mathbb{N}_0$ stands for the set $\mathbb{N} \cup \{0\}$.

$\mathcal{R}$ stands for the Riemann surface of the logarithm.

For any $r > 0$ we denote by D_r an open disc of radius r and center at 0, i.e., $D_r = \{z \in \mathbb{C} : \ |z| < r\}$. Similarly, a closed disc of radius r and center at 0 is denoted by $\bar{D}_r$ and defined as $\{z \in \mathbb{C} : \ |z| \leq r\}$.

For fixed $d \in \mathbb{R}$, $\alpha > 0$ and $R > 0$ an open sector of direction d, opening α and radius R is defined as

$$S_d(\alpha, R) = \left\{ z \in \mathcal{R} : \ 0 < |z| < R, \ d - \frac{\alpha}{2} < \arg z < d + \frac{\alpha}{2} \right\}.$$

For a sector with an infinite radius we simply write $S_d(\alpha)$. If furthermore the opening is not specified, we use notation S_d.

Analogously

$$\bar{S}_d(\alpha, R) = \left\{ z \in \mathcal{R} : \ 0 < |z| \leq R, \ d - \frac{\alpha}{2} \leq \arg z \leq d + \frac{\alpha}{2} \right\}.$$

defines a closed sector and we use notation $\bar{S}_d(\alpha)$ whenever the radius is not given. Please note that in this case the radius cannot be infinite. Moreover, a closed sector is not a closed subset of $\mathbb{C}$, as it never contains the origin. It is however a closed subset of $\mathcal{R}$.

An open and connected set $G \subset \mathcal{R}$ is a sectorial region if $G \subset S_d(\alpha)$ for certain $d \in \mathbb{R}$ and $\alpha > 0$ and for every $0 < \beta < \alpha$ there exists $r > 0$ such that $\bar{S}_d(\beta, r) \subset G$. As is the case with sectors, we call α the opening of G and d its direction. We also use notation $G_d(\alpha)$ to specify these two parameters.

For any fixed complex Banach space $(\mathbb{E}, \|\cdot\|_{\mathbb{E}})$ we denote by $\mathcal{O}(U, \mathbb{E})$ the set of all functions with values in $\mathbb{E}$ that are holomorphic on an open set U. For $\mathbb{E} = \mathbb{C}$ we use a simplified notation $\mathcal{O}(U)$. The set of all formal power series with coefficients in $\mathbb{E}$ will be denoted by $\mathbb{E}[[z]]$.

For any sector $S_d(\alpha)$, $f \in \mathcal{O}(S_d(\alpha), \mathbb{E})$ and $k > 0$ we will say that $f(z)$ is of an exponential growth no greater than k on $S_d(\alpha)$ for $z \to \infty$ if for every $0 < \beta < \alpha$

there exist constants $C_1, C_2 > 0$ such that $\|f(z)\|_{\mathbb{E}} \le C_1 e^{C_2|z|^k}$ for every $z \in S_d(\beta)$. The set of all such functions will be denoted by $\mathcal{O}^k(S_d(\alpha), \mathbb{E})$.

For $\hat{f}(z) = \sum_{n=0}^{\infty} f_n z^n \in \mathbb{E}[[z]]$ we call a series with non-negative real coefficients $\hat{g}(z) = \sum_{n=0}^{\infty} g_n z^n$ a majorant of $\hat{f}$ if $\|f_n\|_{\mathbb{E}} \le g_n$ for every $n \ge 0$. We denote this relation by $\hat{f}(z) \ll \hat{g}(z)$.

3. Moment functions and moment operators

3.1. Kernel functions and moment functions.

DEFINITION 3.1. We call a pair of functions e, E *kernel functions of order* $k > \frac{1}{2}$ if the following conditions are met:

(1) $e \in \mathcal{O}\left(S_0(\frac{\pi}{k})\right)$, $e(x) \in \mathbb{R}_+$ for $x \in \mathbb{R}_+$ and for every $\varepsilon > 0$ there exist constants $A, B > 0$ such that $|e(z)| \le Ae^{-(|z|/B)^k}$ for $z \in S_0\left(\frac{\pi}{k} - \varepsilon\right)$.

(2) There exists $\alpha > 0$ such that for every direction $\theta \in \left(-\frac{\pi}{2k}, \frac{\pi}{2k}\right)$ there exist $C, r > 0$ satisfying

$$|e(z)| \le C|z|^{\alpha} \text{ for every } z \in \tilde{S}_\theta \subset S_0\left(\frac{\pi}{k}\right),\ |z| < r. \tag{3.1}$$

(3) $E \in \mathcal{O}^k(\mathbb{C})$ and there exists $\beta > 0$ such that for every direction $\theta \in \left(\frac{\pi}{2k}, 2\pi - \frac{\pi}{2k}\right)$ there exist constants $K, R > 0$ satisfying

$$|E(z)| \le K|z|^{-\beta} \text{ for every } z \in \tilde{S}_\theta \subset S_\pi\left(2\pi - \frac{\pi}{k}\right),\ |z| > R. \tag{3.2}$$

(4) Functions e and E are connected by *a corresponding moment function* m *of order* $s = \frac{1}{k}$ given by the formula

$$m(u) := \int_0^\infty x^{u-1} e(x)\, dx$$

for all u satisfying $\Re u \ge 0$. Then the kernel function E can be represented as a power series of the form: $E(z) = \sum_{n=0}^{\infty} \frac{z^n}{m(n)}$ for $z \in \mathbb{C}$.

(5) Additionally we assume that $m(0) = 1$.

The aforementioned characterization of kernel functions is based for the most part on the classic definition from [**1**]. However, more general conditions of integrability at zero have been replaced with (3.1) and (3.2) as seen in [**4**], where the authors use the term *strong kernel functions*. Moreover, let us note that, as seen in [**1**], kernel functions of orders $k \le \frac{1}{2}$ can also be defined. For the sake of brevity these definitions will be omitted. For more details on the subject see [**1**].

DEFINITION 3.2. Let $e(z)$, $E(z)$ be a pair of kernel functions of order k and let $m(u)$ be their corresponding moment function of order $s = \frac{1}{k}$. Then we call the sequence $(m(n))_{n\ge 0}$ *a moment sequence of order* s.

The example below presents kernel functions and their corresponding moment function that are used in the classical theory of summability.

EXAMPLE 3.3.

- $e(z) = kz^k e^{-z^k}$;
- $m(u) = \Gamma(1 + \frac{u}{k})$,
- $E(z) = \sum_{j=0}^{\infty} \frac{z^j}{\Gamma(1+\frac{j}{k})} = \mathbf{E}_{1/k}(z)$, where $\mathbf{E}_{1/k}$ denotes the Mittag-Leffler function of index $\frac{1}{k}$.

As shown in [**1**], the set of all moment functions of positive orders is closed with respect to multiplication.

LEMMA 3.4. *Let m_1 and m_2 be two moment functions of positive orders s_1 and s_2, respectively. Then $m_1 \cdot m_2$ is a moment function of order $s_1 + s_2$. If moreover we have $s_1 > s_2$ then $\frac{m_1}{m_2}$ is a moment function of order $s_1 - s_2$.*

Further ahead we will rely on a property of moment sequences that connects all of the moment sequences of the same positive order s with a sequence $(\Gamma_s(n))_{n\geq 0} = (\Gamma(1+sn))_{n\geq 0}$, where $\Gamma(z)$ denotes the Gamma function. For any given moment sequence of order $s > 0$ there exist constants $a, A > 0$ such that

$$a^n \Gamma_s(n) \leq m(n) \leq A^n \Gamma_s(n) \quad \text{for every} \quad n \in \mathbb{N}_0. \tag{3.3}$$

For the proof of this fact please see Section 5.5 in [**1**].

We are also going to consider a class of moment sequences that was first defined in [**11**]. It was later used in [**16**] and [**7**] as well.

DEFINITION 3.5. For any given $s > 0$ we will say that $(m(n))_{n\geq 0}$ is *a regular moment sequence of order* s if it is a moment sequence of order s and moreover there exist constants $a, A > 0$ satisfying

$$a(n+1)^s \leq \frac{m(n+1)}{m(n)} \leq A(n+1)^s \quad \text{for every} \quad n \in \mathbb{N}_0. \tag{3.4}$$

We will denote the set of all regular moment sequences of order s by $\mathcal{M}_s$.

LEMMA 3.6. *Regular moment sequences have the following properties:*

(1) *Let $m_1 \in \mathcal{M}_{s_1}$, $m_2 \in \mathcal{M}_{s_2}$. Then $m_1 \cdot m_2 \in \mathcal{M}_{s_1+s_2}$. If we moreover assume that $s_1 > s_2$ then $\frac{m_1}{m_2} \in \mathbb{M}_{s_1-s_2}$.*
(2) *For any $s > 0$ the sequence $\Gamma_s \in \mathcal{M}_s$.*

3.2. Moment operators. Let us now recall the definitions of m-Borel and m-Laplace operators and their formal analogues as well as a definition of m-differential operator and its inverse. To this end first suppose that $e(z)$, $E(z)$ are kernel functions of order $k > \frac{1}{2}$ and $m(u)$ is their corresponding moment function.

DEFINITION 3.7. Let $S = S_d(\alpha)$ and $f \in \mathcal{O}^k(S, \mathbb{E})$. For any direction θ satisfying $|d - \theta| < \frac{\alpha}{2}$ we call an operator $\mathcal{L}_{e,d}$ given by a formula

$$(\mathcal{L}_{e,d} f)(z) = \int_0^{e^{i\theta}\infty} e\left(\frac{u}{z}\right) f(u) \frac{du}{u} \tag{3.5}$$

a *m-Laplace transform in direction d.*

From the fact that (3.5) is absolutely and locally uniformly convergent it follows that we can integrate along an arbitrary direction θ so long as it is sufficiently close to d.

Let us now consider a sectorial region $G_d(\alpha)$. We choose

$$\theta \in \left(d - \frac{\alpha}{2} + \frac{\pi}{2k}, d + \frac{\alpha}{2} - \frac{\pi}{2k}\right)$$

and ε, $r > 0$ in such a way that $S_\theta(\frac{\varepsilon+\pi}{k}, r) \subset G_d(\alpha)$. Then we denote a negatively oriented path that is the border of $S_\theta(\frac{\varepsilon+\pi}{k}, r)$ by $\gamma_k(\theta)$.

Now we can move on to the definition of m-Borel transform:

DEFINITION 3.8. For any $f \in \mathcal{O}(G, \mathbb{E})$ continuous at zero and $u \in S_\theta\left(\frac{\varepsilon}{k}\right)$ we call an operator of the form

$$(\mathcal{B}_{e,\theta} f)(u) = -\frac{1}{2\pi i} \int_{\gamma_k(\theta)} E\left(\frac{u}{z}\right) f(z) \frac{dz}{z} \tag{3.6}$$

a *m-Borel transform in direction* θ.

If we consider a family of operators $\{\mathcal{B}_{e,\theta}\}_\theta$ such that $2|\theta - d| < \alpha - k^{-1}\pi$ then this family defines a function $\mathcal{B}_{e,d} f$ holomorphic in the sector $S_d(\alpha - k^{-1}\pi)$ such that $\mathcal{B}_{e,d} f(z) \in \mathcal{O}^k(S_d(\alpha - k^{-1}\pi), \mathbb{E})$.

For more details on the second part of the definition please consult Section 5.2 and Theorem 24 from [**1**].

THEOREM 3.9. *Suppose that* $d \in \mathbb{R}$, $\theta > 0$ *and* $k > 0$. *Moreover, let* $f \in \mathcal{O}^k(S_d(\theta), \mathbb{E})$. *Let us define* $g(z) = \mathcal{L}_{e,d} f(z)$ *for* $z \in G(d, \theta + k^{-1}\pi)$. *Then* $f \equiv \mathcal{B}_{e,d} g$.

Please consult [**1**, Theorem 30] for a detailed proof of this fact. For a more general version of the theorem above see [**7**, Theorem 1].

We can also easily define moment analogues of formal Laplace and Borel transforms.

DEFINITION 3.10. Let $\hat{f}(u) = \sum_{n=0}^{\infty} f_n u^n$ be a formal power series. Then we define a *formal m-Laplace transform* as follows:

$$(\hat{\mathcal{L}}_{m,z} \hat{f})(z) = \sum_{n=0}^{\infty} f_n m(n) z^n.$$

DEFINITION 3.11. A *formal m-Borel transform* is an operator $\hat{\mathcal{B}}_{m,z} \colon \mathbb{E}[[z]] \to \mathbb{E}[[z]]$ given by:

$$\hat{\mathcal{B}}_{m,z}\left(\sum_{n=0}^{\infty} u_n z^n\right) := \sum_{n=0}^{\infty} \frac{u_n}{m(n)} z^n.$$

Let us now recall the definition of m-derivative and its inverse.

DEFINITION 3.12. Let $(m(n))_{n\geq 0}$ denote a moment sequence. We define a *m-differential operator* $\partial_{m,z} \colon \mathbb{E}[[z]] \to \mathbb{E}[[z]]$ by the formula:

$$\partial_{m,z}\left(\sum_{n=0}^{\infty} \frac{u_n}{m(n)} z^n\right) := \sum_{n=0}^{\infty} \frac{u_{n+1}}{m(n)} z^n.$$

DEFINITION 3.13. Let $(m(n))_{n\geq 0}$ denote a moment sequence. Then we define a *m-integral operator* $\partial_{m,t}^{-1} \colon \mathbb{E}[[z]] \to \mathbb{E}[[z]]$ by a formula:

$$\partial_{m,z}^{-1}\left(\sum_{n=0}^{\infty} \frac{u_n}{m(n)} z^n\right) := \sum_{n=1}^{\infty} \frac{u_{n-1}}{m(n)} z^n.$$

To describe a m-differential operator we can use integration of kernel functions. This way we receive a formula that can be treated as an analogue of the standard Cauchy Formula.

LEMMA 3.14. *Let* $f \in \mathcal{O}(D_r, \mathbb{E})$ *for a certain* $r > 0$ *and suppose that* $e(z), E(z)$ *are kernel functions of order* k *with* $m(u)$ *denoting their corresponding moment function of order* $\frac{1}{k}$*. Then for every* $|z| < \varepsilon < r$ *and* $n \in \mathbb{N}_0$ *we get*

$$\partial_{m,z}^n f(z) = \frac{1}{2\pi i} \int_{\partial D_\varepsilon} f(w) \int_0^{e^{i\theta}\infty} \zeta^n E(z\zeta) \frac{e(w\zeta)}{w\zeta}\, d\zeta\, dw \tag{3.7}$$

with $\theta \in \left(-\arg w - \frac{\pi}{2k}, -\arg w + \frac{\pi}{2k}\right)$.

PROOF. Since $f(z)$ is holomorphic on D_r, it can be expanded into a power series as seen below:

$$f(z) = \sum_{j=0}^{\infty} f_j z^j = \sum_{j=0}^{\infty} \frac{f^{(j)}(0)}{j!} z^j.$$

Moreover, notice that

$$\partial_{m,z}^n f(z) = \sum_{j=0}^{\infty} \frac{\frac{f^{(j+n)}(0)}{(j+n)!} m(j+n)}{m(j)} z^j,$$

for every $n \in \mathbb{N}$, which means that $\partial_{m,z}^j f(0) = \frac{f^{(j)}(0)}{j!} m(j)$ for every $j \in \mathbb{N}_0$. Using this fact and the Cauchy Formula we get

$$\partial_{m,z}^j f(0) = \frac{m(j)}{j!} \cdot \frac{j!}{2\pi i} \int_{\partial D_\varepsilon} \frac{f(w)}{w^{j+1}}\, dw.$$

Let us now recall that

$$m(j) = \int_0^{\infty} x^{j-1} e(x)\, dx \text{ for any } j \in \mathbb{N}.$$

Hence,

$$\partial_{m,z}^j f(0) = \frac{1}{2\pi i} \int_{\partial D_\varepsilon} \frac{f(w)}{w^{j+1}} \int_0^{\infty} x^{j-1} e(x)\, dx\, dw,$$

which after the change of variable $x = w\zeta$ gives us

$$\begin{aligned} \partial_{m,z}^j f(0) &= \frac{1}{2\pi i} \int_{\partial D_\varepsilon} \frac{f(w)}{w} \int_0^{e^{-i\arg w}\infty} \zeta^{j-1} e(w\zeta)\, d\zeta\, dw \\ &= \frac{1}{2\pi i} \int_{\partial D_\varepsilon} f(w) \int_0^{e^{-i\arg w}\infty} \zeta^j \frac{e(w\zeta)}{w\zeta}\, d\zeta\, dw. \end{aligned}$$

Instead of $\theta = -\arg w$ we can use any given θ from $\left(-\arg w - \frac{\pi}{2k}, -\arg w + \frac{\pi}{2k}\right)$. Then for $z \in D_\varepsilon$ we receive

$$f(z) = \sum_{j=0}^{\infty} \frac{\partial_{m,z}^j f(0)}{m(j)} z^j = \frac{1}{2\pi i} \int_{\partial D_\varepsilon} f(w) \int_0^{e^{i\theta}\infty} \zeta^j \frac{e(w\zeta)}{w\zeta} \sum_{j=0}^{\infty} \frac{z^j}{m(j)}\, d\zeta\, dw.$$

Seeing as $E(z) = \sum_{j=0}^{\infty} \frac{z^j}{m(j)}$, the formula takes the form

$$f(z) = \frac{1}{2\pi i} \int_{\partial D_\varepsilon} f(w) \int_0^{e^{i\theta}\infty} \frac{e(w\zeta)}{w\zeta} E(z\zeta)\, d\zeta\, dw.$$

It is now enough to notice that $\partial_{m,z}^n E(z\zeta) = \zeta^n E(z\zeta)$, which gives us

$$\partial_{m,z}^n f(z) = \frac{1}{2\pi i} \int_{\partial D_\varepsilon} f(w) \int_0^{e^{i\theta}\infty} \frac{e(w\zeta)}{w\zeta} \zeta^n E(z\zeta)\, d\zeta\, dw.$$

□

REMARK 3.15. Similarly to Theorem 3 from [**6**], Lemma 3.14 can be applied in a more general case when $f \in \mathcal{O}^k(S_d(\alpha) \cup D_r, \mathbb{E})$ for any given $k > 0$, $\alpha \in \mathbb{R}$ and $r > 0$. More precisely, if $f \in \mathcal{O}^k(S_d(\alpha) \cup D_r, \mathbb{E})$ then for a certain $0 < \tilde{r} < r$ and any $0 < \beta < \alpha$, $n \in \mathbb{N}_0$ and $z \in S_d(\beta) \cup D_{\tilde{r}}$ it follows that

$$\partial_{m,z}^n f(z) = \frac{1}{2\pi i} \int_\gamma f(w) \int_0^{e^{i\theta}\infty} \zeta^n E(z\zeta) \frac{e(w\zeta)}{w\zeta} \, d\zeta \, dw, \tag{3.8}$$

where $\theta \in \left(-\arg w - \frac{\pi}{2k}, -\arg w + \frac{\pi}{2k}\right)$ and γ denotes the border of a set $S_d(\beta_0) \cup D_\varepsilon$ for certain $\beta < \beta_0 < \alpha$ and $\tilde{r} < \varepsilon < r$.

Using both Lemma 3.14 and Remark 3.15 we can formulate a theorem characterizing functions of exponential growth no greater than k. It is a special case of a more general result given in [**6**] in Theorem 3 and the proof leans heavily on the one that was presented there.

THEOREM 3.16. *Let us consider a pair of kernel functions $e(z)$, $E(z)$ and their corresponding moment function $m(u)$. Let $(m(n))_{n\geq 0}$ be a moment sequence and suppose that $d \in \mathbb{R}$ and α, k, $r > 0$ are given. Consider a function $f \in \mathcal{O}^k(S_d(\alpha) \cup D_r, \mathbb{E})$. Then there exists $0 < \tilde{r} < r$ such that for any $0 < \beta < \alpha$, $z \in S_d(\beta) \cup D_{\tilde{r}}$ and $n \in \mathbb{N}_0$ there exist A, B, $C > 0$ satisfying*

$$\|\partial_{m,z}^n f(z)\|_{\mathbb{E}} \leq AB^n m(n) e^{C|z|^k} \quad \text{for} \quad n \in \mathbb{N}_0, \; z \in S_d(\beta) \cup D_{\tilde{r}}. \tag{3.9}$$

PROOF. From Remark 3.15 it follows that

$$\partial_{m,z}^n f(z) = \frac{1}{2\pi i} \int_{\gamma(z)} f(w) \int_0^{e^{i\theta}\infty} \zeta^n E(z\zeta) \frac{e(w\zeta)}{w\zeta} \, d\zeta \, dw \tag{3.10}$$

for a certain $\theta \in \left(-\arg w - \frac{\pi}{2k}, -\arg w + \frac{\pi}{2k}\right)$ and a path γ that depends on the choice of z. Let us now consider an integral

$$\int_0^{e^{i\theta}\infty} \zeta^n E(z\zeta) \frac{e(w\zeta)}{w\zeta} \, d\zeta,$$

where θ is as given above, z belongs to a certain neighborhood of the origin and w satisfies $|w| = \varepsilon$ for $\varepsilon > 0$ that depends on z. After using a change of variable $\zeta = se^{i\theta}$ with $s \in (0, +\infty)$ we receive

$$\begin{aligned} \left| \int_0^{e^{i\theta}\infty} \zeta^n E(z\zeta) \frac{e(w\zeta)}{w\zeta} \, d\zeta \right| &\leq \frac{1}{\varepsilon} \int_0^\infty s^{n-1} |E(zse^{i\theta}) e(wse^{i\theta})| \, ds \\ &\leq \frac{C_1 C_3}{\varepsilon} \int_0^\infty s^{n-1} \exp\left(C_2 s^k |z|^k - \frac{\varepsilon^k s^k}{C_4^k} \right) ds \end{aligned}$$

for certain constants C_1, C_2, C_3, $C_4 > 0$.

Let now $|z| \leq \tilde{r} = \frac{\varepsilon}{(2C_2)^{\frac{1}{k}} C_4}$. Then $\exp\left(C_2 s^k |z|^k - \frac{\varepsilon^k s^k}{C_4^k}\right) \leq \exp\left(-\frac{\varepsilon^k s^k}{2C_4^k}\right)$ and

$$\begin{aligned} \left| \int_0^{e^{i\theta}\infty} \zeta^n E(z\zeta) \frac{e(w\zeta)}{w\zeta} \, d\zeta \right| &\leq \frac{C_1 C_3}{\varepsilon} \int_0^\infty s^{n-1} \exp\left(-\frac{\varepsilon^k s^k}{2C_4^k} \right) ds \\ &= \frac{C_1 C_3 2^{\frac{n}{k}} C_4^n}{\varepsilon^{n+1} k} \int_0^\infty \xi^{\frac{n}{k}-1} \exp(-\xi) \, d\xi \qquad (3.11) \\ &= \frac{C_1 C_3 2^{\frac{n}{k}} C_4^n}{\varepsilon^{n+1} k} \Gamma\left(\frac{n}{k}\right) \end{aligned}$$

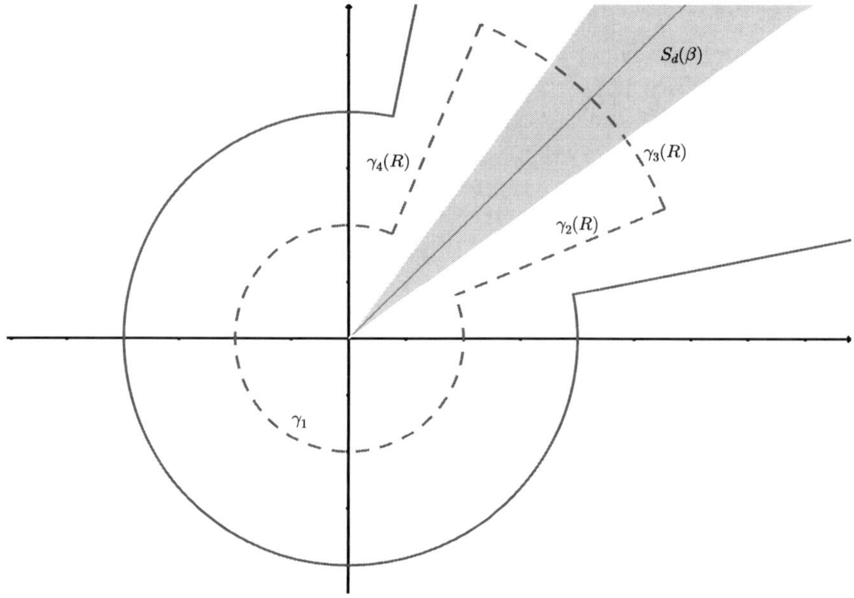

FIGURE 1. Path $\gamma(z)$.

From (3.3) it follows that

$$\left| \int_0^{e^{i\theta}\infty} \zeta^n E(z\zeta) \frac{e(w\zeta)}{w\zeta}\, d\zeta \right| \le A_0 B_0^n m(n) \tag{3.12}$$

and moreover

$$\|\partial_{m,z}^n f(z)\|_{\mathbb{E}} \le \sup_{|w|=\varepsilon} \|f(w)\|_{\mathbb{E}} A_0 B_0^n m(n) \le \tilde{A}_0 \tilde{B}_0^n m(n) e^{C_0|z|^k} \tag{3.13}$$

for $z \in \bar{D}_{\tilde{r}}$.

In the case when $z \notin \bar{D}_{\tilde{r}}$ let us consider $0 < \beta < \alpha$ and a sector $S_d(\beta)$. Then the integration path γ takes a shape as seen on Figure 1.

Let β_0 be such that $0 < \beta < \beta_0 < \alpha$ and let $R(z) = C_4(2C_2)^{\frac{1}{k}}|z|$. We consider a positively oriented path $\gamma(z) = \gamma(R) = \gamma_1 + \gamma_2(R) + \gamma_3(R) + \gamma_4(R)$, where

- γ_1 is an arc connecting $\varepsilon e^{i(d+\frac{\beta_0}{2})}$ and $\varepsilon e^{i(d-\frac{\beta_0}{2})}$,
- $\gamma_2(R)$ is a segment from $\varepsilon e^{i(d-\frac{\beta_0}{2})}$ to $Re^{i(d-\frac{\beta_0}{2})}$,
- $\gamma_3(R)$ is an arc connecting points $Re^{i(d-\frac{\beta_0}{2})}$ and $Re^{i(d+\frac{\beta_0}{2})}$,
- $\gamma_4(R)$ is a segment from $Re^{i(d+\frac{\beta_0}{2})}$ to $\varepsilon e^{i(d+\frac{\beta_0}{2})}$.

On every part of $\gamma(R)$ we are going to compute a separate estimate of the integral $\int_0^{e^{i\theta}\infty} \zeta^n E(z\zeta) \frac{e(w\zeta)}{w\zeta}\, d\zeta$.

For $w \in \gamma_1$ notice that

$$\left(-\arg(w) - \frac{\pi}{2k}, -\arg(w) + \frac{\pi}{2k}\right) \cap \left(-\arg(z) + \frac{\pi}{2k}, -\arg(z) + 2\pi - \frac{\pi}{2k}\right) \ne \emptyset$$

and we can choose a direction θ from this set. After applying substitution $\zeta = se^{i\theta}$ and using a procedure similar to the one used before we can write

$$\int_0^\infty s^{n-1}|E(zse^{i\theta})|\frac{|e(wse^{i\theta})|}{\varepsilon}\,ds = \int_0^{\frac{M}{|z|}} s^{n-1}|E(zse^{i\theta})|\frac{|e(wse^{i\theta})|}{\varepsilon}\,ds + \\ + \int_{\frac{M}{|z|}}^\infty s^{n-1}|E(zse^{i\theta})|\frac{|e(wse^{i\theta})|}{\varepsilon}\,ds$$

for a certain fixed $M > 0$. Then

$$\int_0^{\frac{M}{|z|}} s^{n-1}|E(zse^{i\theta})|\frac{|e(wse^{i\theta})|}{\varepsilon}\,ds \leq \sup_{y\in\bar{D}_M}|E(y)|\varepsilon^{-1}C_3\int_0^{\frac{M}{|z|}} s^{n-1}\exp\left(-\frac{\varepsilon^k s^k}{C_4^k}\right)ds.$$

Like in the previous case, we can bound the last integral using $A_1B_1^n m(n)$. Moreover from Definition 3.1 it follows that

$$\begin{aligned}\int_{\frac{M}{|z|}}^\infty s^{n-1}|E(zse^{i\theta})|\frac{|e(wse^{i\theta})|}{\varepsilon}\,ds &\leq \int_{\frac{M}{|z|}}^\infty s^{n-1}\frac{K}{(s|z|)^\beta}\frac{|e(wse^{i\theta})|}{\varepsilon}\,ds \\ &\leq \frac{K}{\varepsilon M^\beta}\int_{\frac{M}{|z|}}^\infty s^{n-1}|e(wse^{i\theta})|\,ds.\end{aligned}$$

The last integral can be bounded from above by $A_2B_2^n m(n)\exp(C_2|z|^k)$. Hence,

$$\left\|\frac{1}{2\pi i}\int_{\gamma_1} f(w)\int_0^{e^{i\theta}\infty}\zeta^n E(z\zeta)\frac{e(w\zeta)}{w\zeta}\,d\zeta\,dw\right\|_{\mathbb{E}} \leq \tilde{A}_1\tilde{B}_1^n m(n)e^{\tilde{C}_1|z|^k} \tag{3.14}$$

for certain constants $\tilde{A}_1$, $\tilde{B}_1$, $\tilde{C}_1 > 0$.

For $w \in \gamma_2(R)$ we can choose θ from the same set of directions. We use substitution $[\varepsilon, R] \ni \rho \mapsto \rho e^{i(d-\frac{\beta_0}{2})}$ and the parametrization from the previous case to arrive at

$$\left|\int_0^{e^{i\theta}\infty}\zeta^n E(z\zeta)\frac{e(w\zeta)}{w\zeta}\,d\zeta\right| \leq \int_0^\infty s^{n-1}|E(zse^{i\theta})|\frac{|e(\rho se^{i(\theta+d-\frac{\beta_0}{2})})|}{\rho}\,ds.$$

Analogously to the previous case, this integral can be bounded from above by $A_3B_3^n m(n)$. Then we receive

$$\begin{aligned}&\left\|\frac{1}{2\pi i}\int_{\gamma_2(R)} f(w)\int_0^{e^{i\theta}\infty}\zeta^n E(z\zeta)\frac{e(w\zeta)}{w\zeta}\,d\zeta\,dw\right\|_{\mathbb{E}} \\ &\qquad\leq \frac{1}{2\pi}\sup_{w\in\gamma_2(R)}\|f(w)\|_{\mathbb{E}}A_3B_3^n m(n)R \leq \tilde{A}_2\tilde{B}_2^n m(n)e^{\tilde{C}_2|z|^k}\end{aligned} \tag{3.15}$$

for certain positive constants $\tilde{A}_2$, $\tilde{B}_2$, $\tilde{C}_2$. For $\gamma_4(R)$ identical reasoning can be followed to arrive at

$$\left\|\frac{1}{2\pi i}\int_{\gamma_4(R)} f(w)\int_0^{e^{i\theta}\infty}\zeta^n E(z\zeta)\frac{e(w\zeta)}{w\zeta}\,d\zeta\,dw\right\|_{\mathbb{E}} \leq \tilde{A}_3\tilde{B}_3^n m(n)e^{\tilde{C}_3|z|^k} \tag{3.16}$$

for $\tilde{A}_3$, $\tilde{B}_3$, $\tilde{C}_3 > 0$.

For $w \in \gamma_3(R)$ we use the previous parametrization as well as substitution $\left[d - \frac{\beta_0}{2}, d + \frac{\beta_0}{2}\right] \ni t \mapsto Re^{it}$. The direction $\theta \in \left(-\arg(w) - \frac{\pi}{2k}, -\arg(w) + \frac{\pi}{2k}\right)$ can be chosen arbitrarily. Then

$$\left|\int_0^{e^{i\theta}\infty} \zeta^n E(z\zeta)\frac{e(w\zeta)}{w\zeta}\,d\zeta\right| \le \int_0^{\infty} s^{n-1}|E(zse^{i\theta})|\frac{|e(Rse^{i(\theta+t)})|}{R}\,ds$$
$$= \frac{1}{R}\int_0^{\infty} s^{n-1}|E(zse^{i\theta})e(Rse^{i(\theta+t)})|\,ds.$$

Notice that $|w| = R(z)$ and so a condition analogous to (3.11) is met. Hence,

$$\left\|\frac{1}{2\pi i}\int_{\gamma_3(R)} f(w)\int_0^{e^{i\theta}\infty} \zeta^n E(z\zeta)\frac{e(w\zeta)}{w\zeta}\,d\zeta\,dw\right\|_{\mathbb{E}} \le \tilde{A}_4\tilde{B}_4^n m(n)e^{\tilde{C}_4|z|^k} \tag{3.17}$$

for certain positive $\tilde{A}_4$, $\tilde{B}_4$, $\tilde{C}_4$.

Combining results from (3.14), (3.15), (3.16) and (3.17) concludes the proof. □

The following lemma describes the connection between m-derivative and m-moment transform.

Lemma 3.17. *Let m_1, m_2 be moment sequences. Then operators $\hat{\mathcal{B}}_{m_2,t}$, $\partial_{m_1,t}$ satisfy*

$$\hat{\mathcal{B}}_{m_2,t}\partial_{m_1,t}\hat{u} = \partial_{m_1m_2,t}\hat{\mathcal{B}}_{m_2,t}\hat{u}$$

for every formal power series $\hat{u}$.

Proof. Let $\hat{u}(t) = \sum_{n=0}^{\infty} u_n t^n$. Then

$$\hat{\mathcal{B}}_{m_2,t}\partial_{m_1,t}\hat{u}(t) = \hat{\mathcal{B}}_{m_2,t}\sum_{n=0}^{\infty}\frac{u_{n+1}m_1(n+1)}{m_1(n)}t^n = \sum_{n=0}^{\infty}\frac{u_{n+1}m_1(n+1)}{m_1(n)m_2(n)}t^n.$$

On the other hand,

$$\partial_{m_1m_2,t}\hat{\mathcal{B}}_{m_2,t}\hat{u}(t) = \partial_{m_1m_2,t}\sum_{n=0}^{\infty}\frac{u_n}{m_2(n)}t^n = \sum_{n=0}^{\infty}\frac{u_{n+1}m_1(n+1)}{m_1(n)m_2(n)}t^n.$$

□

4. The problem of k-summability

The main focus of this section is the concept of k-summability in the framework of moment sequences and operators. To this end basic facts and definitions related to the topic will be presented.

Definition 4.1. Suppose that G is a sectorial region and $s > 0$. Let $f \in \mathcal{O}(G, \mathbb{E})$. We call a formal power series $\hat{f}(z) = \sum_{n=0}^{\infty} f_n z^n \in \mathbb{E}[[z]]$ an *asymptotic expansion of order $s > 0$ of f on G* if for every closed sector $\bar{S} \subset G$ there exist constants $C, K > 0$ such that

$$|z|^{-N}\left\|f(z) - \sum_{n=0}^{N-1} f_n z^n\right\|_{\mathbb{E}} \le CK^N\Gamma_s(N) \text{ for every } N \in \mathbb{N},\ z \in \bar{S}.$$

This relation will be denoted by $f \cong_s \hat{f}$.

DEFINITION 4.2. Let $\hat{f}$ be a formal power series with coefficients from a given Banach space $\mathbb{E}$. Moreover, assume that $k > 0$ and $d \in \mathbb{R}$. Then we say that $\hat{f}$ is *k-summable in the direction d* if there exist $\alpha > \frac{\pi}{k}$, a sectorial region $G = G_d(\alpha)$ and a function $f \in \mathcal{O}(G, \mathbb{E})$ such that $\hat{f}$ is an asymptotic expansion of Gevrey order $\frac{1}{k}$ of f on G. We call f the k-sum of $\hat{f}$ in direction d and denote this fact by $f(z) = (S_{k,d}\hat{f})(z)$.

The set of all series k-summable in direction d will be denoted by $\mathbb{E}\{z\}_{k,d}$.

Note that from Watson's Lemma it follows that the k-sum of any power series in a given direction d, if it exists, is uniquely determined. For more details on Watson's Lemma and its proof see Section 4.7 of [**1**].

DEFINITION 4.3. Let $(m(n))_{n\geq 0}$ be a moment sequence of order $s > 0$. A formal power series $\hat{f}(z) \in \mathbb{E}[[z]]$ is said to be *of a Gevrey order s* if there exists $r > 0$ such that $\hat{B}_{m,z}\hat{f} \in \mathcal{O}(D_r, \mathbb{E})$. The set of all such formal power series will be denoted by $\mathbb{E}[[z]]_s$.

THEOREM 4.4. *Let $\hat{f}(z) \in \mathbb{E}[[z]]_{1/k}$ and suppose that $\hat{g}(z) = \hat{\mathcal{B}}_{m,z}\hat{f}(z)$ for a certain moment sequence $(m(n))_{n\geq 0}$ of order $\frac{1}{k}$. Let $g(z)$ denote its sum. Then $\hat{f}(z)$ is k-summable in direction $d \in \mathbb{R}$ if and only if there exists a sector $S_d(\varepsilon)$, with sufficiently small $\varepsilon > 0$, such that $g(z)$ can be continued analytically to $S_d(\varepsilon)$ and its continuation there is of exponential growth no greater than k.*

The theorem above is a direct consequence of the properties of the Laplace transform. More details can be found in [**1**], in Section 6.1 and Chapter 5.

Analogously to the way it has been done in Definition 13 from [**7**], we can define m-derivative of a sum of a formal power series.

DEFINITION 4.5. Let $\mathbb{E}$ be an arbitrary Banach space and let us consider a moment sequence $(m(n))_{n\geq 0}$ of order $\frac{1}{k}$ as well as a formal power series $\hat{f}(z) \in \mathbb{E}\{z\}_{k,d}$ for some direction $d \in \mathbb{R}$. Moreover let $f(z)$ denote a k-sum of $\hat{f}(z)$. Then we define the *m-derivative of $f(z)$* as $\partial_{m,z} f(z) = g(z)$, where $g(z) = S_{k,d}\left(\partial_{m,z}\hat{f}(z)\right)$.

THEOREM 4.6. *For fixed $k > 0$ and $d \in \mathbb{R}$ the set $\mathbb{E}\{z\}_{k,d}$ is a differential algebra closed under the operator $\partial_{m,z}$.*

The result above is a conclusion from [**15**, Proposition 6.20.] and [**6**, Corollary 1]. A more general formulation can be found in [**7**, Proposition 1].

LEMMA 4.7. *Let k, $s > 0$ and $(m(n))_{n\geq 0}$ be a moment sequence of order s. If $\hat{f}(z) \in \mathbb{E}_{k,d}\{z\}$ and its k-sum is denoted by $f(z)$ then the formal power series $\partial_{m,z}^{-1}\hat{f}(z)$ is also k-summable in direction d and its k-sum $F(z)$ satisfies:*

$$\partial_{m,z} F(z) = f(z).$$

PROOF. To execute the proof of this lemma we use the Caputo fractional derivative. Assume for the moment that $(m(n)))_{n\geq 0} = (\Gamma(1 + sn))_{n\geq 0}$. The m-differential operator $\partial_{m,z}$ is connected to the aforementioned fractional derivative ∂_z^s as follows:

$$(\partial_{m,z}\hat{f})(z^s) = \partial_z^s(\hat{f}(z^s)).$$

Hence,

$$\partial_z^{-s} z^{sn} = \frac{\Gamma(1 + sn)}{\Gamma(1 + sn + s)} z^{sn+s}.$$

As mentioned in [**7**, Proposition 5], the same property holds for the Riemann-Liouville fractional integral given by

$$I^s_{0+} := \frac{1}{\Gamma(s)} \int_0^z f(t)(z-t)^{s-1}\, dt.$$

Let us fix a sector $S \subset G$. We aim to find a boundary for

$$\left\| \frac{1}{\Gamma(s)} \int_0^z f(t^s)(z-t)^{s-1}\, dt - \sum_{n=1}^{N-1} \frac{f_{n-1}}{\Gamma(1+sn)} z^{sn} \right\|_{\mathbb{E}} \tag{4.1}$$

for any sector $S \subset G$ and integer $N \geq 2$. To this end first let us note that

$$z^{sn} B\left(s,\, sn-s+1\right) = \int_0^z (z-t)^{s-1} t^{sn-s}\, dt,$$

where $B\left(s,\, sn-s+1\right) = \frac{\Gamma(s)\Gamma(1+sn-s)}{\Gamma(1+sn)}$ denotes the Beta function. From this it follows that (4.1) takes form:

$$\left\| \frac{1}{\Gamma(s)} \int_0^z \frac{t^{-sN+s}\left(f(t^s) - \sum_{n=1}^{N-1} \frac{f_{n-1}}{\Gamma(1+sn-s)} t^{sn-s}\right)}{t^{-sN+s}(z-t)^{1-s}}\, dt \right\|_{\mathbb{E}}. \tag{4.2}$$

We know that $\hat{f}(z) \cong_{\frac{1}{k}} f(z)$. As such

$$\left\| t^{-sN+s}\left(f(t^s) - \sum_{n=1}^{N-1} \frac{f_{n-1}}{\Gamma(1+sn-s)} t^{sn-s}\right) \right\|_{\mathbb{E}} \leq CK^{N-1}\Gamma\left(1 + \frac{N-1}{k}\right)$$

for certain positive constants C, K. Moreover, after the change of variable $t = z\xi$ we receive

$$\int_0^z t^{s(N-1)}(z-t)^{s-1}\, dt = z^{sN} \int_0^1 \xi^{s(N-1)}(1-\xi)^{s-1}\, d\xi.$$

The last integral defines the Beta function $B\left(s, 1+sN-s\right) = \frac{\Gamma(s)\Gamma(1+sN-s)}{\Gamma(1+sN)}$.

From this and the properties of the Gamma function the formula (4.2) can be bounded from above by

$$CK^{N-1}\Gamma\left(1 + \frac{N-1}{k}\right) |z|^{sN} \frac{\Gamma(s)\Gamma(1+sN-s)}{\Gamma(1+sN)} \leq \tilde{C}\tilde{K}^N \Gamma\left(1 + \frac{N}{k}\right) |z|^{sN}.$$

To show that the thesis holds for any moment sequence of order s, it suffices to define a moment sequence of order 0 with terms given by $\tilde{m}(n) = \frac{\Gamma(1+sn)}{m(n)}$ for every $n \geq 0$. Then, if $\hat{f}(z) = \sum_{n=0}^{\infty} \frac{f_n}{m(n)} z^n$, it is enough to consider a formal power series $\hat{\mathcal{B}}_{\tilde{m},z}\hat{f}(z)$ instead. □

We can now proceed to define an anti-derivative for a sum of a formal power series.

Definition 4.8. Let us consider a Banach space $\mathbb{E}$ and a moment sequence $(m(n))_{n\geq 0}$ of order $\frac{1}{k}$ as well as a formal power series $\hat{f}(z) \in \mathbb{E}\{z\}_{k,d}$ for some direction $d \in \mathbb{R}$. Suppose that $f(z)$ denotes the k-sum of $\hat{f}(z)$. Then we define the *m-anti-derivative* of $f(z)$ as $\partial_{m,z}^{-1} f(z) = g(z)$, where $g(z) = S_{k,d}(\partial_{m,z}^{-1}\hat{f}(z))$.

It is worth mentioning that the definition given above is correct due to Lemma 4.7. Please note that it does not need to be correct in a more general framework of strongly regular sequences. See [**7**] for more details on the issue. In the cited paper thesis of Lemma 4.7 is referred to as Assumption (A) and does not need to be satisfied in some cases.

LEMMA 4.9. *Let* $\hat{f}(z) \in \mathbb{E}\{z\}_{k,d}$ *for certain* $k > 0$, $d \in \mathbb{R}$ *and let* $f(z) = S_{k,d}(\hat{f}(z))$ *on a sectorial region* $G_d(\alpha)$, *where* $\alpha > \frac{\pi}{k}$. *Then for every closed sector* $\bar{S} \subset G_d(\alpha)$ *there exist positive constants* A, B *such that*

$$\|\partial_{m,z}^n f(z)\|_{\mathbb{E}} \leq AB^n m(n)\Gamma\left(1 + \frac{n}{k}\right) \text{ for any } z \in \bar{S},\ n \in \mathbb{N}_0.$$

PROOF. From the definition of k-summability and Lemma 3.17 we receive

$$\partial_{m,z}^n f(z) = \mathcal{L}_{e,d}(\hat{\mathcal{B}}_{\Gamma_{1/k},z}\partial_{m,z}^n \hat{f}(z)) = \mathcal{L}_{e,d}(\partial_{m\Gamma_{1/k},z}^n \hat{\mathcal{B}}_{\Gamma_{1/k},z}\hat{f}(z)).$$

for any $n \in \mathbb{N}_0$ and certain kernel function $e(z)$ of order k. Moreover, function $\hat{\mathcal{B}}_{\Gamma_{1/k},z}\hat{f}(z)$ is of an exponential growth no greater than k on $S_d(\delta) \cup D_r$ for certain δ, $r > 0$. Therefore, we can apply Theorem 3.16 to this function and the moment sequence $\left(m(n)\Gamma\left(1 + \frac{n}{k}\right)\right)_{n \geq 0}$. For any closed sector $\bar{S} \subset G$ we receive

$$\|\partial_{m\Gamma_{1/k},z}^n \hat{\mathcal{B}}_{\Gamma_{1/k},z}\hat{f}(z)\|_{\mathbb{E}} \leq A_1 B_1^n m(n)\Gamma\left(1 + \frac{n}{k}\right) e^{C_1|z|^k}$$

for certain fixed constants A_1, B_1, $C_1 > 0$ and for every $z \in \bar{S}$.

Let $g(z) = \hat{\mathcal{B}}_{\Gamma_{1/k},z}\hat{f}(z)$. Then $\partial_{m,z}^n f(z) = \mathcal{L}_{e,d}g(z)$. To reach the conclusion it is enough to find a boundary of the integral $\int_0^{e^{i\tau}\infty} e\left(\frac{u}{z}\right) g(u)\frac{du}{u}$ for any direction τ sufficiently close to d, which can be done in a way analogous to the one presented in proof of Theorem 3.16. □

5. Summability of formal solutions for linear moment integro-differential equation

5.1. Banach spaces of holomorphic functions. Let $0 < \tilde{r} < r$. By $(\mathbb{E}, \|\cdot\|_{\mathbb{E}})$ we are henceforth going to denote the space of all functions holomorphic on a disc $\bar{D} = \bar{D}_r \subset \mathbb{C}$ with the norm defined as

$$\left\|\sum_{n=0}^{\infty} f_n z^n\right\|_{\tilde{r}} = \sum_{n=0}^{\infty} |f_n|\,|z|^n, \quad z \in \bar{D},\ |z| = \tilde{r}.$$

LEMMA 5.1. *Suppose that* f, $g \in \mathbb{E}$, $f(z) = \sum_{n=0}^{\infty} f_n z^n$ *and* $g(z) = \sum_{n=0}^{\infty} g_n z^n$. *Then*

$$\|f(z)g(z)\|_{\tilde{r}} \leq \|f(z)\|_{\tilde{r}}\|g(z)\|_{\tilde{r}} \text{ for } z \in \bar{D},\ |z| = \tilde{r} < r.$$

PROOF. For any fixed $\tilde{r} < r$ it is enough to notice that

$$\|f(z)g(z)\|_{\tilde{r}} \leq \sum_{n=0}^{\infty}\sum_{k=0}^{n} |f_k g_{n-k}||z|^n = \|f(z)\|_{\tilde{r}}\|g(z)\|_{\tilde{r}}.$$

□

We will call $(a_n)_{n \geq 0}$ *a logarithmically convex sequence* if inequality $a_n^2 \leq a_{n-1}a_{n+1}$ holds for every $n \in \mathbb{N}$.

LEMMA 5.2. *Let m be a moment function such that the moment sequence $(m(n))_{n\geq 0}$ is logarithmically convex. Moreover suppose that $f\in\mathcal{O}(\bar{D})$. If there exist $C<\infty$ and $n\in\mathbb{N}_0$ satisfying*

$$\|f(z)\|_{\tilde{r}}\leq C\frac{|z|^n}{m(n)}\text{ for every } z\in\bar{D},\ |z|=\tilde{r} \tag{5.1}$$

then we have

$$\|\partial_{m,z}^{-k}f(z)\|_{\tilde{r}}\leq C\frac{|z|^{n+k}}{m(n+k)}\text{ for every } k\in\mathbb{N}_0 \text{ and } z\in\bar{D},\ |z|=\tilde{r}. \tag{5.2}$$

PROOF. Let us note that from (5.1) it follows that $f=O(z^n)$. Hence, we can write $f(z)=\sum_{j=n}^{\infty}f_jz^j$ for every $z\in\bar{D}$. Let us define a function $g(z)=\sum_{j=0}^{\infty}|f_{j+n}|m(n)z^j$. Using (5.1) we conclude that

$$\|g(z)\|_{\tilde{r}}=\frac{m(n)}{|z|^n}\|f(z)\|_{\tilde{r}}\leq C.$$

Moreover, $f(z)\ll\frac{|z|^n}{m(n)}g(z)$. Hence we get

$$\partial_{m,z}^{-k}f(z)=\sum_{j=n+k}^{\infty}\frac{m(j-k)}{m(j)}f_{j-k}z^j=\frac{z^{n+k}}{m(n+k)}\sum_{j=0}^{\infty}\frac{m(j+n)m(n+k)}{m(j+n+k)}f_{j+n}z^j.$$

We can use an equivalent definition of logarithmic convexity to write that

$$\frac{m(n)}{m(n+1)}\leq\frac{m(n-1)}{m(n)}\qquad\text{for every } n\geq 1.$$

In particular, after using the inequality above j times we receive

$$\frac{m(j+n)}{m(j+n+k)}\leq\frac{m(j+n-1)}{m(j+n+k-1)}\leq\frac{m(j+n-2)}{m(j+n+k-2)}\leq\dots\leq\frac{m(n)}{m(n+k)}.$$

Hence, it follows that

$$\frac{m(j+n)m(n+k)}{m(j+n+k)}\leq m(n).$$

Using this inequality and the connection between f and g, we arrive at

$$\partial_{m,z}^{-k}f(z)\ll\frac{z^{n+k}}{m(n+k)}\sum_{j=0}^{\infty}m(n)|f_{j+n}|z^j=\frac{z^{n+k}}{m(n+k)}g(z).$$

Hence,

$$\|\partial_{m,z}^{-k}f(z)\|_{\tilde{r}}\leq\frac{|z|^{n+k}}{m(n+k)}\|g(z)\|_{\tilde{r}}\leq\frac{|z|^{n+k}}{m(n+k)}C,$$

which concludes the proof. □

It is easy to see that Lemma 5.2 can be expanded to a finite sum of holomorphic functions.

LEMMA 5.3. *Suppose that $f(z)=f_1(z)+\dots+f_p(z)$ with $f_j\in\mathcal{O}(\bar{D})$ for $j=1$, $\dots$, p and that*

$$\|f_j(z)\|_{\tilde{r}}\leq C_j\frac{|z|^{n_j}}{m(n_j)}\text{ for every } z\in\bar{D},\ |z|=\tilde{r},\ j=1,\dots,p.$$

Moreover, assume that $\{m(n)\}_{n\geq 0}$ is a logarithmically convex moment sequence. Then we receive

$$\|\partial_{m,z}^{-k} f(z)\|_{\tilde{r}} \leq \sum_{j=1}^{p} C_j \frac{|z|^{n_j+k}}{m(n_j+k)} \quad \textit{for every } k \in \mathbb{N}_0,\ z \in \bar{D},\ |z| = \tilde{r}.$$

5.2. The Newton polygon. The Newton polygon, introduced in [**18**], has been used to estimate the Gevrey order of formal solutions for standard partial differential equations. Since then it has been utilized, sometimes in a slightly modified form, as a tool for geometric representation of various types of partial differential equations (see for example [**17**]). The standard definition of a Newton polygon was then further generalized for use in the case of moment partial differential equations with constant coefficients in [**8**] and with variable coefficients in [**11**] and [**16**].

Let us now recall a general definition of the Newton polygon.

Consider moment functions m_1, m_2 of orders s_1, s_2, respectively, $t \in \mathbb{C}$ and $z \in \mathbb{C}^N$ and a linear moment differential operator defined by the formula:

$$P(t, z, \partial_{m_1,t}, \partial_{m_2,z}) = \sum_{(\sigma,j,\alpha)\in\Sigma\times J\times A} a_{\sigma,j,\alpha}(z) t^{\sigma} \partial_{m_1,t}^{j} \partial_{m_2,z}^{\alpha}, \tag{5.3}$$

where $J \subset \mathbb{N}_0, A \subset \mathbb{N}_0^N$ are finite and $\Sigma \subset \mathbb{N}_0$. Then for the operator (5.3) the Newton polygon is defined as follows:

DEFINITION 5.4. *The Newton polygon* for an operator P given by (5.3) as a set

$$N(P) = \operatorname{conv}\left\{ \bigcup_{(\sigma,j,\alpha)\in\Sigma\times J\times A} \Delta\left(s_1 j + s_2\alpha,\ \sigma - j\right) \right\} \tag{5.4}$$

with $\Delta(a, b) = \{(x, y) \in \mathbb{R}^2 \mid x \leq a,\ y \geq b\}$ and conv$F$ denoting a convex hull of a given set F.

5.3. The main problem. As mentioned before, the following section contains a special case of results presented in [**7**]. We are restricting ourselves to operators defined in Section 2, whereas in the aforementioned paper more general moment differentiation is considered.

Let us consider a moment integro-differential operator of the form:

$$\left(1 - \sum_{i\in\mathcal{K}} \sum_{q=0}^{p_i} a_{iq}(z) \partial_{m_1,t}^{-i} \partial_{m_2,z}^{q}\right) u(t,z) = \hat{f}(t,z), \tag{5.5}$$

where $\mathcal{K} \subset \{1, \ldots, \kappa\} \subset \mathbb{N}$ for a certain integer $\kappa \geq 1$, $\mathcal{K} \neq \emptyset$, $p_i \geq 0$ for every $i \in \mathcal{K}$ and m_1, m_2 are regular moment sequences of positive orders s_1, s_2, respectively. We moreover assume that m_2 is logarithmically convex. Furthermore, let us suppose that all $a_{iq}(z)$ and $\frac{1}{a_{\kappa p_\kappa}(z)}$ are holomorphic for all $z \in D_r$, $r < 1$.

We further assume that the Newton polygon for

$$P(\partial_{m_1,t}, \partial_{m_2,z}) = 1 - \sum_{i\in\mathcal{K}} \sum_{q=0}^{p_i} a_{iq}(z) \partial_{m_1,t}^{-i} \partial_{m_2,z}^{q}$$

contains exactly one non-horizontal segment, as seen in Figure 2. The slope of this segment, denoted by k, is given by the formula

$$\frac{1}{k} = \frac{s_2 p_\kappa - s_1 \kappa}{\kappa} = \frac{s_2 p_\kappa}{\kappa} - s_1. \tag{5.6}$$

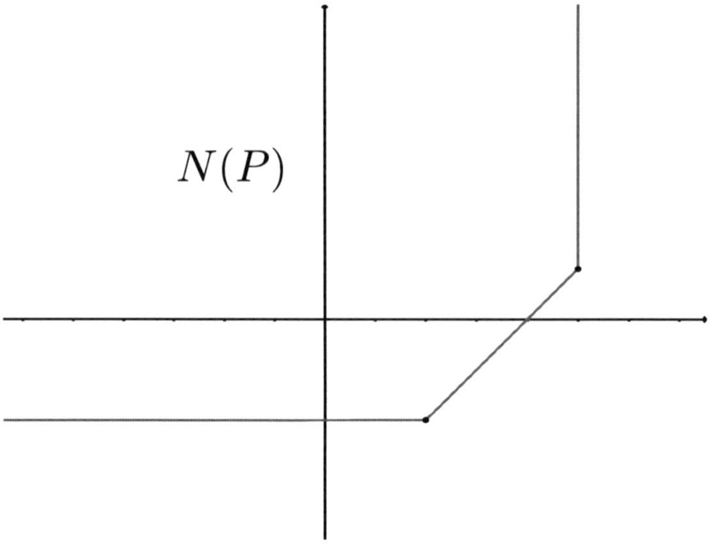

FIGURE 2. The Newton polygon for operator P.

REMARK 5.5. Notice that for any fixed $i \in \mathcal{K}$ we have $q \leq p_i \leq p_\kappa$ and moreover $p_i = p_\kappa$ if and only if $i = \kappa$. Furthermore, inequality $\frac{p_i}{i} \leq \frac{p_\kappa}{\kappa}$ holds for every $i \in \mathcal{K}$.

PROOF. The first part of the thesis follows directly from the definition of the operator P. To prove the last inequality, first let us fix $i \in \mathcal{K}$. Then

$$\frac{s_2 p_i - s_1 i}{i} \leq \frac{s_2 p_\kappa - s_1 \kappa}{\kappa}.$$

Hence, $s_2 \kappa p_i - s_1 \kappa i \leq s_2 i p_\kappa - s_1 \kappa i$, from which it follows that $\kappa p_i \leq i p_\kappa$. □

LEMMA 5.6. *Equation* (5.5) *admits a unique formal solution* $\hat{u}(t,z)$.

PROOF. Let $\hat{f}(t,z) = \sum_{n=0}^{\infty} f_n(t) z^n$ and $a_{iq}(z) = \sum_{n=0}^{\infty} a_{iq,n} z^n$ for every $i \in \mathcal{K}$, $0 \leq q \leq p_i$. Moreover, let $f_n(t) \in \mathbb{C}[[t]]$ for every $n \in \mathbb{N}_0$. Fix $\hat{u}(t,z) = \sum_{n=0}^{\infty} u_n(t) \frac{z^n}{m_2(n)}$ such that $u_n(t) \in \mathbb{C}[[t]]$ for every $n \in \mathbb{N}_0$. After substituting $\hat{u}$, $\hat{f}$ and a_{iq} in (5.5), we receive:

$$\frac{u_n(t)}{m_2(n)} = \sum_{i \in \mathcal{K}} \sum_{q=0}^{p_i} \sum_{p=0}^{n} a_{iq,p} \frac{\partial_{m_1,t}^{-i} u_{n-p+q}(t)}{m_2(n-p)} + f_n(t). \tag{5.7}$$

Let us introduce the following notation:

$$Q_\kappa = \{0, 1, \ldots, p_\kappa - 1\} \quad \text{and} \quad Q_i = \{0, 1, \ldots, p_i\} \text{ for } i < \kappa.$$

From (5.7) it follows that

$$a_{\kappa p_\kappa, 0} \partial_{m_1,t}^{-\kappa} u_{n+p_\kappa}(t) = u_n(t) - f_n(t) m_2(n) - \sum_{p=1}^{n} a_{\kappa p_\kappa, p} \partial_{m_1,t}^{-\kappa} u_{n-p+p_\kappa}(t) \frac{m_2(n)}{m_2(n-p)}$$
$$- \sum_{i \in \mathcal{K}} \sum_{q \in Q_i} \sum_{p=0}^{n} a_{iq,p} \partial_{m_1,t}^{-i} u_{n-p+q}(t) \frac{m_2(n)}{m_2(n-p)}$$

for every $n \geq 0$. Note that, since $a_{\kappa p_\kappa,0} \neq 0$, the formula above describes every coefficient $u_n(t)$ uniquely as long as $u_0(t)$, $u_1(t)$, ..., $u_{p_\kappa-1}(t)$ are known. □

Under all the assumptions introduced at the beginning of this section, the following theorem holds true:

THEOREM 5.7. *Let us consider equation* (5.5) *and let* k *be given by* (5.6). *Moreover, suppose that* $\hat{f}(t,z) \in \mathbb{E}[[t]]$ *and* $\partial^q_{m_2,z}\hat{u}(t,0)$ *for* $0 \leq q \leq p_\kappa - 1$ *are* k*-summable in direction* $d \in \mathbb{R}$. *Then the formal solution* $\hat{u}(t,z)$ *of* (5.5) *is also* k*-summable in direction* d.

Note that it is enough to assume the k-summability of $\partial^q_{m_2,z}\hat{u}(t,0)$ for every $0 \leq q \leq p_\kappa - 1$ in light of Lemma 4.7. It is however worth mentioning that it is not so if we consider more general cases with moment sequences derived from strongly regular sequences like in [**6, 7**].

Consider $z \in \bar{D}_r$ for a certain sufficiently small $r > 0$. For $\hat{v}(t,z) = \partial^{p_\kappa}_{m_2,z}\hat{u}(t,z)$ we receive the following formula:

$$\hat{u}(t,z) = \sum_{n=0}^{p_\kappa-1} u_n(t)z^n + \partial^{-p_\kappa}_{m_2,z}\hat{v}(t,z).$$

Next let us substitute $\hat{v}$ with $\hat{v}(t,z) = \partial^{\kappa}_{m_1,t}\hat{w}(t,z)$ for a certain $\hat{w}(t,z)$. Then (5.5) takes form

$$\left(1 - b_{\kappa p_\kappa}(z)\partial^{\kappa}_{m_1,t}\partial^{-p_\kappa}_{m_2,z} + \sum_{i\in\mathcal{K}}\sum_{q\in Q_i} b_{iq}(z)\partial^{\kappa-i}_{m_1,t}\partial^{q-p_\kappa}_{m_2,z}\right)\hat{w}(t,z) = \hat{g}(t,z)$$

for

$$b_{iq}(z) = \begin{cases} \frac{1}{a_{\kappa p_\kappa}(z)} & \text{for } i = \kappa,\ q = p_\kappa \\ \frac{a_{iq}(z)}{a_{\kappa p_\kappa}(z)} & \text{otherwise} \end{cases}$$

and

$$\hat{g}(t,z) = \frac{1}{a_{\kappa p_\kappa}(z)}\left(\sum_{n=0}^{p_\kappa-1} u_n(t)z^n - \sum_{i\in\mathcal{K}}\sum_{q=0}^{p_i}\sum_{n=0}^{p_\kappa-1} a_{iq}(z)\partial^{-i}_{m_1,t}\partial^{q}_{m_2,z}u_n(t)z^n - \hat{f}(t,z)\right),$$

which is k-summable under the assumptions from Theorem 5.7. Let $g(t,z)$ denote the k-sum of $\hat{g}(t,z)$ in direction d holomorphic for t from a certain sectorial region $G = G_d(\alpha)$, with $\alpha > \frac{\pi}{k}$, and $z \in \bar{D}_r$. Note that assumptions concerning $a_{iq}(z)$ guarantee the holomorphy of all $b_{iq}(z)$.

Notation $B = \max_{i,q}\max_{\tilde{r}\leq r}\|b_{iq}(z)\|_{\tilde{r}}$ will be used alongside the symbols introduced in Lemma 5.6 in the following calculations for the sake of clarity.

We can now use the classical fixed point method that has been utilized many times previously, for example in [**2**] and [**12**]. To this end first let us write $\hat{w}(t,z) = \sum_{p\geq 0}\hat{w}_p(t,z)$ with $\hat{w}_0(t,z) = \hat{g}(t,z)$ and all the other coefficients given by a recursive formula:

$$\hat{w}_{p+1}(t,z) = \left(b_{\kappa p_\kappa}(z)\partial^{\kappa}_{m_1,t}\partial^{-p_\kappa}_{m_2,z} + \sum_{i\in\mathcal{K}}\sum_{q\in Q_i} b_{iq}(z)\partial^{\kappa-i}_{m_1,t}\partial^{q-p_\kappa}_{m_2,z}\right)\hat{w}_p(t,z) \text{ for } p \geq 0.$$

Analogously we define $w(t,z)=\sum_{p\geq 0} w_p(t,z)$ with $w_0(t,z)=g(t,z)$ and the subsequent coefficients given by

$$w_{p+1}(t,z)=\left(b_{\kappa p_\kappa}(z)\partial^{\kappa}_{m_1,t}\partial^{-p_\kappa}_{m_2,z}+\sum_{i\in\mathcal{K}}\sum_{q\in Q_i} b_{iq}(z)\partial^{\kappa-i}_{m_1,t}\partial^{q-p_\kappa}_{m_2,z}\right)w_p(t,z) \text{ for } p\geq 0.$$

Seeing as $w_0(t,z)\in\mathcal{O}(G\times\bar{D}_r)$, observe that also $w_p(t,z)\in\mathcal{O}(G\times\bar{D}_r)$ for every $p\geq 1$. Moreover, $\hat{w}_p(t,z)$ is an asymptotic expansion of $w_p(t,z)$ with respect to t for every $p\geq 0$. Hence, $w_p(t,z)$ is a k-sum of $\hat{w}_p(t,z)$ for every $p\geq 0$.

Let us also notice that for every $p\geq 1$ functions $w_p(t,z)$ can be bounded from above using Lemma 5.3. Indeed, $w_1(t,z)$ can be described as a sum of holomorphic functions of the form $b_{iq}(z)\partial^{\kappa-i}_{m_1,t}\partial^{q-p_\kappa}_{m_2,z}w_0(t,z)$. More precisely, using the recursive formula we arrive at

$$w_1(t,z)=b_{\kappa p_\kappa}(z)\partial^{\kappa}_{m_1,t}\partial^{-p_\kappa}_{m_2,z}w_0(t,z)-\sum_{i\in\mathcal{K}}\sum_{q\in Q_i} b_{iq}(z)\partial^{\kappa-i}_{m_1,t}\partial^{q-p_\kappa}_{m_2,z}w_0(t,z).$$

It is now possible to bound the norm of any m-derivative of $w_0(t,z)$ with respect to t as follows:

$$\|\partial^n_{m_1,t}w_0(t,z)\|_{\tilde{r}}\leq CK^n m_1(n)\Gamma\left(1+\frac{n}{k}\right)\frac{|z|^0}{m_2(0)} \text{ for } n\in\mathbb{N}_0,\ z\in\bar{D},\ |z|=\tilde{r},\ t\in G.$$

From this fact and Lemma 5.2 we conclude that

$$\left\|b_{iq}(z)\partial^{\kappa-i}_{m_1,t}\partial^{q-p_\kappa}_{m_2,z}w_0(t,z)\right\|_{\tilde{r}}\leq BC'K'^{\kappa-i}m_1(\kappa-i)\Gamma\left(1+\frac{\kappa-i}{k}\right)\frac{|z|^{p_\kappa-q}}{m_2(p_\kappa-q)},$$

for $|z|=\tilde{r}$ and $t\in G$.

Hence,

$$\begin{aligned}\|w_1(t,z)\|_{\tilde{r}}&\leq\sum_{i\in\mathcal{K}\cup\{0\}}\sum_{q\in Q_i}\|b_{iq}(z)\partial^{\kappa-i}_{m_1,t}\partial^{q-p_\kappa}_{m_2,z}w_0(t,z)\|_{\tilde{r}}\\&\leq\sum_{i\in\mathcal{K}\cup\{0\}}\sum_{q\in Q_i}BC'K'^{\kappa-i}m_1(\kappa-i)\Gamma\left(1+\frac{\kappa-i}{k}\right)\frac{|z|^{p_\kappa-q}}{m_2(p_\kappa-q)}\end{aligned}$$

and

$$\begin{aligned}\|\partial^n_{m_1,t}w_1(t,z)\|_{\tilde{r}}\leq\sum_{i\in\mathcal{K}\cup\{0\}}\sum_{q\in Q_i}BCK^{n+\kappa-i}m_1(n+\kappa-i)\Gamma\left(1+\frac{n+\kappa-i}{k}\right)\\ \cdot\frac{|z|^{p_\kappa-q}}{m_2(p_\kappa-q)}.\end{aligned}$$

Analogous procedure can be repeated for every $p\geq 2$.

To conclude the proof of Theorem 5.7 two more technical Lemmas are required. Please note that they are modified versions of properties described at length in [**13**].

LEMMA 5.8. *Let $z\in D(0,r)$, where $r<1$. Then there exist positive constants C,K,B' such that*

$$\left\|\partial^n_{m_1,t}w_p(t,z)\right\|_{\tilde{r}}\leq B'^pCK^{n+\kappa p}m_1(n+\kappa p)\Gamma\left(1+\frac{n+\kappa p}{k}\right)P_p(|z|) \tag{5.8}$$

for every $n \in \mathbb{N}_0$, $\tilde{r} = |z|$. *By* $P_p(z)$ *we denote a polynomial*

$$\begin{cases} P_0(z) &= 1 \\ P_{p+1}(z) &= \left[\partial_{m_2,z}^{-p_\kappa} + \sum_{i\in\mathcal{K}'}\sum_{q\in Q_i} p_\kappa^{s_2 p_\kappa}\frac{(p_\kappa p)!^{s_2}}{(p_\kappa p+p_i)!^{s_2}}\partial_{m_2,z}^{q-p_\kappa}\right] P_p(z) \text{ for } p \geq 0. \end{cases}$$

PROOF. Inequality (5.8) holds true for $p = 0$. Let us then suppose that it holds true for a fixed $p \geq 0$. Let $Q_0 = \{0\}$. Using this notation we can write

$$\left\|\partial_{m_1,t}^n w_{p+1}(t,z)\right\|_{\tilde{r}} \leq B'^p C K^{n+\kappa p+\kappa} B$$
$$\cdot \sum_{i\in\mathcal{K}\cup\{0\}}\sum_{q\in Q_i} K^{-i} m_1(n+\kappa-i+\kappa p)\Gamma\left(1+\frac{n+\kappa-i+\kappa p}{k}\right)\partial_{m_2,z}^{q-p_\kappa}(P_p)(|z|).$$

Notice that $\Gamma\left(1+\frac{n}{k}\right)$ is a regular moment function of order $\frac{1}{k}$, and so for certain constants $0 < a$, $A < \infty$ we have

$$a(n+1)^{\frac{1}{k}} \leq \frac{\Gamma\left(1+\frac{n+1}{k}\right)}{\Gamma\left(1+\frac{n}{k}\right)} \leq A(n+1)^{\frac{1}{k}} \text{ for every } n \in \mathbb{N}_0.$$

Using this fact we arrive at

$$m_1(n+\kappa-i+\kappa p)\Gamma\left(1+\frac{n+\kappa-i+\kappa p}{k}\right)$$
$$\leq \frac{a_1^{-i}a^{-i}}{\prod_{l=1}^{i}(n+\kappa-i+\kappa p+l)^{\frac{s_2 p_\kappa}{\kappa}}} m_1(n+\kappa+\kappa p)\Gamma\left(1+\frac{n+\kappa+\kappa p}{k}\right).$$

Hence, for $K \geq \frac{1}{aa_1}$, we receive

$$\text{(5.9)}\quad \left\|\partial_{m_1,t}^n w_{p+1}(t,z)\right\|_{\tilde{r}} \leq B'^p C K^{n+\kappa p+\kappa} B m_1(n+\kappa+\kappa p)\Gamma\left(1+\frac{n+\kappa+\kappa p}{k}\right)$$
$$\cdot \sum_{i\in\mathcal{K}\cup\{0\}}\sum_{q\in Q_i}\frac{1}{\prod_{l=1}^{i}(n+\kappa-i+\kappa p+l)^{\frac{s_2 p_\kappa}{\kappa}}}\partial_{m_2,z}^{q-p_\kappa}(P_p)(|z|).$$

Moreover, let us note that

$$\prod_{l=1}^{i}(n+\kappa-i+\kappa p+l)^{\frac{s_2 p_\kappa}{\kappa}} \geq \prod_{l=1}^{i}(\kappa p+l)^{\frac{s_2 p_\kappa}{\kappa}} = \left[\frac{(\kappa p+i)!}{(\kappa p)!}\right]^{\frac{s_2 p_\kappa}{\kappa}}.$$

We can proceed to further bound the right-hand side of (5.9) as follows:

$$\left\|\partial_{m_1,t}^n w_{p+1}(t,z)\right\|_{\tilde{r}} \leq B'^p C K^{n+\kappa p+\kappa} B m_1(n+\kappa+\kappa p)\Gamma\left(1+\frac{n+\kappa+\kappa p}{k}\right)$$
$$\cdot \sum_{i\in\mathcal{K}\cup\{0\}}\sum_{q\in Q_i}\left[\frac{(\kappa p)!}{(\kappa p+i)!}\right]^{\frac{s_2 p_\kappa}{\kappa}}\partial_{m_2,z}^{q-p_\kappa}(P_p)(|z|).$$

Let us introduce notation $\mathcal{K}' = \{i \in \mathcal{K} : p_i \geq 1\}$. Then

$$\sum_{i\in\mathcal{K}\cup\{0\}}\sum_{q\in Q_i}\left[\frac{(\kappa p)!}{(\kappa p+i)!}\right]^{\frac{s_2 p_\kappa}{\kappa}} \leq \left[(1+\kappa)\partial_{m_2,z}^{-p_\kappa} + \sum_{i\in\mathcal{K}'}\sum_{q\in Q_i}\left[\frac{(\kappa p)!}{(\kappa p+i)!}\right]^{\frac{s_2 p_\kappa}{\kappa}}\right]$$
$$\leq (1+\kappa)\left[\partial_{m_2,z}^{-p_\kappa} + \sum_{i\in\mathcal{K}'}\sum_{q\in Q_i}\left[\frac{(\kappa p)!}{(\kappa p+i)!}\right]^{\frac{s_2 p_\kappa}{\kappa}}\right].$$

Moreover, after applying Remark 5.5 we receive

$$\frac{(\kappa p)!^{\frac{p_\kappa}{\kappa}}(p_\kappa p+p_i)!}{(\kappa p+i)!^{\frac{p_\kappa}{\kappa}}(p_\kappa p)!}=\frac{(p_\kappa p+p_i)\dots(p_\kappa p+1)}{(\kappa p+i)^{\frac{p_\kappa}{\kappa}}\dots(\kappa p+1)^{\frac{p_\kappa}{\kappa}}}=\frac{\left(\frac{p_\kappa}{\kappa}\right)^{p_i}\prod_{\sigma=1}^{p_i}\left(\kappa p+\frac{\kappa\sigma}{p_\kappa}\right)}{(\kappa p+i)^{\frac{p_\kappa}{\kappa}}\dots(\kappa p+1)^{\frac{p_\kappa}{\kappa}}}$$
$$\leq\left(\frac{p_\kappa}{\kappa}\right)^{p_i}\frac{(\kappa p+\kappa)^{p_i}}{(\kappa p+1)^{\frac{p_\kappa}{\kappa}i}}\leq\left(\frac{p_\kappa}{\kappa}\right)^{p_i}\frac{(\kappa p+\kappa)^{p_i}}{(\kappa p+1)^{p_i}}$$
$$\leq\left(\frac{p_\kappa}{\kappa}\right)^{p_\kappa}\left(1+\frac{\kappa-1}{1}\right)^{p_i}\leq p_\kappa^{p_\kappa}.$$

Using the estimations above, we arrive at:

(5.10)

$$\left\|\partial_{m_1,t}^n w_{p+1}(t,z)\right\|_{\tilde r}\leq B'^pCK^{n+\kappa p+\kappa}B(1+\kappa)m_1(n+\kappa+\kappa p)\Gamma\left(1+\frac{n+\kappa+\kappa p}{k}\right)$$
$$\cdot\left[\partial_{m_2,z}^{-p_\kappa}+\sum_{i\in\mathcal{K}'}\sum_{q\in Q_i}p_\kappa^{s_2p_\kappa}\frac{(p_\kappa p)!^{s_2}}{(p_\kappa p+p_i)!^{s_2}}\partial_{m_2,z}^{q-p_\kappa}\right](P_p)(|z|).$$

Hence, it is enough to put $B'\geq B(1+\kappa)$ to finish the proof. □

LEMMA 5.9. *Let $0<r<1$. There exists $F>0$ such that*

$$P_p(|z|)\leq F^p\frac{|z|^p}{m_2(p_\kappa p)}. \tag{5.11}$$

for every $z\in D_r$ and $p\in\mathbb{N}_0$.

PROOF. Inequality (5.11) is of course true for $p=0$. Let us then fix $p\in\mathbb{N}$. Then $P_p(z)$ takes the form

$$\left(\partial_{m_2,z}^{-p_\kappa}+\sum_{i\in\mathcal{K}'}\sum_{q\in Q_i}p_\kappa^{s_2p_\kappa}\frac{(p_\kappa p-p_\kappa)!^{s_2}}{(p_\kappa p-p_\kappa+p_i)!^{s_2}}\partial_{m_2,z}^{q-p_\kappa}\right)\dots$$
$$\dots\left(\partial_{m_2,z}^{-p_\kappa}+\sum_{i\in\mathcal{K}'}\sum_{q\in Q_i}p_\kappa^{s_2p_\kappa}\frac{1}{p_i!^{s_2}}\partial_{m_2,z}^{q-p_\kappa}\right)(P_0)(z).$$

We can simplify the formula above to arrive at

$$P_p(z)=\left[\partial_{m_2,z}^{-p_\kappa p}+\sum_{j=1}^{p}\sum_{1\leq l_1<l_2<\dots<l_j\leq p}\sum_{(i_1,\dots,i_j)\in\mathcal{K}'^j}\sum_{q_{i_1}\in Q_{i_1},\dots,q_{i_j}\in Q_{i_j}}p_\kappa^{s_2jp_\kappa}\right.$$
$$\left.\cdot\prod_{s=1}^{j}\frac{(p_\kappa(l_s-1))!^{s_2}}{(p_\kappa(l_s-1)+p_{i_s})!^{s_2}}\partial_{m_2,z}^{q_{i_1}+\dots+q_{i_j}-p_\kappa p}\right]P_0(z).$$

We further estimate $P_p(|z|)$ as follows:

$$P_p(|z|)\leq\frac{|z|^{p_\kappa p}}{m_2(p_\kappa p)}+\sum_{j=1}^{p}\sum_{1\leq l_1<l_2<\dots<l_j\leq p}\sum_{(i_1,\dots,i_j)\in\mathcal{K}'^j}\sum_{q_{i_1}\in Q_{i_1},\dots,q_{i_j}\in Q_{i_j}}p_\kappa^{s_2jp_\kappa}$$
$$\cdot\prod_{s=1}^{j}\frac{(p_\kappa(l_s-1))!^{s_2}}{(p_\kappa(l_s-1)+p_{i_s})!^{s_2}}\frac{|z|^{p_\kappa p-(q_{i_1}+\dots+q_{i_j})}}{m_2\left(p_\kappa p-\left(q_{i_1}+\dots+q_{i_j}\right)\right)}.$$

From Remark 5.5 it follows that $p_i < p_\kappa$ for $i < \kappa$. Hence,

$$p_\kappa p - \left(q_{i_1} + \cdots + q_{i_j}\right) \geq p_\kappa p - \left(p_{i_1} + \cdots + p_{i_j}\right) \geq p_\kappa p - \left(p_\kappa - 1\right) p = p.$$

Seeing as $|z| < r < 1$, we further receive:

$$P_p(|z|) = \left(1 + \sum_{j=1}^{p} \sum_{1 \leq l_1 < l_2 < \cdots < l_j \leq p} \sum_{(i_1,\ldots,i_j) \in \mathcal{K}'^j} \sum_{q_{i_1} \in Q_{i_1},\ldots,q_{i_j} \in Q_{i_j}} p_\kappa^{s_2 j p_\kappa} \right.$$
$$\left. \cdot \prod_{s=1}^{j} \frac{(p_\kappa(l_s - 1))!^{s_2}}{(p_\kappa(l_s - 1) + p_{i_s})!^{s_2}} \frac{m_2\left(p_\kappa p\right)}{m_2\left(p_\kappa p - \left(q_{i_1} + \cdots + q_{i_j}\right)\right)} \right) \frac{|z|^p}{m_2(p_\kappa p)}.$$

Let us also note that by (3.3) there exists a constant A_2 satisfying

$$\frac{m_2\left(p_\kappa p\right)}{m_2\left(p_\kappa p - \left(q_{i_1} + \cdots + q_{i_j}\right)\right)} \leq A_2^{q_{i_1} + \cdots + q_{i_j}} \frac{(p_\kappa p)!^{s_2}}{\left(p_\kappa p - \left(q_{i_1} + \cdots + q_{i_j}\right)\right)!^{s_2}}$$
$$\leq A_2^{p_\kappa p} \frac{(p_\kappa p)!^{s_2}}{\left(p_\kappa\left(p - j\right) + \left(p_\kappa - q_{i_1}\right) + \cdots + \left(p_\kappa - q_{i_j}\right)\right)!^{s_2}}.$$

After applying a notation

$$\alpha_s = \begin{cases} 1 & \text{for } i_s = \kappa \\ p_\kappa - p_{i_s} & \text{for } i_s < \kappa \end{cases}$$

with $s = 1, 2, \ldots, j$, we receive

$$\frac{(p_\kappa p)!^{s_2}}{\left(p_\kappa\left(p - j\right) + \left(p_\kappa - q_{i_1}\right) + \cdots + \left(p_\kappa - q_{i_j}\right)\right)!^{s_2}}$$
$$\leq \frac{(p_\kappa p)!^{s_2}}{\left(p_\kappa\left(p - j\right)\right)!^{s_2}} \prod_{s=1}^{j} \prod_{\sigma=1}^{\alpha_s} \frac{1}{\left(\alpha_1 + \ldots + \alpha_{s-1} + \sigma\right)^{s_2}}.$$

At the same time we also have

$$\prod_{s=1}^{j} \frac{(p_\kappa(l_s - 1))!^{s_2}}{(p_\kappa(l_s - 1) + p_{i_s})!^{s_2}} = \prod_{s=1}^{j} \prod_{\sigma=1}^{p_{i_s}} \frac{1}{\left(p_\kappa(l_s - 1) + \sigma\right)^{s_2}}$$
$$\leq \prod_{s=1}^{j} \prod_{\sigma=1}^{p_{i_s}} \frac{1}{\left(p_\kappa(s - 1) + \sigma\right)^{s_2}} = \frac{1}{(p_\kappa j)!^{s_2}} \prod_{s=1}^{j} B_s,$$

where

$$B_s = \begin{cases} 1 & \text{for } i_s = \kappa \\ \prod_{\sigma=1}^{\alpha_s} \left(p_\kappa\left(s - 1\right) + p_{i_s} + \sigma\right)^{s_2} & \text{for } i_s < \kappa \end{cases}.$$

Using both properties listed above, we arrive at

$$\prod_{s=1}^{j} \frac{(p_\kappa(l_s - 1))!^{s_2}}{(p_\kappa(l_s - 1) + p_{i_s})!^{s_2}} \frac{m_2\left(p_\kappa p\right)}{m_2\left(p_\kappa p - \left(q_{i_1} + \cdots + q_{i_j}\right)\right)}$$
$$\leq \frac{(p_\kappa p)!^{s_2}}{\left(p_\kappa\left(p - j\right)\right)!^{s_2}} \frac{A_2^{p_\kappa p}}{(p_\kappa j)!^{s_2}} \prod_{s=1}^{j} \left[B_s \prod_{\sigma=1}^{\alpha_s} \frac{1}{\left(\alpha_1 + \ldots + \alpha_{s-1} + \sigma\right)^{s_2}}\right].$$

For $i_s = \kappa$ inequality

$$B_s \prod_{\sigma=1}^{\alpha_s} \frac{1}{(\alpha_1 + \ldots + \alpha_{s-1} + \sigma)^{s_2}} = \frac{1}{(\alpha_1 + \ldots + \alpha_{s-1} + 1)^{s_2}} \leq 1.$$

holds. For $i_s < \kappa$ we have

$$B_s \prod_{\sigma=1}^{\alpha_s} \frac{1}{(\alpha_1 + \ldots + \alpha_{s-1} + \sigma)^{s_2}} = \prod_{\sigma=1}^{\alpha_s} \frac{(p_\kappa (s-1) + p_{i_s} + \sigma)^{s_2}}{(\alpha_1 + \ldots + \alpha_{s-1} + \sigma)^{s_2}}$$
$$= \prod_{\sigma=1}^{\alpha_s} \frac{(\alpha_1 + \ldots + \alpha_{s-1} + p_\kappa (s-1) + p_{i_s} + \sigma - \alpha_1 - \ldots - \alpha_{s-1})^{s_2}}{(\alpha_1 + \ldots + \alpha_{s-1} + \sigma)^{s_2}}.$$

Hence,

$$B_s \prod_{\sigma=1}^{\alpha_s} \frac{1}{(\alpha_1 + \ldots + \alpha_{s-1} + \sigma)^{s_2}} = \prod_{\sigma=1}^{\alpha_s} \left(1 + \frac{p_\kappa (s-1) + p_{i_s} - \alpha_1 - \ldots - \alpha_{s-1}}{\alpha_1 + \ldots + \alpha_{s-1} + \sigma}\right)^{s_2}$$
$$\leq \prod_{\sigma=1}^{\alpha_s} \left(1 + \frac{p_\kappa s}{s}\right)^{s_2} \leq (1 + p_\kappa)^{s_2 (p_\kappa - p_{i_s})}.$$

Using the estimations above, we finally arrive at

$$\prod_{s=1}^{j} \frac{(p_\kappa(l_s - 1))!^{s_2}}{(p_\kappa(l_s - 1) + p_{i_s})!^{s_2}} \frac{m_2 (p_\kappa p)}{m_2 \left(p_\kappa p - \left(q_{i_1} + \cdots + q_{i_j}\right)\right)}$$
$$\leq A_2^{p_\kappa p} \binom{p_\kappa p}{p_\kappa j}^{s_2} \prod_{s=1}^{j} (1 + p_\kappa)^{s_2 (p_\kappa - p_{i_s})},$$

and so

$$P_p(|z|) \leq \left(1 + \sum_{j=1}^{p} \sum_{1 \leq l_1 < l_2 < \cdots < l_j \leq p} \sum_{(i_1, \ldots, i_j) \in \mathcal{K}'^j} \sum_{q_{i_1} \in Q_{i_1}, \ldots, q_{i_j} \in Q_{i_j}} p_\kappa^{s_2 j p_\kappa} A_2^{p_\kappa p} \right.$$
$$\left. \cdot \binom{p_\kappa p}{p_\kappa j}^{s_2} \prod_{s=1}^{j} (1 + p_\kappa)^{s_2 (p_\kappa - p_{i_s})} \right) \frac{|z|^p}{m_2(p_\kappa p)} \leq$$
$$\leq \left(1 + \sum_{j=1}^{p} \sum_{1 \leq l_1 < l_2 < \cdots < l_j \leq p} \sum_{(i_1, \ldots, i_j) \in \mathcal{K}'^j} \sum_{q_{i_1} \in Q_{i_1}, \ldots, q_{i_j} \in Q_{i_j}} p_\kappa^{s_2 j p_\kappa} A_2^{p_\kappa p} \right.$$
$$\left. \cdot 2^{s_2 p_\kappa p} (1 + p_\kappa)^{s_2 (p_\kappa j - p_{i_1} - \cdots - p_{i_j})} \right) \frac{|z|^p}{m_2(p_\kappa p)} \leq$$
$$\leq \left(p_\kappa^{s_2 p_\kappa} A_2^{p_\kappa} 2^{s_2 p_\kappa} (1 + p_\kappa)^{s_2 (p_\kappa - 1)}\right)^p$$
$$\cdot \left(1 + \sum_{j=1}^{p} \sum_{1 \leq l_1 < l_2 < \cdots < l_j \leq p} \sum_{(i_1, \ldots, i_j) \in \mathcal{K}'^j} \sum_{q_{i_1} \in Q_{i_1}, \ldots, q_{i_j} \in Q_{i_j}} 1 \right) \frac{|z|^p}{m_2(p_\kappa p)}.$$

All that remains is to find an upper bound for

$$1 + \sum_{j=1}^{p} \sum_{1 \leq l_1 < l_2 < \cdots < l_j \leq p} \sum_{(i_1, \ldots, i_j) \in \mathcal{K}'^j} \sum_{q_{i_1} \in Q_{i_1}, \ldots, q_{i_j} \in Q_{i_j}} 1.$$

To this end first let us notice that $|\mathcal{K}'^j| \leq \kappa^j$ and $|Q_i| \leq p_\kappa$ for every $i \in \mathcal{K}'$. Thus we conclude that

$$1+\sum_{j=1}^{p} \sum_{1\leq l_1<l_2<\cdots<l_j\leq p} \sum_{(i_1,\ldots,i_j)\in\mathcal{K}'^j} \sum_{q_{i_1}\in Q_{i_1},\ldots,q_{i_j}\in Q_{i_j}} 1 \leq 1+\sum_{j=1}^{p}\binom{p}{j}\kappa^j p_\kappa^j = (1+\kappa p_\kappa)^p .$$

From the last inequality it follows that

$$P_p(|z|) \leq \left(p_\kappa^{s_2 p_\kappa} A_2^{p_\kappa} 2^{s_2 p_\kappa}\left(1+p_\kappa\right)^{s_2(p_\kappa-1)}\left(1+\kappa p_\kappa\right)\right)^p \frac{|z|^p}{m_2(p_\kappa p)}.$$

□

It is now possible to conduct the proof of Theorem 5.7.

Proof of Theorem 5.7. Using previously proven lemmas we arrive at

$$\left\|\partial_{m_1,t}^n w_p(t,z)\right\|_{\tilde r} \leq B'^p C K^{n+\kappa p} m_1(n+\kappa p)\Gamma\left(1+\frac{n+\kappa p}{k}\right) F^p \frac{|z|^p}{m_2(p_\kappa p)}$$

for every $p \in \mathbb{N}_0,\ n \in \mathbb{N}_0,\ \tilde r = |z|$. Hence,

$$\sum_{p\geq 0}\left\|\partial_{m_1,t}^n w_p(t,z)\right\|_{\tilde r} \leq C K^n m_1(n)\Gamma\left(1+\frac{n}{k}\right)\sum_{p\geq 0} B'^p K^{\kappa p} \cdot \frac{m_1(n+\kappa p)\Gamma\left(1+\frac{n+\kappa p}{k}\right)}{m_1(n)\Gamma\left(1+\frac{n}{k}\right)m_2(p_\kappa p)} F^p |z|^p .$$

To find a boundary for

$$\frac{m_1(n+\kappa p)\Gamma\left(1+\frac{n+\kappa p}{k}\right)}{m_1(n)\Gamma\left(1+\frac{n}{k}\right)m_2(p_\kappa p)}$$

let us note that

$$\frac{m_1(n+\kappa p)\Gamma\left(1+\frac{n+\kappa p}{k}\right)}{m_1(n)\Gamma\left(1+\frac{n}{k}\right)m_2(p_\kappa p)} \sim \frac{(n+\kappa p)!^{\frac{s_2 p_\kappa}{\kappa}}}{n!^{\frac{s_2 p_\kappa}{\kappa}}\,(\kappa p)!^{\frac{s_2 p_\kappa}{\kappa}}} \leq \binom{n+\kappa p}{n}^{\frac{s_2 p_\kappa}{\kappa}} .$$

Hence, there exists a constant $\tilde A > 0$ satisfying

$$\frac{m_1(n+\kappa p)\Gamma\left(1+\frac{n+\kappa p}{k}\right)}{m_1(n)\Gamma\left(1+\frac{n}{k}\right)m_2(p_\kappa p)} \leq \tilde A^{n+\kappa p}. \tag{5.12}$$

Thus,

$$\sum_{p\geq 0}\left\|\partial_{m_1,t}^n w_p(t,z)\right\|_{\tilde r} \leq C K^n \tilde A^n m_1(n)\Gamma\left(1+\frac{n}{k}\right)\sum_{p\geq 0}\left(B' K^\kappa \tilde A^\kappa F|z|\right)^p . \tag{5.13}$$

Let us take $\rho = \min\left\{r,\ (B' K^\kappa \tilde A^\kappa F)^{-1}\right\}$. Then for every $z \in D_\rho$ the last sum on the right-hand side of (5.13) is finite. Moreover, there exists a constant $L > 0$ such that

$$\sum_{p\geq 0}\left(B' K^\kappa \tilde A^\kappa F|z|\right)^p \leq L \text{ for } |z| < \rho.$$

Hence, we receive

$$\sum_{p\geq 0} \left\|\partial^n_{m_1,t} w_p(t,z)\right\|_{\tilde{r}} \leq CLK^n\tilde{A}^n m_1(n)\Gamma\left(1+\frac{n}{k}\right) \quad \text{for } z\in D_\rho,\ |z|=\tilde{r}.$$

If we now put $\tilde{C}=CL$ and $\tilde{K}=K\tilde{A}$ then for every $n\in\mathbb{N}_0$ it follows that

$$\left\|\partial^n_{m_1,t} w(t,z)\right\|_{\tilde{r}} \leq \tilde{C}\tilde{K}^n m_1(n)\Gamma\left(1+\frac{n}{k}\right) \quad \text{for } z\in D_\rho,\ |z|=\tilde{r}. \tag{5.14}$$

To conclude the proof of Theorem 5.7 it remains to show that $w(t,z)$ is the k-sum of $\hat{w}(t,z)$ in direction d. To this end let us fix a kernel function $e(z)$ of order k and its corresponding moment function $m_e(u)$. Seeing as for any $p\geq 0$ we have $w_p(t,z)=\mathcal{L}_{e,d}\hat{\mathcal{B}}_{m_e,t}\hat{w}_p(t,z)$, we conclude that $w(t,z)=\sum_{p\geq 0}\mathcal{L}_{e,d}\hat{\mathcal{B}}_{m_e,t}\hat{w}_p(t,z)$. From (5.14), for $n=0$, it follows that $w(t,z)$ is holomorphic for $t\in G$ and $z\in\tilde{D}_\rho$. Moreover, the limit $\lim_{G\ni t\to 0}\partial^n_{m_1,t}w(t,z)$ exists for every $z\in D_\rho$, so a Taylor series at $t_0=0$ exists for $w(t,z)$ for every such z. As reference, see for example, [**2**] or [**12**]. Using this observation, Definitions 3.8 and 4.2, and Theorem 4.4, we further conclude that $\mathcal{B}_{e,d}w(t,z)\in\mathcal{O}(G\times\bar{D}_\rho)$ and it moreover is of exponential growth no greater than k on G (see Theorem 3.9). Furthermore, from (5.14) and the previous observations about individual $\hat{w}_p(t,z)$'s we conclude that $\hat{\mathcal{B}}_{m_e,t}\hat{w}_p(t,z)$ are uniformly convergent power series for all $p\geq 0$. Hence, we may write

$$\mathcal{B}_{e,d}w(t,z)=\mathcal{B}_{e,d}\sum_{p\geq 0}\mathcal{L}_{e,d}\hat{\mathcal{B}}_{m_e,t}\hat{w}_p(t,z)=\mathcal{B}_{e,d}\mathcal{L}_{e,d}\sum_{p\geq 0}\hat{\mathcal{B}}_{m_e,t}\hat{w}_p(t,z)=\hat{\mathcal{B}}_{m_e,t}\hat{w}(t,z).$$

Seeing as $w(t,z)$ is of exponential growth no greater than k, we conclude that $\hat{w}(t,z)$ is k-summable in direction d with $w(t,z)$ being its k-sum. k-summability of $\hat{u}(t,z)$ follows from the fact that

$$\hat{u}(t,z)=\sum_{n=0}^{p_\kappa-1}u_n(t)z^n+\partial^{-p_\kappa}_{m_2,z}\partial^\kappa_{m_1,t}\hat{w}(t,z).$$

□

It is easy to prove the converse theorem as well, which leads to the main result of this paper. It is an analogue of the results presented in [**2**] and [**13**] for standard partial differential equations. A more general case is touched on in [**7**].

Theorem 5.10. *Let us consider equation* (5.5) *with all previous assumptions. Then the following conditions are equivalent:*

(1) $\hat{f}(t,z)\in\mathbb{E}[[t]]$ *and* $u_n(t)\in\mathbb{C}[[t]]$ *for* $0\leq n\leq p_\kappa-1$ *are* k*-summable in direction* $d\in\mathbb{R}$.
(2) *The formal solution* $\hat{u}(t,z)\in\mathbb{E}[[t]]$ *of* (5.5) *is* k*-summable in direction* d *and its* k*-sum* $u(t,z)$ *is an analytic solution of* (5.5) *with* $\hat{f}(t,z)$ *substituted with its* k*-sum* $f(t,z)$.

Proof. Implication (2) $\Longrightarrow$ (1) is obvious.

The second implication for the most part has been proven with Theorem 5.7. To conclude the proof, let us consider a mapping

$$t\mapsto u(t,z)-\sum_{i\in\mathcal{K}}\sum_{q=0}^{p_i}a_{iq}(z)\partial^{-i}_{m_1,t}\partial^q_{m_2,z}u(t,z)-f(t,z).$$

The asymptotic expansion of such a mapping in a sectorial region $G = G_d(\alpha)$, where $\alpha > \frac{\pi}{k}$, is given by a null series. From Watson's Lemma it follows that this mapping has to be constant and equal to 0. □

6. Possible generalizations and open problems

As seen in [**6**] and [**7**], the problem described in this paper can be further generalized to include a broader scope of differential operators. To this end it suffices to utilize the theory of strongly regular sequences. In this case some results demand the additional assumptions about the m-differential operator and its inverse.

As for now, it remains to be seen whether the results from [**13**] for linear integro-differential equations with coefficients dependent on both t and z variables can be extended to similar moment differential equations. The main issue is the fact that we don't have an analogue of the Leibniz product rule for moment differential operators. Because of this fact the reasoning from [**13**] cannot be directly generalized.

Many other problems concerning moment differential equations still remain open. It is as of yet unknown, whether the classical results concerning multisummability seen for example in [**10**] and [**17**] can be further generalized. We similarly lack results in regards to behavior of solutions of nonlinear moment partial differential equations.

References

[1] Werner Balser, *Formal power series and linear systems of meromorphic ordinary differential equations*, Universitext, Springer-Verlag, New York, 2000. MR1722871

[2] Werner Balser and Michèle Loday-Richaud, *Summability of solutions of the heat equation with inhomogeneous thermal conductivity in two variables*, Adv. Dyn. Syst. Appl. **4** (2009), no. 2, 159–177. MR2599531

[3] Werner Balser and Masafumi Yoshino, *Gevrey order of formal power series solutions of inhomogeneous partial differential equations with constant coefficients*, Funkcial. Ekvac. **53** (2010), no. 3, 411–434, DOI 10.1619/fesi.53.411. MR2761091

[4] Javier Jiménez-Garrido, Shingo Kamimoto, Alberto Lastra, and Javier Sanz, *Multisummability in Carleman ultraholomorphic classes by means of nonzero proximate orders*, J. Math. Anal. Appl. **472** (2019), no. 1, 627–686, DOI 10.1016/j.jmaa.2018.11.043. MR3906392

[5] Alberto Lastra, Sławomir Michalik, and Maria Suwińska, *Estimates of formal solutions for some generalized moment partial differential equations*, J. Math. Anal. Appl. **500** (2021), no. 1, Paper No. 125094, 18, DOI 10.1016/j.jmaa.2021.125094. MR4224842

[6] Alberto Lastra, Sławomir Michalik, and Maria Suwińska, *Summability of formal solutions for some generalized moment partial differential equations*, Results Math. **76** (2021), no. 1, Paper No. 22, 27, DOI 10.1007/s00025-020-01324-y. MR4198409

[7] Alberto Lastra, Sławomir Michalik, and Maria Suwińska, *Summability of formal solutions for a family of generalized moment integro-differential equations*, Fract. Calc. Appl. Anal. **24** (2021), no. 5, 1445–1476, DOI 10.1515/fca-2021-0061. MR4331893

[8] Sławomir Michalik, *Analytic solutions of moment partial differential equations with constant coefficients*, Funkcial. Ekvac. **56** (2013), no. 1, 19–50, DOI 10.1619/fesi.56.19. MR3099032

[9] Sławomir Michalik, *Summability of formal solutions of linear partial differential equations with divergent initial data*, J. Math. Anal. Appl. **406** (2013), no. 1, 243–260, DOI 10.1016/j.jmaa.2013.04.062. MR3062417

[10] Sławomir Michalik, *Analytic and summable solutions of inhomogeneous moment partial differential equations*, Funkcial. Ekvac. **60** (2017), no. 3, 325–351, DOI 10.1619/fesi.60.325. MR3701264

[11] Sławomir Michalik and Maria Suwińska, *Gevrey estimates for certain moment partial differential equations*, Complex differential and difference equations, De Gruyter Proc. Math., De Gruyter, Berlin, [2020] ©2020, pp. 391–407. MR4401556

[12] Pascal Remy, *Gevrey order and summability of formal series solutions of some classes of inhomogeneous linear partial differential equations with variable coefficients*, J. Dyn. Control Syst. **22** (2016), no. 4, 693–711, DOI 10.1007/s10883-015-9301-8. MR3538953

[13] Pascal Remy, *Gevrey order and summability of formal series solutions of certain classes of inhomogeneous linear integro-differential equations with variable coefficients*, J. Dyn. Control Syst. **23** (2017), no. 4, 853–878, DOI 10.1007/s10883-017-9371-x. MR3688898

[14] Pascal Remy, *Gevrey properties and summability of formal power series solutions of some inhomogeneous linear Cauchy-Goursat problems*, J. Dyn. Control Syst. **26** (2020), no. 1, 69–108, DOI 10.1007/s10883-019-9428-0. MR4042940

[15] Javier Sanz, *Asymptotic analysis and summability of formal power series*, Analytic, algebraic and geometric aspects of differential equations, Trends Math., Birkhäuser/Springer, Cham, 2017, pp. 199–262, DOI 10.1007/978-3-319-52842-7_4. MR3706197

[16] Maria Suwińska, *Gevrey estimates of formal solutions for certain moment partial differential equations with variable coefficients*, J. Dyn. Control Syst. **27** (2021), no. 2, 355–370, DOI 10.1007/s10883-020-09504-3. MR4231675

[17] Hidetoshi Tahara and Hiroshi Yamazawa, *Multisummability of formal solutions to the Cauchy problem for some linear partial differential equations*, J. Differential Equations **255** (2013), no. 10, 3592–3637, DOI 10.1016/j.jde.2013.07.061. MR3093376

[18] Akiyoshi Yonemura, *Newton polygons and formal Gevrey classes*, Publ. Res. Inst. Math. Sci. **26** (1990), no. 1, 197–204, DOI 10.2977/prims/1195171666. MR1053912

Faculty of Mathematics and Natural Sciences, College of Science, Cardinal Stefan Wyszyński University, Wóycickiego 1/3, 01-938 Warszawa, Poland

Email address: `m.suwinska@op.pl`

Contemporary Mathematics
Volume **782**, 2023
https://doi.org/10.1090/conm/782/15729

Uniqueness of the solution of some nonlinear singular partial differential equations of the second order

Hidetoshi Tahara

ABSTRACT. In this paper, we consider a nonlinear singular second order partial differential equation of the form

$$\Big(t\frac{\partial}{\partial t}\Big)^2 u = F\Big(t, x, \Big\{\Big(t\frac{\partial}{\partial t}\Big)^i \Big(\frac{\partial}{\partial x}\Big)^\alpha u\Big\}_{i+|\alpha|\le 2, i<2}\Big)$$

in the complex domain. If $F(t,x,z)$ (with $z = \{z_{i,\alpha}\}_{i+|\alpha|\le 2, i<2}$) is a holomorphic function satisfying $F(0,x,0) \equiv 0$ and $(\partial F/\partial z_{i,\alpha})(0,x,0) \equiv 0$ (if $|\alpha| > 0$), then this equation is called a nonlinear Fuchsian type partial differential equation in t. Under a very weak assumption, we show the uniqueness of the solution. The result is applied to the problem of analytic continuation of local holomorphic solutions of this equation.

1. Introduction

To study the uniqueness of the solution is one of the most fundamental problems in the theory of partial differential equations. In this paper, we consider the case of nonlinear singular partial differential equations (1.1) given below.

Let $m \in \mathbb{N}^*$ $(= \{1, 2, \ldots\})$. Let $(t, x) = (t, x_1, \ldots, x_n) \in \mathbb{C}_t \times \mathbb{C}_x^n$ be a complex variable and let $z = \{z_{i,\alpha}\}_{i+|\alpha|\le m, i<m} \in \mathbb{C}^N$ be also a complex variable, where we used the notations: $i \in \mathbb{N}$ $(= \{0, 1, 2, \ldots\})$, $\alpha = (\alpha_1, \ldots, \alpha_n) \in \mathbb{N}^n$, $|\alpha| = \alpha_1 + \cdots + \alpha_n$, and $N = \#\{(i, \alpha) \in \mathbb{N} \times \mathbb{N}^n\,;\, i + |\alpha| \le m, i < m\}$. We write $(\partial/\partial x)^\alpha = (\partial/\partial x_1)^{\alpha_1} \cdots (\partial/\partial x_n)^{\alpha_n}$.

Let $F(t, x, z)$ be a function in a neighborhood Δ of the origin of $\mathbb{C}_t \times \mathbb{C}_x^n \times \mathbb{C}_z^N$. Set $\Delta_0 = \Delta \cap \{t = 0, z = 0\}$. In this paper, we consider a nonlinear singular partial differential equation

$$\Big(t\frac{\partial}{\partial t}\Big)^m u = F\Big(t, x, \Big\{\Big(t\frac{\partial}{\partial t}\Big)^i \Big(\frac{\partial}{\partial x}\Big)^\alpha u\Big\}_{i+|\alpha|\le m, i<m}\Big) \tag{1.1}$$

(in the germ sense at $(0, 0) \in \mathbb{C}_t \times \mathbb{C}_x^n$) under the following assumptions:

A$_1$) $F(t, x, z)$ is a holomorphic function in Δ.
A$_2$) $F(0, x, 0) \equiv 0$ in Δ_0.
A$_3$) $(\partial F/\partial z_{i,\alpha})(0, x, 0) \equiv 0$ in Δ_0, if $|\alpha| > 0$.

2020 *Mathematics Subject Classification.* Primary 35A02; Secondary 35G20, 35B60.
Key words and phrases. Uniqueness of the solution, nonlinear singular PDEs.

Then, (1.1) is called a nonlinear Fuchsian type partial differential equation with respect to t, and the roots $\lambda_1(x), \dots, \lambda_m(x)$ of

$$\lambda^m - \sum_{i<m} (\partial F/\partial z_{i,0})(0,x,0)\lambda^i = 0$$

are called the characteristic exponents of (1.1).

Equations of this type were first studied by Gérard-Tahara [**3**, **4**], and then by Tahara-Yamazawa [**7**]. The most fundamental result on (1.1) is:

THEOREM 1 (Gérard-Tahara [**3**]). *Suppose that* $\lambda_i(0) \notin \mathbb{N}^*$ *holds for* $i = 1, \dots, m$. *Then, the equation* (1.1) *has a unique holomorphic solution* $u_0(t,x)$ *in a neighborhood of* $(0,0) \in \mathbb{C}_t \times \mathbb{C}_x^n$ *satisfying* $u_0(0,x) \equiv 0$.

In this paper, we consider the uniqueness of the solution of (1.1) under the assumption

$$\operatorname{Re}\lambda_i(0) < 0, \quad i = 1, \dots, m. \tag{1.2}$$

We denote by $\mathcal{R}(\mathbb{C}_t \setminus \{0\})$ the universal covering space of $\mathbb{C}_t \setminus \{0\}$. For a nonempty open interval $I = (\theta_1, \theta_2)$ and $r > 0$ we write $S_I = \{t \in \mathcal{R}(\mathbb{C}\setminus\{0\})\,;\, \theta_1 < \arg t < \theta_2\}$ and $S_I(r) = \{t \in S_I\,;\, 0 < |t| < r\}$. For $R > 0$ we write $D_R = \{x \in \mathbb{C}^n\,;\, |x| < R\}$, where $|x| = \max_{1 \le j \le n} |x_j|$.

We know:

(1) (Gérard-Tahara [**3**]). If (1.2) is satisfied, and if $u(t,x)$ is a holomorphic solution of (1.1) on $S_I(r_0) \times D_{R_0}$ (for some nonempty open interval I, $r_0 > 0$ and $R_0 > 0$) satisfying

$$\sup_{x \in D_{R_0}} |u(t,x)| = O(|t|^a) \quad (\text{as } S_I \ni t \longrightarrow 0) \tag{1.3}$$

for some $a > 0$, we have $u(t,x) = u_0(t,x)$ on $S_I(r) \times D_R$ for some $r > 0$ and $R > 0$.

(2) In Tahara [**5**], this condition (1.3) was weakened to the following one: there is an $\epsilon > 0$ such that

$$\sup_{x \in D_{R_0}} |u(t,x)| = O\Big(\frac{1}{|\log t|^\epsilon}\Big) \quad (\text{as } S_I \ni t \longrightarrow 0).$$

(3) After that, the author has been trying to prove the following conjecture as a working hypothesis.

CONJECTURE. *Suppose* (1.2). *If* $u(t,x)$ *is a holomorphic solution of* (1.1) *on* $S_I(r_0) \times D_{R_0}$ *(for some nonempty open interval* I, $r_0 > 0$ *and* $R_0 > 0$*) satisfying*

$$\sup_{x \in D_{R_0}} |u(t,x)| = o(1) \quad (\text{as } S_I \ni t \longrightarrow 0),$$

we have $u(t,x) = u_0(t,x)$ *on* $S_I(r) \times D_R$ *for some* $r > 0$ *and* $R > 0$.

The purpose of this paper is to show that the above conjecture is true in the case $m = 2$. The case $m = 1$ has already been proved in Tahara [**6**] by a method similar to the Cauchy's characteristics method. In this paper, we will modify its argument so that it works also in the case $m = 2$. Since the argument here works only in the case $m = 2$, the above conjecture is still open in the case $m \ge 3$.

2. Main result

From now, we consider the case $m = 2$. Our equation is

$$\left(t\frac{\partial}{\partial t}\right)^2 u = F\left(t, x, \left\{\left(t\frac{\partial}{\partial t}\right)^i \left(\frac{\partial}{\partial x}\right)^\alpha u\right\}_{i+|\alpha|\le 2, i<2}\right). \tag{2.1}$$

Let $\lambda_1(x), \lambda_2(x)$ be the characteristic exponents of (2.1). We suppose the conditions A_1), A_2), A_3) and

$$\operatorname{Re}\lambda_i(0) < 0, \quad i = 1, 2. \tag{2.2}$$

The following result is the main theorem of this paper.

THEOREM 2. *Let $u(t,x)$ be a holomorphic solution of (2.1) on $S_I(r_0) \times D_{R_0}$ for some nonempty open interval I, $r_0 > 0$ and $R_0 > 0$. If $u(t,x)$ satisfies*

$$\overline{\lim_{R\to+0}}\left[\lim_{r\to+0}\left(\frac{1}{R^4}\sup_{S_I(r)\times D_R}|u(t,x)|\right)\right] = 0, \tag{2.3}$$

we have $u(t,x) = u_0(t,x)$ on $S_I(\sigma) \times D_\delta$ for some $\sigma > 0$ and $\delta > 0$, where $u_0(t,x)$ is the unique holomorphic solution in Theorem 1.

If

$$\sup_{x\in D_R}|u(t,x)| = o(1), \quad (\text{as } S_I \ni t \longrightarrow 0) \tag{2.4}$$

holds for some $R > 0$ we have (2.3), and so we have

COROLLARY 1. *If $u(t,x)$ satisfies (2.4), we have $u(t,x) = u_0(t,x)$ on $S_I(\sigma)\times D_\delta$ for some $\sigma > 0$ and $\delta > 0$.*

This proves the conjecture posed in §1 in the case $m = 2$.

REMARK. (1) If $\operatorname{Re}\lambda_1(0) > 0$ or $\operatorname{Re}\lambda_2(0) > 0$ holds, we have many solutions satisfying (2.4) and so the uniqueness of the solution is not valid. See [**3, 7**].

(2) In the case $\operatorname{Re}\lambda_1(0) = 0$ or $\operatorname{Re}\lambda_2(0) = 0$, we have the following counter example: the equation

$$\left(t\frac{\partial}{\partial t}\right)^2 u = -\left(t\frac{\partial}{\partial t}\right)u + \left(\frac{\partial u}{\partial x}\right)^2 + 8u\left(\frac{\partial^2 u}{\partial x^2}\right)^2$$

has a trivial solution $u(t,x) \equiv 0$ and a nontrivial solution

$$u(t,x) = \frac{-x^2}{4\log t}$$

which satisfies (2.4). In this case, the characteristic exponents are 0 and -1.

(3) The following example shows that the condition (2.3) will be reasonable: the equation

$$\left(t\frac{\partial}{\partial t}\right)^2 u = -3\left(t\frac{\partial}{\partial t}\right)u - 2u + \left(\frac{\partial^2 u}{\partial x^2}\right)^2$$

has a trivial solution $u(t,x) \equiv 0$ and a nontrivial solution $u = x^4/72$. We note that for $u = x^4/72$ we have

$$\overline{\lim_{R\to+0}}\left[\lim_{r\to+0}\left(\frac{1}{R^4}\sup_{S_I(r)\times D_R}|u(t,x)|\right)\right] = \frac{1}{72}.$$

In this case, the characteristic exponents are -1 and -2.

3. Some preparatory discussion

Before the proof of Theorem 2, let us present some preparatory discussion. Let $u(t,x)$ be a holomorphic solution of (2.1) on $S_I(\sigma_0^*) \times D_{R_0^*}$ (for some nonempty open interval I, $\sigma_0^* > 0$ and $R_0^* > 0$) satisfying the condition (2.3). For simplicity, we set

$$\Lambda = \{(i,\alpha) \in \mathbb{N} \times \mathbb{N}^n \,;\, i + |\alpha| \leq 2, i < 2\},$$
$$N = \#\Lambda \ \text{(the cardinal of } \Lambda),$$
$$z = \{z_{i,\alpha}\}_{(i,\alpha)\in\Lambda} \quad \text{and} \quad z' = \{z_{0,\alpha}\}_{|\alpha|=2}.$$

For $\nu = \{\nu_{i,\alpha}\}_{(i,\alpha)\in\Lambda} \in \mathbb{N}^N$ we write

$$|\nu| = \sum_{(i,\alpha)\in\Lambda} \nu_{i,\alpha}, \quad z^\nu = \prod_{(i,\alpha)\in\Lambda} (z_{i,\alpha})^{\nu_{i,\alpha}}.$$

Suppose the conditions A$_1$), A$_2$), A$_3$): then $F(t,x,z)$ is a holomorphic function in a neighborhood of the origin of $\mathbb{C}_t \times \mathbb{C}_x^n \times \mathbb{C}_z^N$ having the expansion of the form

$$F(t,x,z) = a(x)t + \beta_0^*(x)z_{0,0} + \beta_1^*(x)z_{1,0} + R(t,x,z)$$

with

$$R(t,x,z) = \sum_{i+|\nu|\geq 2} \gamma_{i,\nu}^*(x)t^i z^\nu.$$

STEP 1. We set $w(t,x) = u(t,x) - u_0(t,x)$. Then, we have

$$\overline{\lim_{R\to+0}}\Big[\lim_{r\to+0}\Big(\frac{1}{R^4}\sup_{S_I(r)\times D_R}|w(t,x)|\Big)\Big] = 0, \tag{3.1}$$

and $w(t,x)$ is a holomorphic solution of the equation

$$\Big(t\frac{\partial}{\partial t}\Big)^2 w = H\Big(t,x,\Big\{\Big(t\frac{\partial}{\partial t}\Big)^i\Big(\frac{\partial}{\partial x}\Big)^\alpha w\Big\}_{(i,\alpha)\in\Lambda}\Big) \tag{3.2}$$

on $S_I(\sigma_0^*) \times D_{R_0^*}$, where

$$\begin{aligned} H(t,x,z) =& F\big(t,x,z + \{u_{0,i,\alpha}(t,x)\}_{(i,\alpha)\in\Lambda}\big) \\ &- F\big(t,x,\{u_{0,i,\alpha}(t,x)\}_{(i,\alpha)\in\Lambda}\big) \end{aligned}$$

under the notation $u_{0,i,\alpha}(t,x) = (t\partial/\partial t)^i(\partial/\partial x)^\alpha u_0(t,x)$ $((i,\alpha)\in\Lambda)$. It is easy to see that $H(t,x,z)$ can be expressed in the form

$$\begin{aligned} H(t,x,z) =& \beta_0^*(x)z_{0,0} + \beta_1^*(x)z_{1,0} + t\sum_{(i,\alpha)\in\Lambda} a_{i,\alpha}^*(t,x,z)z_{i,\alpha} \\ &+ \sum_{(i,\alpha)\in\Lambda,|\alpha|\leq 1} b_{i,\alpha}^*(x,z)z_{i,\alpha} + \sum_{|\alpha|=|\beta|=2} c_{\alpha,\beta}^*(x,z')z_{0,\alpha}z_{0,\beta} \end{aligned}$$

for some holomorphic functions $\beta_i^*(x)$ $(i=0,1)$, $a_{i,\alpha}^*(t,x,z)$ $((i,\alpha)\in\Lambda)$, $b_{i,\alpha}^*(x,z)$ $((i,\alpha)\in\Lambda, |\alpha|\leq 1)$, and $c_{\alpha,\beta}^*(x,z')$ $(|\alpha|=|\beta|=2)$ in a common neighborhood of $(0,0,0) \in \mathbb{C}_t \times \mathbb{C}_x^n \times \mathbb{C}_z^N$ satisfying $b_{i,\alpha}^*(x,0) \equiv 0$ $((i,\alpha)\in\Lambda, |\alpha|\leq 1)$. In addition, we see that the roots of $\lambda^2 - \beta_1^*(x)\lambda - \beta_0^*(x) = 0$ are just the same as $\lambda_1(x), \lambda_2(x)$.

From now, we consider the equation (3.2). Our purpose is to show that $w(t,x) \equiv 0$ holds on $S_I(\sigma) \times D_\delta$ for some $\sigma > 0$ and $\delta > 0$.

STEP 2. We set $\Theta_0 = 1$ and

$$\Theta_1 = \Big(t\frac{\partial}{\partial t} - \lambda_1(0)\Big),$$
$$\Theta_2 = \Big(t\frac{\partial}{\partial t} - \lambda_2(0)\Big)\Big(t\frac{\partial}{\partial t} - \lambda_1(0)\Big).$$

We write $Dw = \{D_{i,\alpha}w\}_{(i,\alpha)\in\Lambda}$ and $D'w = \{D_{0,\alpha}w\}_{|\alpha|=2}$ with

$$D_{i,\alpha}w = \Theta_i\Big(\frac{\partial}{\partial x}\Big)^{\alpha} w, \quad (i,\alpha)\in\Lambda.$$

Then, (3.2) is expressed in the form

$$\begin{aligned}(3.3)\qquad \Theta_2 w &= \beta_0(x)w + \beta_1(x)\Theta_1 w + t\sum_{(i,\alpha)\in\Lambda} a_{i,\alpha}(t,x,Dw)\Theta_i\Big(\frac{\partial}{\partial x}\Big)^{\alpha} w \\ &+ \sum_{(i,\alpha)\in\Lambda,|\alpha|\le 1} b_{i,\alpha}(x,Dw)\Theta_i\Big(\frac{\partial}{\partial x}\Big)^{\alpha} w \\ &+ \sum_{|\alpha|=|\beta|=2} c_{\alpha,\beta}(x,D'w)\Big[\Big(\frac{\partial}{\partial x}\Big)^{\alpha} w\Big]\times\Big[\Big(\frac{\partial}{\partial x}\Big)^{\beta} w\Big]\end{aligned}$$

for some holomorphic functions $\beta_i(x)$ $(i=0,1)$, $a_{i,\alpha}(t,x,z)$ $((i,\alpha)\in\Lambda)$, $b_{i,\alpha}(x,z)$ $((i,\alpha)\in\Lambda, |\alpha|\le 1)$ and $c_{\alpha,\beta}(x,z')$ $(|\alpha|=|\beta|=2)$ in a common neighborhood of $(0,0,0)\in\mathbb{C}_t\times\mathbb{C}_x^n\times\mathbb{C}_z^N$ satisfying $\beta_i(0)=0$ $(i=0,1)$ and $b_{i,\alpha}(x,0)\equiv 0$ $((i,\alpha)\in\Lambda, |\alpha|\le 1)$.

STEP 3. By a rotation, we may suppose that $0\in I$ holds: then we have $(0,\sigma_0^*)\subset S_I(\sigma_0^*)$. From now, we consider the equation (3.3) only on $(0,\sigma_0^*)\times D_{R_0^*}$. By (3.1) and Nagumo type lemma in a sector (for example, see Lemma 4.2 in [**1**]) we have the condition

$$(3.4)\qquad \overline{\lim_{R\to+0}}\Big[\lim_{\sigma\to+0}\Big(\frac{1}{R^4}\sup_{(0,\sigma]\times D_R}\Big|\Big(t\frac{\partial}{\partial t}\Big)^{i} w(t,x)\Big|\Big)\Big] = 0, \quad i=0,1,2.$$

For a formal power series $f(t,x)=\sum_{|\alpha|\ge 0} f_\alpha(t)x^\alpha$ in x with coefficients in $C^0((0,T))$ we write

$$\|f(t)\|_\rho = \sum_{|\alpha|\ge 0}|f_\alpha(t)|\frac{\alpha!}{|\alpha|!}\rho^{|\alpha|}$$

(which is a formal power series in ρ with coefficients in $C^0((0,T))$). When $\|f(t)\|_\rho$ converges, we regard it as a function in (t,ρ). We can easily see:

$$\|f(t)g(t)\|_\rho \ll \|f(t)\|_\rho\|g(t)\|_\rho,$$
$$\|(\partial/\partial x_j)f(t)\|_\rho \ll (\partial/\partial\rho)\|f(t)\|_\rho, \quad j=1,\dots,n.$$

where $\sum_{i\ge 0} a_i\rho^i \ll \sum_{i\ge 0} b_i\rho^i$ means that $|a_i|\le b_i$ holds for all $i\ge 0$. For a holomorphic function $f(t,x,z)=\sum_{i+|\nu|\ge 0} f_{i,\nu}(x)t^i z^\nu$ we write

$$\|f\|_\rho(t,z) = \sum_{i+|\nu|\ge 0}\|f_{i,\nu}\|_\rho t^i z^\nu.$$

By (3.4) we have

LEMMA 1. *Under the above situation, for $i = 0, 1, 2$ we have*

$$\lim_{\sigma \to +0} \Big(\sup_{(0,\sigma]\times[0,R]} \Big\| \Big(t\frac{\partial}{\partial t} \Big)^i w(t) \Big\|_\rho \Big) = o(R^4) \quad (\text{as } R \longrightarrow +0). \tag{3.5}$$

STEP 4. Set $J = \{(i,j) \in \mathbb{N}^2 \,;\, i+j \le 2, i < 2\}$: actually, we have

$$J = \{(0,0), (1,0), (0,1), (1,1), (0,2)\}.$$

For $(i,j) \in J$ we set

$$\begin{aligned} \phi_{0,0}(t,\rho) &= \int_0^t \Big(\frac{\tau}{t}\Big)^{-\operatorname{Re}\lambda_1(0)} \|\Theta_1 w(\tau)\|_\rho \frac{d\tau}{\tau}, \\ \phi_{1,0}(t,\rho) &= \int_0^t \Big(\frac{\tau}{t}\Big)^{-\operatorname{Re}\lambda_2(0)} \|\Theta_2 w(\tau)\|_\rho \frac{d\tau}{\tau}, \\ \phi_{0,1}(t,\rho) &= \frac{\partial}{\partial\rho}\phi_{0,0}(t,\rho), \\ \phi_{1,1}(t,\rho) &= \frac{\partial}{\partial\rho}\phi_{1,0}(t,\rho), \\ \phi_{0,2}(t,\rho) &= \frac{\partial}{\partial\rho}\phi_{0,1}(t,\rho) = \Big(\frac{\partial}{\partial\rho}\Big)^2 \phi_{0,0}(t,\rho). \end{aligned}$$

By making $\sigma_0^* > 0$ and $R_0^* > 0$ smaller if necessary, we may suppose that $\phi_{i,j}(t,\rho)$ ($(i,j) \in J$) are convergent on $(0,\sigma_0^*] \times [0, R_0^*]$. By (3.5) we have

LEMMA 2. (1) *For any $(i,j) \in J$ we have*

$$\lim_{\sigma \to +0} \Big(\sup_{(0,\sigma]\times[0,R]} \phi_{i,j}(t,\rho) \Big) = o(R^{4-j}) \quad (\text{as } R \longrightarrow +0).$$

(2) *For any $\epsilon > 0$ we can find $\sigma > 0$ and $R > 0$ such that $|\phi_{i,j}(t,\rho)| \le \epsilon$ holds on $(0,\sigma] \times [0,R]$ for any $(i,j) \in J$.*

STEP 5. By (2.2) we can take an $h > 0$ such that

$$\operatorname{Re}\lambda_i(0) < -2h, \quad i = 1, 2.$$

LEMMA 3. (1) *For any $(i,\alpha) \in \Lambda$ we have*

$$\|D_{i,\alpha} w(t)\|_\rho = \Big\| \Theta_i \Big(\frac{\partial}{\partial x}\Big)^\alpha w(t) \Big\|_\rho \ll \phi_{i,|\alpha|}(t,\rho) \quad \text{on } (0,\sigma_0^*].$$

(2) *We set $\Phi = (\phi_{i,|\alpha|}(t,\rho)\,;\, (i,\alpha) \in \Lambda)$ and $\Phi' = (\phi_{0,|\alpha|}(t,\rho)\,;\, |\alpha| = 2)$: we have*

$$\begin{aligned} &\Big(t\frac{\partial}{\partial t} + 2h\Big)\phi_{0,0}(t,\rho) \ll \phi_{1,0}(t,\rho), \\ &\Big(t\frac{\partial}{\partial t} + 2h\Big)\phi_{1,0}(t,\rho) \\ &\qquad \ll \|\beta_0\|_\rho \phi_{0,0}(t,\rho) + \|\beta_1\|_\rho \phi_{1,0}(t,\rho) + t \sum_{(i,\alpha)\in\Lambda} \|a_{i,\alpha}\|_\rho(t,\Phi)\phi_{i,|\alpha|}(t,\rho) \\ &\qquad + \sum_{(i,\alpha)\in\Lambda, |\alpha|\le 1} \|b_{i,\alpha}\|_\rho(\Phi)\phi_{i,|\alpha|}(t,\rho) + \sum_{|\alpha|=|\beta|=2} \|c_{\alpha,\beta}\|_\rho(\Phi')(\phi_{0,2}(t,\rho))^2 \end{aligned}$$

on $(0,\sigma_0^*]$.

(3) *By taking $R_0^* > 0$ a smaller one if necessary, we may assume that* $\|\beta_i\|_\rho$ $(i = 0, 1)$, $\|a_{i,\alpha}\|_\rho(t, z)$ $((i, \alpha) \in \Lambda)$, $\|b_{i,\alpha}\|_\rho(z)$ $((i, \alpha) \in \Lambda, |\alpha| \le 1)$ *and* $\|c_{\alpha,\beta}\|_\rho(z')$ *are convergent on* $(0, \sigma_0^*] \times [0, R_0^*] \times D_L$ *(where* $D_L = \{z \in \mathbb{C}^N\,;\, |z_{i,\alpha}| < L\ ((i, \alpha) \in \Lambda)\}$*) for some* $L > 0$. *In addition, we have*

$$\|\beta_i\|_\rho \le H_i\rho \quad \text{on } [0, R_0^*], \quad i = 0, 1,$$
$$\|b_{i,\alpha}\|_\rho(z) \le B_{i,\alpha}|z| \quad \text{on } [0, R_0^*] \times D_L, \quad (i, \alpha) \in \Lambda,\ |\alpha| \le 1$$

for some constants $H_i > 0$ $(i = 0, 1)$ *and* $B_{i,\alpha} > 0$ $((i, \alpha) \in \Lambda, |\alpha| \le 1)$, *where* $|z| = \sum_{(i,\alpha)\in\Lambda} |z_{i,\alpha}|$.

(4) *For any* $(i, \alpha) \in \Lambda$ *with* $|\alpha| \le 1$ *we have*

$$\lim_{\sigma\to+0}\Big(\sup_{(0,\sigma]\times[0,R]} \frac{\partial\|b_{i,\alpha}\|_\rho(\Phi)}{\partial\rho}\Big) = o(R) \quad (\text{as } R \longrightarrow +0).$$

Proof. Let us show (1). Since

$$\Big(t\frac{\partial}{\partial t} - \lambda_{i+1}(0)\Big)\Theta_i w = (\Theta_{i+1}w)(t, x), \quad i = 0, 1$$

hold, by integrating this we have

$$\Theta_i w(t, x) = \int_0^t \Big(\frac{\tau}{t}\Big)^{-\lambda_{i+1}(0)} (\Theta_{i+1}w)(\tau, x)\frac{d\tau}{\tau}$$

and so by taking the norm

$$\|\Theta_i w(t)\|_\rho \ll \int_0^t \Big(\frac{\tau}{t}\Big)^{-\operatorname{Re}\lambda_{i+1}(0)} \|\Theta_{i+1}w(\tau)\|_\rho\frac{d\tau}{\tau} = \phi_{i,0}(t, \rho), \quad i = 0, 1.$$

Hence, for any $(i, \alpha) \in \Lambda$ we have

$$\Big\|\Theta_i\Big(\frac{\partial}{\partial x}\Big)^\alpha w(t)\Big\|_\rho \ll \Big(\frac{\partial}{\partial\rho}\Big)^{|\alpha|}\|\Theta_i w(t)\|_\rho \ll \Big(\frac{\partial}{\partial\rho}\Big)^{|\alpha|}\phi_{i,0}(t, \rho) = \phi_{i,|\alpha|}(t, \rho).$$

Let us show (2). We have

$$\begin{aligned}\Big(t\frac{\partial}{\partial t} + 2h\Big)\phi_{0,0}(t, \rho) &\ll \Big(t\frac{\partial}{\partial t} + \operatorname{Re}\lambda_1(0)\Big)\phi_{0,0}(t, \rho)\\ &= \|\Theta_1 w(\tau)\|_\rho \ll \phi_{1,0}(t, \rho) \quad \text{on } (0, \sigma_0^*].\end{aligned}$$

This proves the first inequality of (2). Since

$$\Big(t\frac{\partial}{\partial t} + 2h\Big)\phi_{1,0}(t, \rho) \ll \Big(t\frac{\partial}{\partial t} + \operatorname{Re}\lambda_2(0)\Big)\phi_{1,0}(t, \rho) = \|\Theta_2 w(\tau)\|_\rho,$$

by applying (3.3) and by using (1) we can easily see the second inequality of (2). The condition (3) is clear. Since $\|b_{j,\alpha}\|_\rho(\Phi) \le B_{j,\alpha}|\Phi|$ $((j, \alpha) \in \Lambda)$ hold, by (1) of Lemma 2 we have

$$\lim_{\sigma\to+0}\Big(\sup_{(0,\sigma]\times[0,R]} \|b_{i,\alpha}\|_\rho(\Phi)\Big) = o(R^2) \quad (\text{as } R \longrightarrow +0).$$

This leads us to the condition (4). □

Step 6. Let $\varepsilon_{0,0} > 0$, $\varepsilon_{1,0} = 1$, $\varepsilon_{0,1} > 0$, $\varepsilon_{1,1} > 0$ and $0 < \kappa < 1$. We set

$$\begin{aligned}q(t, \rho) =&\varepsilon_{0,0}\phi_{0,0}(t, \rho) + \phi_{1,0}(t, \rho) + t^\kappa\phi_{0,2}(t, \rho)\\ &+ \varepsilon_{0,1}\phi_{0,1}(t, \rho) + \varepsilon_{1,1}\phi_{1,1}(t, \rho) + (\phi_{0,2}(t, \rho))^{3/2}.\end{aligned}$$

LEMMA 4. *We have the following inequality:*

$$\Big(t\frac{\partial}{\partial t}+2h\Big)q\le A(t,\rho)\,q+B(t,\rho)\frac{\partial}{\partial\rho}q \tag{3.6}$$

on $(0,\sigma_0^*]\times[0,R_0^*]$, *where*

$$\begin{aligned}
A(t,\rho)={}&\varepsilon_{0,0}+\Big(\frac{1}{\varepsilon_{0,0}}\|\beta_0\|_\rho+\|\beta_1\|_\rho\Big)\\
&+\sum_{(i,\alpha)\in\Lambda,|\alpha|\le 1}\frac{1}{\varepsilon_{i,|\alpha|}}\times t\,\|a_{i,\alpha}\|_\rho(t,\Phi)+\sum_{|\alpha|=2}t^{1-\kappa}\|a_{0,\alpha}\|_\rho(t,\Phi)\\
&+\sum_{(i,\alpha)\in\Lambda,|\alpha|\le 1}\frac{1}{\varepsilon_{i,|\alpha|}}\times\|b_{i,\alpha}\|_\rho(\Phi)+\sum_{|\alpha|=|\beta|=2}\|c_{\alpha,\beta}\|_\rho(\Phi')(\phi_{0,2})^{1/2}\\
&+\kappa+\frac{\varepsilon_{0,1}}{\varepsilon_{1,1}}+\varepsilon_{1,1}\Big(\frac{1}{\varepsilon_{0,0}}\frac{\partial\|\beta_0\|_\rho}{\partial\rho}+\frac{\partial\|\beta_1\|_\rho}{\partial\rho}\Big)\\
&+\varepsilon_{1,1}\Big(\frac{\|\beta_0\|_\rho}{\varepsilon_{0,1}}+\frac{\|\beta_1\|_\rho}{\varepsilon_{1,1}}\Big)+\sum_{(i,\alpha)\in\Lambda,|\alpha|\le 1}\frac{\varepsilon_{1,1}}{\varepsilon_{i,|\alpha|}}\times t\,\frac{\partial\|a_{i,\alpha}\|_\rho(t,\Phi)}{\partial\rho}\\
&+\varepsilon_{1,1}\sum_{|\alpha|=2}t^{1-\kappa}\frac{\partial\|a_{0,\alpha}\|_\rho(t,\Phi)}{\partial\rho}+\sum_{(i,\alpha)\in\Lambda,|\alpha|\le 1}\frac{\varepsilon_{1,1}}{\varepsilon_{i,|\alpha|}}\times\frac{\partial\|b_{i,\alpha}\|_\rho(\Phi)}{\partial\rho}\\
&+\varepsilon_{1,1}\sum_{|\alpha|=|\beta|=2}\frac{\partial\|c_{\alpha,\beta}\|_\rho(\Phi')}{\partial\rho}(\phi_{0,2})^{1/2},\\
B(t,\rho)={}&\frac{t^\kappa}{\varepsilon_{1,1}}+\sum_{(i,\alpha)\in\Lambda,|\alpha|\le 1}\frac{\varepsilon_{1,1}}{\varepsilon_{i,|\alpha|}}\times t\,\|a_{i,\alpha}\|_\rho(t,\Phi)\\
&+\varepsilon_{1,1}\sum_{|\alpha|=2}t^{1-\kappa}\|a_{0,\alpha}\|_\rho(t,\Phi)+\sum_{(i,\alpha)\in\Lambda,|\alpha|\le 1}\frac{\varepsilon_{1,1}}{\varepsilon_{i,|\alpha|}}\times\|b_{i,\alpha}\|_\rho(\Phi)\\
&+\frac{4\varepsilon_{1,1}}{3}\sum_{|\alpha|=|\beta|=2}\|c_{\alpha,\beta}\|_\rho(\Phi')(\phi_{0,2})^{1/2}+\frac{3}{2\varepsilon_{1,1}}(\phi_{0,2})^{1/2}.
\end{aligned}$$

PROOF. By the definition of $q(t,\rho)$ we have

$$\phi_{i,j}\le\frac{1}{\varepsilon_{i,j}}q\quad(i,j=0,1),\qquad \phi_{0,2}\le q^{2/3},\qquad t^\kappa\phi_{0,2}\le q \tag{3.7}$$

on $(0,\sigma_0^*]\times[0,R_0^*]$. Since

$$\begin{aligned}
\frac{\partial}{\partial\rho}q={}&\varepsilon_{0,0}\frac{\partial}{\partial\rho}\phi_{0,0}+\frac{\partial}{\partial\rho}\phi_{1,0}+t^\kappa\frac{\partial}{\partial\rho}\phi_{0,2}\\
&+\varepsilon_{0,1}\frac{\partial}{\partial\rho}\phi_{0,1}+\varepsilon_{1,1}\frac{\partial}{\partial\rho}\phi_{1,1}+\frac{3}{2}(\phi_{0,2})^{1/2}\frac{\partial}{\partial\rho}\phi_{0,2}
\end{aligned}$$

we have also

$$\begin{aligned}
&\frac{\partial}{\partial\rho}\phi_{i,j}\le\frac{1}{\varepsilon_{i,j}}\frac{\partial}{\partial\rho}q\quad(i,j=0,1),\\
&(\phi_{0,2})^{1/2}\frac{\partial}{\partial\rho}\phi_{0,2}\le\frac{2}{3}\frac{\partial}{\partial\rho}q,\quad t^\kappa\frac{\partial}{\partial\rho}\phi_{0,2}\le\frac{\partial}{\partial\rho}q
\end{aligned} \tag{3.8}$$

on $(0,\sigma_0^*]\times[0,R_0^*]$. By using these inequality, let us do a calculation.

1) About $\varepsilon_{0,0}\phi_{0,0}(t,\rho)$, by (2) of Lemma 3 and (3.7) we have

$$\Big(t\frac{\partial}{\partial t}+2h\Big)(\varepsilon_{0,0}\phi_{0,0}) \ll \varepsilon_{0,0}\phi_{1,0} \le \varepsilon_{0,0}q.$$

2) About $\phi_{1,0}$, by (2) of Lemma 3 and (3.7) we have

$$\begin{aligned}
&\Big(t\frac{\partial}{\partial t}+2h\Big)\phi_{1,0}\\
&\ll \|\beta_0\|_\rho\phi_{0,0}+\|\beta_1\|_\rho\phi_{1,0}+t\sum_{(i,\alpha)\in\Lambda}\|a_{i,\alpha}\|_\rho(t,\Phi)\phi_{i,|\alpha|}\\
&\qquad+\sum_{(i,\alpha)\in\Lambda,|\alpha|\le 1}\|b_{i,\alpha}\|_\rho(\Phi)\phi_{i,|\alpha|}+\sum_{|\alpha|=|\beta|=2}\|c_{\alpha,\beta}\|_\rho(\Phi')(\phi_{0,2})^2\\
&\le \Big(\frac{1}{\varepsilon_{0,0}}\|\beta_0\|_\rho+\|\beta_1\|_\rho\Big)q\\
&\qquad+\sum_{(i,\alpha)\in\Lambda,|\alpha|\le 1}\frac{1}{\varepsilon_{i,|\alpha|}}t\,\|a_{i,\alpha}\|_\rho(t,\Phi)\,q+t^{1-\kappa}\sum_{|\alpha|=2}\|a_{0,\alpha}\|_\rho(t,\Phi)q\\
&\qquad+\sum_{(i,\alpha)\in\Lambda,|\alpha|\le 1}\frac{1}{\varepsilon_{i,|\alpha|}}\|b_{i,\alpha}\|_\rho(\Phi)\,q+\sum_{|\alpha|=|\beta|=2}\|c_{\alpha,\beta}\|_\rho(\Phi')(\phi_{0,2})^{1/2}q.
\end{aligned}$$

3) Let us consider $t^\kappa\phi_{0,2}(t,\rho)$. Since $(\partial/\partial\rho)\phi_{1,1}\ll(1/\varepsilon_{1,1})(\partial/\partial\rho)q$ holds, we have

$$\Big(t\frac{\partial}{\partial t}+2h\Big)\phi_{0,2}=\Big(\frac{\partial}{\partial\rho}\Big)^2\Big(t\frac{\partial}{\partial t}+2h\Big)\phi_{0,0}\ll\Big(\frac{\partial}{\partial\rho}\Big)^2\phi_{1,0}=\frac{\partial}{\partial\rho}\phi_{1,1}\le\frac{1}{\varepsilon_{1,1}}\frac{\partial}{\partial\rho}q$$

and so

$$\Big(t\frac{\partial}{\partial t}+2h\Big)t^\kappa\phi_{0,2}=t^\kappa\Big(t\frac{\partial}{\partial t}+2h\Big)\phi_{0,2}+\kappa t^\kappa\phi_{0,2}\le\frac{t^\kappa}{\varepsilon_{1,1}}\frac{\partial}{\partial\rho}q+\kappa q.$$

4) About $\varepsilon_{0,1}\phi_{0,1}(t,\rho)$, we have

$$\begin{aligned}
\Big(t\frac{\partial}{\partial t}+2h\Big)(\varepsilon_{0,1}\phi_{0,1})&=\varepsilon_{0,1}\frac{\partial}{\partial\rho}\Big(t\frac{\partial}{\partial t}+2h\Big)\phi_{0,0}\\
&\ll\varepsilon_{0,1}\frac{\partial}{\partial\rho}\phi_{1,0}=\varepsilon_{0,1}\phi_{1,1}\le\frac{\varepsilon_{0,1}}{\varepsilon_{1,1}}q.
\end{aligned}$$

5) About $\varepsilon_{1,1}\phi_{1,1}(t,\rho)$, by (2) of Lemma 3 we have

$$\begin{aligned}
&\Big(t\frac{\partial}{\partial t}+2h\Big)(\varepsilon_{1,1}\phi_{1,1})=\varepsilon_{1,1}\frac{\partial}{\partial\rho}\Big(t\frac{\partial}{\partial t}+2h\Big)\phi_{1,0}\\
&\ll\varepsilon_{1,1}\frac{\partial}{\partial\rho}\Big[\|\beta_0\|_\rho\phi_{0,0}+\|\beta_1\|_\rho\phi_{1,0}+t\sum_{(i,\alpha)\in\Lambda}\|a_{i,\alpha}\|_\rho(t,\Phi)\phi_{i,|\alpha|}\\
&\qquad+\sum_{(i,\alpha)\in\Lambda,|\alpha|\le 1}\|b_{i,\alpha}\|_\rho(\Phi)\phi_{i,|\alpha|}+\sum_{|\alpha|=|\beta|=2}\|c_{\alpha,\beta}\|_\rho(\Phi')(\phi_{0,2})^2\Big].
\end{aligned}$$

By calculating the right side of the above formula and then by using (3.7) and (3.8) we have

$$\begin{aligned}
&\Big(t\frac{\partial}{\partial t}+2h\Big)(\varepsilon_{1,1}\phi_{1,1}) \\
&\le \varepsilon_{1,1}\Big(\frac{1}{\varepsilon_{0,0}}\frac{\partial\|\beta_0\|_\rho}{\partial\rho}+\frac{\partial\|\beta_1\|_\rho}{\partial\rho}\Big)q+\varepsilon_{1,1}\Big(\frac{\|\beta_0\|_\rho}{\varepsilon_{0,1}}+\frac{\|\beta_1\|_\rho}{\varepsilon_{1,1}}\Big)q \\
&+\sum_{(i,\alpha)\in\Lambda,|\alpha|\le 1}\frac{\varepsilon_{1,1}}{\varepsilon_{i,|\alpha|}}\times t\frac{\partial\|a_{i,\alpha}\|_\rho(t,\Phi)}{\partial\rho}\times q \\
&+\sum_{(i,\alpha)\in\Lambda,|\alpha|\le 1}\frac{\varepsilon_{1,1}}{\varepsilon_{i,|\alpha|}}\times t\|a_{i,\alpha}\|_\rho(t,\Phi)\frac{\partial}{\partial\rho}q \\
&+\varepsilon_{1,1}t^{1-\kappa}\sum_{|\alpha|=2}\frac{\partial\|a_{0,\alpha}\|_\rho(t,\Phi)}{\partial\rho}q+\varepsilon_{1,1}t^{1-\kappa}\sum_{|\alpha|=2}\|a_{0,\alpha}\|_\rho(t,\Phi)\frac{\partial}{\partial\rho}q \\
&+\sum_{(i,\alpha)\in\Lambda,|\alpha|\le 1}\frac{\varepsilon_{1,1}}{\varepsilon_{i,|\alpha|}}\times\frac{\partial\|b_{i,\alpha}\|_\rho(\Phi)}{\partial\rho}q+\sum_{(i,\alpha)\in\Lambda,|\alpha|\le 1}\frac{\varepsilon_{1,1}}{\varepsilon_{i,|\alpha|}}\times\|b_{i,\alpha}\|_\rho(\Phi)\frac{\partial}{\partial\rho}q \\
&+\varepsilon_{1,1}\sum_{|\alpha|=|\beta|=2}\frac{\partial\|c_{\alpha,\beta}\|_\rho(\Phi')}{\partial\rho}(\phi_{0,2})^{1/2}q \\
&+\frac{4\varepsilon_{1,1}}{3}\sum_{|\alpha|=|\beta|=2}\|c_{\alpha,\beta}\|_\rho(\Phi')(\phi_{0,2})^{1/2}\frac{\partial}{\partial\rho}q.
\end{aligned}$$

6) Let us consider $(\phi_{0,2})^{3/2}$. Since

$$\Big(t\frac{\partial}{\partial t}+2h\Big)\phi_{0,2}=\frac{\partial}{\partial\rho}\Big(t\frac{\partial}{\partial t}+2h\Big)\phi_{0,1}\ll\frac{\partial}{\partial\rho}\phi_{1,1}\le\frac{1}{\varepsilon_{1,1}}\frac{\partial}{\partial\rho}q$$

we have

$$\begin{aligned}
\Big(t\frac{\partial}{\partial t}+2h\Big)(\phi_{0,2})^{3/2}&=\frac{3}{2}(\phi_{0,2})^{1/2}t\frac{\partial}{\partial t}\phi_{0,2}+2h(\phi_{0,2})^{3/2} \\
&\le\frac{3}{2}(\phi_{0,1})^{1/2}\Big(t\frac{\partial}{\partial t}+2h\Big)\phi_{0,2}\le\frac{3}{2\varepsilon_{1,1}}(\phi_{0,2})^{1/2}\frac{\partial}{\partial\rho}q.
\end{aligned}$$

7) Thus, by taking the summation from 1) to 6) we have the result (3.6). □

STEP 7. Let us estimate $A(t,\rho)$ and $B(t,\rho)$. We have

LEMMA 5. *By taking* $\varepsilon_{0,0}>0$, $\varepsilon_{0,1}>0$, $\varepsilon_{1,1}>0$, $0<\kappa<1/2$, $\sigma_0>0$ *and* $R_0>0$ *suitably, we have*

(3.9) $$A(t,\rho)\le h,$$

(3.10) $$B(t,\rho)\le C_1t^\kappa+C_2q+C_3q^{2/3}+C_4q^{1/3}$$

on $(0,\sigma_0]\times[0,R_0]$ *for some* $C_i>0$ $(i=1,2,3,4)$.

PROOF. 1) First, we take $\varepsilon_{0,0}$ so that $0<\varepsilon_{0,0}\le h/4$.

2) Since $\varepsilon_{0,0}$ is a fixed constant, by taking $\varepsilon_{1,1}>0$ sufficiently small we have the condition

$$\varepsilon_{1,1}\Big(\frac{1}{\varepsilon_{0,0}}\frac{\partial\|\beta_0\|_\rho}{\partial\rho}+\frac{\partial\|\beta_1\|_\rho}{\partial\rho}\Big)\le\frac{h}{4}\quad\text{on } (0,\sigma_0^*]\times[0,R_0^*].$$

3) Thirdly, we take $\kappa>0$ and $\varepsilon_{0,1}>0$ so that $\kappa+\varepsilon_{0,1}/\varepsilon_{1,1}\le h/4$.

4) Now, $\varepsilon_{i,|\alpha|} > 0$ ($(i, \alpha) \in \Lambda, |\alpha| \leq 1$) are fixed. Since $\|\beta_i\|_\rho \leq H_i\rho$ ($i = 0, 1$) and $\|b_{i,\alpha}\|_\rho(\Phi) \leq B_{i,\alpha}|\Phi|$ ($(i, \alpha) \in \Lambda, |\alpha| \leq 1$) are known (by (3) of Lemma 3), by the conditions (2) of Lemma 2 and (4) of Lemma 3 we can take $\sigma_0 > 0$ and $R_0 > 0$ suitably sufficiently small so that the condition (3.9) is valid on $(0, \sigma_0] \times [0, R_0]$.

5) Since $0 < \kappa < 1/2$ is supposed, we have $t = O(t^\kappa)$ and $t^{1-\kappa} = O(t^\kappa)$. Hence, by the condition $\|b_{i,\alpha}\|_\rho(\Phi) \leq B_{i,\alpha}|\Phi|$ we have

$$B(t, \rho) \leq K_1 t^\kappa + K_2|\Phi| + K_3(\phi_{0,2})^{1/2}$$

for some $K_i > 0$ ($i = 1, 2, 3$). By applying (3.7) to this estimate we have the condition (3.10). □

4. Proof of Theorem 2

In order to prove Theorem 2, it is enough to show

PROPOSITION 1. *Under the situation in section 3, there are $\sigma > 0$ and $\delta > 0$ such that $q(t, \rho) = 0$ holds on $(0, \sigma] \times [0, \delta)$.*

PROOF. Let us show this, step by step.

STEP 1. By Lemmas 4 and 5 we have

$$\left(t\frac{\partial}{\partial t} + h\right)q \leq B(t, \rho)\frac{\partial}{\partial \rho}q, \tag{4.1}$$

$$B(t, \rho) \leq C_1 t^\kappa + C_2 q + C_3 q^{2/3} + C_4 q^{1/3} \tag{4.2}$$

on $(0, \sigma_0] \times [0, R_0]$. For $\sigma > 0$ and $R > 0$ we set

$$r = \sup_{(0,\sigma]\times[0,R]} q(t, \rho).$$

By (1) of Lemma 2 and the definition of $q(t, \rho)$ we have

$$\lim_{\sigma \to +0} r = o(R^3) \quad (\text{as } R \longrightarrow +0).$$

This shows

LEMMA 6. *By taking $\sigma > 0$ and $R > 0$ sufficiently small, we have the condition*

$$\frac{C_1}{\kappa}\sigma^\kappa + \frac{C_2}{h}r + \frac{3C_3}{2h}r^{2/3} + \frac{3C_4}{h}r^{1/3} < \frac{R}{2} \quad \text{on } (0, \sigma] \times [0, R].$$

STEP 2. Let $\sigma > 0$ and $R > 0$ be as in Lemma 6. Take any $t_0 \in (0, \sigma]$ and $\xi \in (0, R)$; for a while we fix them.

Let us consider the equation

$$\frac{d\rho}{dt} = -\frac{B(t, \rho)}{t}, \quad \rho(t_0) = \xi \tag{4.3}$$

in the region $(0, t_0] \times (0, R)$. Since $B(t, \rho)/t$ is a continuous function on $(0, t_0] \times (0, R)$ we have a local solution (not necessarily unique). We take a maximally extended solution $\rho(t)$ and we denote by $(t_\xi, t_0]$ the interval of existence of this maximally extended solution. Set

$$q^*(t) = q(t, \rho(t)), \quad t_\xi < t \leq t_0.$$

LEMMA 7. *Under the above situation, we have the following inequality for any (t_1, τ) satisfying $t_\xi < t_1 < \tau \leq t_0$:*

$$q^*(\tau) \leq \left(\frac{t_1}{\tau}\right)^h q^*(t_1). \tag{4.4}$$

PROOF. By (4.1) we have

$$\begin{aligned}\Big(t\frac{d}{dt}+h\Big)q^*(t) &= \Big(t\frac{\partial}{\partial t}+h\Big)q(t,\rho)\Big|_{\rho=\rho(t)} + \frac{\partial q}{\partial \rho}(t,\rho(t))\times t\frac{d\rho(t)}{dt} \\ &\le \Big[B(t,\rho)\frac{\partial q}{\partial \rho}+\frac{\partial q}{\partial \rho}(-B(t,\rho))\Big]\Big|_{\rho=\rho(t)} = 0,\end{aligned}$$

that is,

$$\Big(t\frac{d}{dt}+h\Big)q^*(t)\le 0 \quad \text{on } (t_\xi,t_0].$$

Since this is equivalent to

$$\frac{d}{dt}(t^h q^*(t))\le 0 \quad \text{on } (t_\xi,t_0],$$

by integrating this from t_1 to τ we have $\tau^h q^*(\tau)\le {t_1}^h q^*(t_1)$. This proves (4.4). □

STEP 3. By (4.2) and Lemma 7 we have

LEMMA 8. *Under the above situation, we have the following inequalities for any t_1 satisfying $t_\xi<t_1<t_0$:*

$$\xi\le\rho(t_1)\le\xi+\frac{C_1}{\kappa}({t_0}^\kappa-{t_1}^\kappa)+\frac{C_2}{h}r+\frac{3C_3}{2h}r^{2/3}+\frac{3C_4}{h}r^{1/3}. \tag{4.5}$$

PROOF. Let $t_\xi<t_1<t_0$. By (4.3) we have

$$\rho(t_1)=\xi+\int_{t_1}^{t_0}B(\tau,\rho(\tau))\frac{d\tau}{\tau}.$$

Since $B(t,\rho)\ge 0$ we have $\rho(t_1)\ge\xi$. Since $q(\tau,\rho(\tau))=q^*(\tau)$, by (4.2) and (4.4) we have

$$\begin{aligned}&B(\tau,\rho(\tau))\\ &\le C_1\tau^\kappa+C_2q^*(\tau)+C_3(q^*(\tau))^{2/3}+C_4(q^*(\tau))^{1/3}\\ &\le C_1\tau^\kappa+C_2\Big(\frac{t_1}{\tau}\Big)^h q^*(t_1)+C_3\Big(\frac{t_1}{\tau}\Big)^{2h/3}q^*(t_1)^{2/3}+C_4\Big(\frac{t_1}{\tau}\Big)^{h/3}q^*(t_1)^{1/3}\\ &\le C_1\tau^\kappa+C_2\Big(\frac{t_1}{\tau}\Big)^h r+C_3\Big(\frac{t_1}{\tau}\Big)^{2h/3}r^{2/3}+C_4\Big(\frac{t_1}{\tau}\Big)^{h/3}r^{1/3}\end{aligned}$$

for $t_1<\tau\le t_0$, and so

$$\begin{aligned}\rho(t_1)\le\xi+\int_{t_1}^{t_0}\Big(&C_1\tau^\kappa+C_2\Big(\frac{t_1}{\tau}\Big)^h r\\ &+C_3\Big(\frac{t_1}{\tau}\Big)^{2h/3}r^{2/3}+C_4\Big(\frac{t_1}{\tau}\Big)^{h/3}r^{1/3}\Big)\frac{d\tau}{\tau}.\end{aligned} \tag{4.6}$$

Since

$$\int_{t_1}^{t_0}\Big(\frac{t_1}{\tau}\Big)^a\frac{d\tau}{\tau}=\frac{1}{a}\Big(1-\frac{{t_1}^a}{{t_0}^a}\Big)\le\frac{1}{a}$$

holds for any $a>0$, by applying this to (4.6) we have (4.5). □

COROLLARY 2. *If $\xi\in(0,R/2)$ we have $t_\xi=0$.*

PROOF. Let $\xi \in (0, R/2)$. Let us show that the condition $t_\xi > 0$ leads us to a contradiction.

Suppose that $t_\xi > 0$ holds. Then, by Lemmas 6 and 8 we have

$$0 < \xi \leq \rho(t_1) \leq \xi + \frac{C_1}{\kappa}\sigma^\kappa + \frac{C_2}{h}r + \frac{3C_3}{2h}r^{2/3} + \frac{3C_4}{h}r^{1/3} = R_1 < R$$

for any t_1 satisfying $t_\xi < t_1 < t_0$. Since $K = \{\rho \in (0, R)\,;\, \xi \leq \rho \leq R_1\}$ is a compact subset of $(0, R)$ and since $\rho(t_1) \in K$ holds for any $t_1 \in (t_\xi, t_0]$, by a general theory of ordinary differential equations (for example, by Theorem 4.1 in [**2**]) we can extend $\rho(t)$ to $(t_\xi - \delta, t_0]$ for some $\delta > 0$. This contradicts the assumption that $(t_\xi, t_0]$ is the interval of existence of a maximally extended solution $\rho(t)$. □

STEP 4. Since $t_\xi = 0$, by (4.4) with $\tau = t_0$ we have

$$0 \leq q^*(t_0) \leq \left(\frac{t_1}{t_0}\right)^h q^*(t_1) \leq \left(\frac{t_1}{t_0}\right)^h r$$

for any $0 < t_1 < t_0$. Since $r > 0$ is independent of t_1, by letting $t_1 \longrightarrow 0$ we have $q^*(t_0) = 0$. Since $q^*(t_0) = q(t_0, \xi)$ we have $q(t_0, \rho) = 0$ for any $\rho \in (0, R/2)$. By the continuity we have $q(t_0, \rho) = 0$ for any $\rho \in [0, R/2)$. Since $t_0 \in (0, \sigma]$ is taken arbitrarily we have $q(t, \rho) = 0$ on $(0, \sigma] \times [0, R/2)$.

This completes the proof of Proposition 1. □

5. Application

As an application of Theorem 2, we have

THEOREM 3 (Analytic continuation). *Let $u(t, x)$ be a holomorphic solution of* (2.1) *on $S_I(r_0) \times D_{R_0}$ for some nonempty open interval I, $r_0 > 0$ and $R_0 > 0$. If $u(t, x)$ satisfies* (2.3)*, $u(t, x)$ can be continued holomorphically up to a neighborhood of $(0, 0) \in \mathbb{C}_t \times \mathbb{C}_x^n$.*

References

[1] Dennis B. Bacani and Hidetoshi Tahara, *Unique solvability of some nonlinear partial differential equations with Fuchsian and irregular singularities*, J. Math. Soc. Japan **66** (2014), no. 3, 1017–1042, DOI 10.2969/jmsj/06631017. MR3238327

[2] Earl A. Coddington and Norman Levinson, *Theory of ordinary differential equations*, McGraw-Hill Book Co., Inc., New York-Toronto-London, 1955. MR0069338

[3] Raymond Gérard and Hidetoshi Tahara, *Solutions holomorphes et singulières d'équations aux dérivées partielles singulières non linéaires* (French), Publ. Res. Inst. Math. Sci. **29** (1993), no. 1, 121–151, DOI 10.2977/prims/1195167545. MR1208031

[4] Raymond Gérard and Hidetoshi Tahara, *Singular nonlinear partial differential equations*, Aspects of Mathematics, Friedr. Vieweg & Sohn, Braunschweig, 1996, DOI 10.1007/978-3-322-80284-2. MR1757086

[5] Hidetoshi Tahara, *Uniqueness of the solution of non-linear singular partial differential equations*, J. Math. Soc. Japan **48** (1996), no. 4, 729–744, DOI 10.2969/jmsj/04840729. MR1404820

[6] Hidetoshi Tahara, *Uniqueness of the solution of nonlinear singular first order partial differential equations*, J. Math. Soc. Japan **72** (2020), no. 4, 1259–1282, DOI 10.2969/jmsj/82788278. MR4165932

[7] Hidetoshi Tahara and Hiroshi Yamazawa, *Structure of solutions of nonlinear partial differential equations of Gérard-Tahara type*, Publ. Res. Inst. Math. Sci. **41** (2005), no. 2, 339–373. MR2138028

SOPHIA UNIVERSITY, TOKYO 102-8554, JAPAN
Current address: Yatsu 3-1-21-403, Narashino-shi, Chiba 275-0026, Japan
Email address: h-tahara@sophia.ac.jp

Contemporary Mathematics
Volume **782**, 2023
https://doi.org/10.1090/conm/782/15730

Solution with movable singular points of some Hamiltonian system

Masafumi Yoshino

ABSTRACT. This paper studies the movable singular points of a solution of the Hamiltonian system with the Hamiltonian $H := H_0 + H_1$. Here H_0 is integrable and the Hamiltonian system of H_0 has a singular solution v, while H_1 is a perturbation. A singular solution for H is given as $F(v)$ for some transformation F being defined in the domain of the phase space which contains the orbit of v tending to the infinity. The map F is given by the homology equation via the global Borel summability.

1. Introduction

Let (q_2, p_2) be the variable in $\mathbb{C}^2$. Let H_0 be a polynomial of q_2 and p_2. Let H_1 be an entire function of q_2 and p_2 analytic in q_1 in the domain $U_1 \subset \mathbb{C}$. Define $H := H_0(q_2, p_2) + H_1(q_1, q_2, p_2)$. Consider the Hamiltonian system

$$\frac{dq_2}{dq_1} = \frac{\partial H}{\partial p_2}, \quad \frac{dp_2}{dq_1} = -\frac{\partial H}{\partial q_2}. \tag{1.1}$$

By the movable singular point of (1.1) we mean the singular point of a solution of the initial value problem of (1.1) which does not appear in (1.1) and that depends on an initial value. (cf. [**FH**]). This paper studies the movable singular points of a solution of (1.1).

Assume that the Hamiltonian system for H_0 has a solution v with a movable singular point. To construct a singular solution of (1.1) we transform the Hamiltonian vector field of H_0 to that of H. The transformation is given by the first order system of partial differential equations, a homology equation. We construct the transformation as a formal power series in a certain variable. Then, by taking the global Borel sum, we obtain a diffeomorphism on some domain of the phase space containing the orbit of v about the singular point. (cf. [**KT**] for a Painlevé equation). The transformation is closely related with the symplectic transformation. (cf. Remark 3.7).

2020 *Mathematics Subject Classification.* Primary 35C10; Secondary 45E10.

Key words and phrases. Movable singular point, Hamiltonian system, homology equation, global Borel summability.

This work was supported by JSPS KAKENHI Grant Number 20K03683 and the Research Institute for Mathematical Sciences, an International Joint Usage/Research Center located in Kyoto University.

This paper is organized as follows. In §2 we state the main result. In §3 we show the solvability of the homology equation via the global Borel summability. In §4 we consider the reduced Hamiltonian system. In §5 we prove the main theorem. In §6 we give examples related to the generalized Emden-Fowler equation. In §7 we show the solvability of the reduced homology equation via the Borel summability. In the appendix we state the properties of the global Borel summability of a formal series solution of the reduced system of the homology equation in the domain of a phase space.

2. Statement of result

Let (q,p) be the variable in $\mathbb{C}^{2n}$. Set $q=(q_1,\ldots,q_n)$ and $p=(q_1,\ldots,p_n)$. Let $H\equiv H(q,p)$ be an entire function of q and p. Consider the Hamiltonian system

$$\frac{dq_j}{dz}=\frac{\partial H}{\partial p_j},\ \frac{dp_j}{dz}=-\frac{\partial H}{\partial q_j},\quad j=1,2,\ldots,n. \tag{2.1}$$

Let a function F be given. Define the Hamiltonian vector field χ_F by

$$\chi_F:=\sum_{j=1}^{n}\left(\frac{\partial F}{\partial p_j}\frac{\partial}{\partial q_j}-\frac{\partial F}{\partial q_j}\frac{\partial}{\partial p_j}\right). \tag{2.2}$$

The Poisson bracket of functions F and G is defined by $\{F,G\}:=\chi_F G$. We say that a function F is the first integral of χ_H if $\chi_H F=0$. We say that the Hamiltonian system is integrable if there exist n first integrals ϕ_j in involution, $\{\phi_i,\phi_j\}=0$ such that ϕ_j's are almost everywhere functionally independent.

In what follows we consider Hamiltonian systems with $n=2$. Let $i>j\geq 1$ be integers. Define

$$H_0:=\lambda q_2p_2+q_2^ip_2^j, \tag{2.3}$$

where $\lambda\neq 0$ is a constant. Let $H_1\equiv H_1(q_1,q_2,p_2)$ be an entire function of q_2 and p_2 analytic in $q_1\in\Omega_2$ for some domain $\Omega_2\subset\mathbb{C}$ containing the origin. Assume that the Taylor expansion of H_1 at the origin $q_2=0$, $p_2=0$ vanishes up to order 1. Define H by $H:=H_0+H_1$. Let $K_0>0$, $\epsilon_0>0$ and $r_0>0$ be constants. Define Σ_0 by

$$\begin{aligned}\Sigma_0:=&\{(q_2,p_2)|\ q_2^{-1}=\alpha r, p_2=\beta r,\ |\alpha|<\epsilon_0,\ |\beta|<\epsilon_0,\\ &0<r<r_0,|q_2^{-1}|\leq K_0,|p_2|\leq K_0\}.\end{aligned} \tag{2.4}$$

We assume

$$\frac{\partial^2 H}{\partial p_2^2}(q_1,q_2,p_2)\neq 0,\quad \forall q_1\in\Omega_2,\ \forall(q_2,p_2)\in\Sigma_0. \tag{2.5}$$

Set $\nu=i-j$. Consider

$$\frac{\partial H}{\partial u_4}(v_1,u_3,u_4)=0, \tag{2.6}$$

$$\frac{\partial H}{\partial u_3}(v_1,u_3,u_4)=0. \tag{2.7}$$

We assume

(A.1) The domain Ω_2 contains the segment $[0,\frac{2\pi i}{\lambda\nu}]$. (2.6) and (2.7) have a solution $(u_3,u_4)=(w_0(v_1),\tilde{w}_0(v_1))$ holomorphic in $v_1\in\Omega_2$ and $w_0\neq 0$, $\tilde{w}_0\neq 0$.

(A.2) $w_0(0)=w_0(\frac{2\pi i}{\lambda\nu})=0,\quad \tilde{w}_0(0)=\tilde{w}_0(\frac{2\pi i}{\lambda\nu})=0.$

Then we have

THEOREM 2.1. *Assume that the conditions* (2.5), *(A.1) and (A.2) are satisfied. Moreover, suppose that* $j = 1$ *in* (2.3). *Then there exist* $\delta_0 > 0$, $K_1 > 0$ *and a neighborhood* $\Omega_1(\subset \Omega_2)$ *of* $[0, \frac{2\pi i}{\lambda\nu}]$ *such that, if*

$$\sup_{q_1\in\Omega_2, |q_2|\le K_1, |p_2|\le K_1} (|w_0(q_1)| + |\tilde{w}_0(q_1)| + |H_1(q_1, q_2, p_2)|) < \delta_0, \tag{2.8}$$

then (1.1) *has a solution* (q_2, p_2) *having at least two movable singular points in* Ω_1.

3. Homology equation

Let $x = (\tilde{q}_1, \tilde{p}_1, \tilde{q}_2, \tilde{p}_2)$ and $y = (q_1, p_1, q_2, p_2)$ be the variables. Write $x = (x_1, \ldots, x_4)$ and $y = (y_1, \ldots, y_4)$, for simplicity. Let H_0 be given by (2.3). Let H_1 be given in Theorem 2.1. Consider $p_1 + H(q_1, q_2, p_2)$, where $H = H_0 + H_1$. We transform $\chi_{p_1+H_0}$ to χ_{p_1+H} by $x = u(y)$, $u = (u_1, \ldots, u_4)$.

For $\phi \equiv \phi(q_1, q_2, p_2)$, define

$$X_0 := \chi_{p_1+\lambda q_2 p_2} \quad Y := \chi_\phi = \sum_j u_j \frac{\partial}{\partial y_j}. \tag{3.1}$$

The component vector of X_0 with respect to the basis $\partial/\partial q_1$, $\partial/\partial p_1$, $\partial/\partial q_2$, $\partial/\partial p_2$ is given by $(1, 0, \lambda q_2, -\lambda p_2) =: \Lambda(y)$. Set

$$f(x) := H(\tilde{q}_1, \tilde{q}_2, \tilde{p}_2) - \lambda\tilde{q}_2\tilde{p}_2, \tag{3.2}$$

(f is independent of $x_2 = \tilde{p}_1$) and define

$$R := \chi_f, \qquad S := \chi_{q_2^i p_2^j}. \tag{3.3}$$

Let $r(x)$ and $s(y)$ be the component vectors of R and S, respectively. The first and the second components of $s(y)$ vanish. The term $r(x) = \Lambda(u(y)))$ is independent of x_2. We have

LEMMA 3.1. *Let* $u = u(y)$ *satisfy*

$$\Lambda(y)\nabla u + s(y)\nabla u = r(u) + \Lambda(u), \tag{3.4}$$

such that $(\nabla u)^{-1}$ *exists. Then,* $x = u(y)$ *transforms* $(\Lambda(y) + s(y))\frac{\partial}{\partial y}$ *to* $(\Lambda(x) + r(x))\frac{\partial}{\partial x}$. *Here* ∇u *is the nambla of* u *with respect to* $(y_1, \ldots, y_4)$ *and* $\Lambda(u) = (1, 0, \lambda u_3, -\lambda u_4)$.

REMARK 3.2. (3.4) is referred as a homology equation.

PROOF.

$$(\Lambda(x) + r(x))\frac{\partial}{\partial x} = (\Lambda(u) + r(u))(\nabla u)^{-1}\frac{\partial}{\partial y} = (\Lambda(y) + s(y))\frac{\partial}{\partial y}.$$

□

Deduction of the homology equation. Consider the third row of (3.4). Set $w = u_3$, $u = (u_1, \cdots, u_4)$. By $\Lambda(y) = (1, 0, \lambda q_2, -\lambda p_2)$ we have

$$\frac{\partial w}{\partial q_1} + \delta w - iq_2^{i-1}p_2^j\frac{\partial w}{\partial p_2} + jq_2^i p_2^{j-1}\frac{\partial w}{\partial q_2} - \lambda w - \mathcal{R}(u) = 0, \tag{3.5}$$

where $\delta = \lambda(q_2\frac{\partial}{\partial q_2} - p_2\frac{\partial}{\partial p_2})$ and $\mathcal{R}(u) = (\partial f/\partial\tilde{p}_2)(u)$.

We consider the first order terms of the left-hand side of (3.5) except for the first term. They are given by the Hamiltonian vector field of H_0. By taking t as the time variable we write it as the system of ordinary differential equations

$$\dot{q}_2 = \lambda q_2 + j q_2^i p_2^{j-1}, \quad \dot{p}_2 = -\lambda p_2 - i p_2^j q_2^{i-1}. \tag{3.6}$$

Set $\nu = i - j$. By the change of the unknown functions

$$q_2 = 1/r, \quad p_2 = rv, \tag{3.7}$$

we have

$$\dot{v} = -\nu r^{-\nu} v^j, \quad \dot{r} = -\lambda r - j v^{j-1} r^{1-\nu}. \tag{3.8}$$

Because $\lambda q_2 p_2 + q_2^i p_2^j$ is the first integral of (3.6), there exists a constant c_1 such that $\lambda q_2 p_2 + q_2^i p_2^j = c_1$. Then, by (3.7) we have $\lambda v + v^j / r^\nu = c_1$. By (3.8) we obtain $\dot{v} - \lambda \nu v + c_1 \nu = 0$, from which we have

$$v = c_1 \lambda^{-1} + c_2 e^{\lambda \nu t}, \tag{3.9}$$

where c_2 is a constant.

By (3.8) and $\dot{v} - \lambda \nu v + c_1 \nu = 0$ we write (3.5) in the form

$$\frac{\partial w}{\partial q_1} + \nu(\lambda v - c_1)\frac{\partial w}{\partial v} - (\lambda r + j v^{j-1} r^{1-\nu})\frac{\partial w}{\partial r} - \lambda w - \mathcal{R}_3(u) = 0, \tag{3.10}$$

where $\mathcal{R}_3(u)$ is the third component of $R(u)$.

Consider the curve $(q_1, v(q_1))$ in the (q_1, v)- space, where $v(q_1)$ is given by (3.9). Then the vector field $\frac{\partial}{\partial q_1} + \nu(\lambda v - c_1)\frac{\partial}{\partial v}$ is written as the vector filed $\frac{\partial}{\partial \tilde{q}_1}$ with respect to the new time variable $\tilde{q}_1$ given by the curve. If there is no fear of confusion we write the new time variable by q_1 instead of $\tilde{q}_1$ and write the corresponding vector field by $\frac{\partial}{\partial q_1}$. Then, by inserting the relation into (3.10) and by multiplying the equation with $r^{\nu-1}$ we obtain

$$r^{\nu-1}\frac{\partial w}{\partial \tilde{q}_1} - (\lambda r^\nu + j v^{j-1})\frac{\partial w}{\partial r} - r^{\nu-1}(\lambda w + \mathcal{R}_3(u)) = 0. \tag{3.11}$$

Define

$$h = -j v^{j-1}. \tag{3.12}$$

Set $c_0 = h^{-1}$. Then we have

$$h\left(r^{\nu-1} c_0 \frac{\partial u_3}{\partial \tilde{q}_1} + (1 - \lambda c_0 r^\nu)\frac{\partial u_3}{\partial r}\right) - r^{\nu-1}\left(\lambda u_3 + \mathcal{R}_3(u)\right) = 0, \tag{3.13}$$

where $u = (u_1, 0, u_3, u_4)$. Similarly, $w = u_4$ satisfies

$$h\left(r^{\nu-1} c_0 \frac{\partial u_4}{\partial \tilde{q}_1} + (1 - \lambda c_0 r^\nu)\frac{\partial u_4}{\partial r}\right) + r^{\nu-1}\left(\lambda u_4 - \mathcal{R}_4(u)\right) = 0, \tag{3.14}$$

where $\mathcal{R}_4(u)$ is the fourth component of $R(u)$.

We consider the solvability of the reduced homology equation. Consider the equation of u_1. We have

LEMMA 3.3. *u_1 is given by*

$$u_1 = -\tilde{u_1} - \frac{2}{\lambda \nu}\log(1 - \lambda c_0 r^\nu), \quad \tilde{u_1} = \frac{q_1}{2} + \frac{1}{2\lambda \nu}\log(\nu \lambda v - c_1 \nu). \tag{3.15}$$

PROOF. Because the first component of $r(u)$ vanishes, we replace $\lambda w + \mathcal{R}(u)$ in (3.13) with 1. Then the first coumn of (3.4) gives the equation of $u_1 =: w$

$$r^{\nu-1}c_0 \frac{\partial w}{\partial \tilde{q}_1} + (1 - \lambda c_0 r^\nu)\frac{\partial w}{\partial r} - r^{\nu-1}c_0 = 0,$$

from which the assertion follows. □

In u_1 we set $-c_1\nu = 1$. Then we have $\tilde{u}_1 = q_1/2$ when $v = 0$. Insert u_1 into $\mathcal{R}_3(u)$ and $\mathcal{R}_4(u)$, respectively. We assume

(A.1)' $1 - \lambda c_0 r^\nu \neq 0$ for every $r \in U_0$. U_0 is a bounded set in the universal covering space of the logarithm, $\log(1 - \lambda c_0 r^\nu)$.

Let $I_0 := [0, \frac{2\pi i}{\lambda\nu}]$. Then the solvability of (3.13)-(3.14) follows from the next theorem.

THEOREM 3.4. *Assume that the conditions (A.1)', (A.1) and (A.2) are satisfied. Then there exist $\delta_0 > 0$, $L_0 > 0$, $K_1 > 0$, $h_0 > 0$, $0 < h_1 < h_0$, a neighborhood Ω_1 of I_0, a neighborhood U_1 of $v(I_0)$ and a sector S_0 in h-variable with vertex at the origin such that, if the conditions $h_1 < |h| < h_0$, $h \in S_0$ and* (2.8) *hold, then* (3.13)-(3.14) *has an analytic solution, (u_3, u_4) when $(q_1, v, r) \in \Omega_1 \times U_1 \times (U_0 \cap \{r| \, |c_0 r| < L_0\})$.*

Theorem 3.4 follows from Theorem 7.1 which follows.

REMARK 3.5. We can choose h_0 in Theorem 3.4 arbitrarily large.

Consider the solvability of (3.4). Note that u_2 is given by the integration of the second equation of (3.4) if we know u_3 and u_4. By the analytic change of variables and the reduction argument it is sufficient to consider the solvability of the reduced homology equation (3.13) - (3.14). By Theorem 3.4 we have

THEOREM 3.6. *Let $p_1^{(0)} \in \mathbb{C}$ be given. Assume that the conditions (A.1)', (A.1) and (A.2) are satisfied. Then there exist $\delta_0 > 0$, $L_0 > 0$, $K_1 > 0$, $\delta_1 > 0$, $h_0 > 0$, $0 < h_1 < h_0$, a neighborhood Ω_1 of I_0, a neighborhood U_1 of $v(I_0)$ and a sector S_0 in h-variable with vertex at the origin such that, if the conditions $h_1 < |h| < h_0$, $h \in S_0$, $|c_0 + 1| < \delta_0$ and (2.8) are satisfied, then (3.4) has an analytic solution, u when $(q_1, v, r) \in \Omega_1 \times U_1 \times (U_0 \cap \{r| \, |c_0 r| < L_0\})$, $|p_1 - p_1^{(0)}| < \delta_1$.*

We note that, in the application, we use Theorem 3.6 when $c_0 = h^{-1} = -1$. Since we can take h_0 sufficiently large, h also satisfies $|h| < h_0$.

REMARK 3.7. Let $u = (u_1, \ldots, u_4)$ be the solution of the homology equation (3.4). We show that (u_3, u_4) is the pull-back of a certain symplectic transformation if (2.5) holds. Especially, this implies that $(\nabla u)^{-1}$ exists. In order to see this, divide (3.13) with $s^{\nu-1}$. Consider the characteristic equation corresponding to the resultant equation

$$\frac{dq_1}{1} = \frac{c_0 s^{\nu-1} ds}{1 - \lambda c_0 s^\nu}. \tag{3.16}$$

Let t be the independent variable of the solution of (3.16). Then we have

$$\frac{du_3}{dt} = \frac{dq_1}{dt}\frac{\partial u_3}{\partial q_1} + \frac{ds}{dt}\frac{\partial u_3}{\partial s} = \frac{\partial u_3}{\partial q_1} + \frac{1 - \lambda c_0 s^\nu}{c_0 s^{\nu-1}}\frac{\partial u_3}{\partial s}. \tag{3.17}$$

Multiply (3.13) with $s^{1-\nu}$ and substitute (3.17). Make the same calculation as for u_4 with (3.13) replaced by (3.14). The reduced system of equations is the

Hamiltonian system with the Hamiltonian $\lambda\tilde{q_2}\tilde{p_2} + f = H$. Let (u_3, u_4) be the solution with the initial value $(u_3, u_4) = (\xi, \eta)$ at $t = 0$. By Hamilton-Jacobi's theory the map $(\xi, \eta) \mapsto (u_3, u_4)$ is symplectic if the non degeneracy condition $(\partial u_3/\partial\eta)(t, \xi, \eta) \neq 0$ holds. Changing the coordinate from (t, ξ, η) to (q_1, q_2, p_2), u_3 and u_4 give the transformation in Lemma 3.1.

By (2.5) we choose $s \in \Omega_1$, $|\xi| < K_0$, $|\eta| < K_0$ such that $(\partial R_3/\partial u_4)(s, \xi, \eta) = \partial^2 H/\partial p_2^2 \neq 0$. Then we verify $(\partial u_3/\partial\eta)(s, \xi, \eta) \neq 0$. Since (u_3, u_4) is the solution of the Hamiltonian system, it holds that $u_3 = \xi + \tilde{u_3}$, $\tilde{u_3} = O(t)$ and $u_4 = \eta + \tilde{u_4}$, $\tilde{u_4} = O(t)$ as $t \to 0$. Here ξ (resp. η) is the initial value of u_3 (resp. u_4). On the other hand, since they are equal to the Borel sum (cf. Theorems 7.1 and 3.4 which follow) we have

$$u_3 = u_3^{(0)} + v_3,\ v_3 = O(h),\ u_4 = u_4^{(0)} + v_4,\ v_4 = O(h), \tag{3.18}$$

for h in (3.12). $u_3^{(0)}$ and $u_4^{(0)}$ are the analytic solution of the equation (cf. (7.6))

$$\lambda u_3^{(0)} + \mathcal{R}_3(u_1, u_3^{(0)}, u_4^{(0)}) = 0. \tag{3.19}$$

Differentiating (3.19) with respect to η we have

$$\lambda\frac{\partial u_3^{(0)}}{\partial\eta} + \frac{\partial\mathcal{R}_3}{\partial u_3}(u_1, u_3^{(0)}, u_4^{(0)})\frac{\partial u_3^{(0)}}{\partial\eta} + \frac{\partial\mathcal{R}_3}{\partial u_4}(u_1, u_3^{(0)}, u_4^{(0)})\frac{\partial u_4^{(0)}}{\partial\eta} = 0. \tag{3.20}$$

We have that $|u_3^{(0)} - \xi|$ and $|u_4^{(0)} - \eta|$ are arbitrarily small if h is sufficiently small. Then we have $(\partial\mathcal{R}_3/\partial u_4)(u_1, u_3^{(0)}, u_4^{(0)}) \neq 0$ by assumption. Since $\partial\mathcal{R}_3/\partial u_3$ is small, it follows from (3.20) that $\partial u_3^{(0)}/\partial\eta \neq 0$. To show the assertion by (3.18) we prove that $(\partial v_3/\partial\eta)$ is sufficiently small. Indeed, the smallness follows (2.8).

4. Study of reduced equation

Let H_0 be given by (2.3). Consider the Hamiltonian system with Hamiltonian $p_1 + H_0$ given by (3.6). Set $\nu = i - j$. By the change of unknown functions (3.7), we obtain (3.8). Then we have

LEMMA 4.1. *The solution of* (3.6) *is given by* (3.7), (3.9) *and* $r = \left(-\nu v^j/\dot{v}\right)^{1/\nu}$. *The solution* (q_2, p_2) *of* (3.6) *has infinitely many movable poles (resp. branch points) at the point given by* $v = c_1\lambda^{-1} + c_2 e^{\lambda\nu t} = 0$ *if* $\nu = 1$ *(resp.* $\nu \geq 2$*). As* t *tends to the singular point from some sector,* q_2 *tends to infinity while* p_2 *tends to zero.*

5. Proof of Theorem 2.1

PROOF. The proof consists of three steps.

STEP 1. Consider $\chi_{p_1+H_0}$. By Lemma 4.1 the Hamiltonian system of H_0 has a solution (q_2, p_2) given by

$$q_2 = 1/r,\ p_2 = rv,\ v = v(t) = c_1/\lambda + c_2 e^{\lambda\nu t},\quad r = (-\nu v^j/\dot{v})^{1/\nu}, \tag{5.1}$$

where $\dot{v} = dv/dt$ and $t = q_1$. By Lemma 4.1 q_2 is singular at the point t_0 such that $v(t_0) = 0$. Hence t_0 is given by

$$e^{\lambda\nu t_0} = -c_1/(\lambda c_2),\ t_0 = (\lambda\nu)^{-1}\log(-c_1/(c_2\lambda)),\ \dot{v}(t_0) = -\nu c_1. \tag{5.2}$$

Let $c_1 \neq 0$ and $c_2 \neq 0$ such that $-c_1/(c_2\lambda)$ is in a small neighborhood of 1. Fix the branch of the logarithm such that t_0 is in a sufficiently small neighborhood of the origin. Then q_2 has the singular points t_0 and $t_0 + 2\pi i(\lambda\nu)^{-1} =: t_1$ in some neighborhoods of the origin and $2\pi i(\lambda\nu)^{-1}$, respectively. Let $\epsilon_0 > 0$ be a small constant. Define the path C_0 by $C_0 = \{t| \ \lambda\nu t = \lambda\nu t_0 + i\theta, \ \epsilon_0 \leq \theta \leq 2\pi - \epsilon_0\}$. Then (5.1) is analytically continued in some neighborhood of C_0. By replacing Ω_1 with a smaller one, if necessary, we assume that (5.1) is analytically continued in Ω_1.

STEP 2. Suppose that the conditions of Theorem 3.6 are satisfied. Then the homology equation (3.4) is solvable. By Lemma 3.1, (2.5) and Remark 3.7, χ_{p_1+H} is the pull back of $\chi_{p_1+H_0}$. Hence we have a solution of the Hamiltonian system of H in Ω_1. By (A.2) the diffeomorphism preserves the infinity at the points $q_1 = 0$ and $q_1 = 2\pi i/(\lambda\nu)$. Since the flow of $\chi_{p_1+H_0}$ goes to infinity at the singular points, the solution of the Hamiltonian system of $p_1 + H$ has the singular points in some neighborhood of $q_1 = 0$ and $q_1 = 2\pi i/(\lambda\nu)$, respectively. Hence, we verify the conditions $|c_0 + 1| < \delta_0$, $|c_0 r| < L_0$, $r \in U_0$, $h_1 < |h| < h_0$ and $h \in S_0$ in Theorem 3.6.

STEP 3. We set $j = 1$ and $\nu = 1$. By (5.1) we have

$$r = \frac{v}{\lambda c_2 e^{\lambda t}} = \frac{1}{\lambda}\left(\frac{c_1}{\lambda c_2 e^{\lambda t}} + 1\right). \tag{5.3}$$

Since $c_1/\lambda c_2$ is close to -1 and $t = q_1$ moves on the bounded set, there exists K_1 such that $|\lambda r| \leq K_1$. Hence, by (5.3) we have

$$|v| \leq |\lambda||c_2 e^{\lambda t}||r| \leq K_1 |c_2 e^{\lambda t}| \leq K_1 |c_1| K_2, \tag{5.4}$$

for some constant K_2 independent of t. Hence $|v|$ is uniformly bounded in t. If we make the symplectic transformation $q_2 \mapsto \epsilon q_2$, $p_2 \mapsto \epsilon^{-1} p_2$ for small $\epsilon > 0$, then we see that r is replaced by ϵr, while v is unchanged. Hence we may assume that q_2^{-1} and p_2 are sufficiently small.

In view of Step 1 we see that q_2^{-1} and p_2 are given by r and rv, respectively. By (5.1) the orbit $(q_2^{-1}, p_2) = (r, rv)$ of $\chi_{p_1+H_0}$ stays in Σ_0 for t in some neighborhood of the path of analytic continuation. Let h be in some small neighborhood of -1. Then we have $|c_0 r| < L_0$ and $r \in U_0$ since U_0 contains some neighborhood of the origin. In order to show $h \in S_0$ we choose the bisecting direction of S_0 such that $-1 \in S_0$. The condition $|h| < h_0$ follows from Remark 3.5 since h_0 can be chosen arbitrarily large. □

6. Example

EXAMPLE 6.1. Consider the generalized Emden-Fowler equation

$$\frac{d^2 v}{dt^2} + A(t)\frac{dv}{dt} + v^2 = 0, \quad t \in \mathbb{C}, \tag{6.1}$$

where $A(t)$ is given by

$$A(t) = \frac{1}{t} + t. \tag{6.2}$$

Set $v = wu$ with $w = \exp(-2^{-1} \int^t A(x)dx$. Then u satisfies

$$\frac{d^2 u}{dt^2} - \frac{(A^2 + 2A')u}{4} + \exp\left(-\frac{1}{2}\int^t A\right) u^2 = 0. \tag{6.3}$$

Set $q := u$ and $p := u_t$, and write (6.3) in the Hamiltonian system with Hamiltonian

$$H := \frac{1}{2}\left(p^2 - \frac{A^2+2A'}{4}q^2\right) + \exp\left(-2^{-1}\int^t A\right)\frac{q^3}{3}. \tag{6.4}$$

By a simple calculation we have

$$-\frac{A^2+2A'}{4} = \frac{1}{4t^2} - \frac{t^2}{4} - 1, \quad -2^{-1}\int^t A = t^{-1/2}\exp(-t^2/4).$$

Let $t_0 \in \mathbb{C}$, $t_0 \neq 0$. Write

$$\begin{aligned} H = {} & \frac{1}{2}\left(p^2 + \left(\frac{1}{4t_0^2} - \frac{t_0^2}{4} - 1\right)q^2\right) - \frac{1}{2}\left(\frac{1}{4t_0^2} - \frac{t_0^2}{4}\right)q^2 \\ & + \frac{1}{2}\left(\frac{1}{4t^2} - \frac{t^2}{4}\right)q^2 + t^{-1/2}\exp(-t^2/4)\frac{q^3}{3}. \end{aligned} \tag{6.5}$$

We take the linear change of variables $(q,p) \mapsto (q_2, p_2)$ such that

$$\frac{1}{2}\left(p^2 + \left(\frac{1}{4t_0^2} - \frac{t_0^2}{4} - 1\right)q^2\right) = \lambda q_2 p_2, \quad q = c_1 q_2 + c_2 p_2, \tag{6.6}$$

for some nonzero constants λ, c_1 and c_2. By (6.5) we have

$$H = \lambda q_2 p_2 + B_1(t)(c_1 q_2 + c_2 p_2)^2 + B_2(t)\frac{(c_1 q_2 + c_2 p_2)^3}{3}, \tag{6.7}$$

where

$$B_1(t) = \frac{1}{8}\left(\frac{1}{t^2} - t^2 - \frac{1}{t_0^2} + t_0^2\right), \quad B_2(t) = t^{-1/2}\exp(-t^2/4).$$

Set $q_2 = u_3$, $p_2 = u_4$ and $t = v_1$. Then (2.6) and (2.7) are given, respectively, by

$$\lambda u_3 + B_1(v_1)2c_2(c_1u_3 + c_2u_4) + B_2(v_1)c_2(c_1u_3 + c_2u_4)^2 = 0, \tag{6.8}$$

$$\lambda u_4 + B_1(v_1)2c_1(c_1u_3 + c_2u_4) + B_2(v_1)c_1(c_1u_3 + c_2u_4)^2 = 0. \tag{6.9}$$

By subtraction we have $c_1u_3 = c_2u_4$. Eliminate u_4 from (6.8). Since $u_3 \neq 0$ we obtain

$$\lambda + 4c_1c_2B_1(v_1) + 4c_1^2c_2B_2(v_1)u_3 = 0. \tag{6.10}$$

Hence u_3 is holomorphic if $v_1 \neq 0$. As for the zero of u_3, we consider $\lambda + 4c_1c_2B_1(v_1) = 0$. It is the algebraic equation of v_1^2 of degree 2. Choose a pair of zero points in the v_1-plane which are not symmetric about the origin. By a suitable translation and a linear change of variables in t we may assume that they are given by 0 and $2\pi i/(\lambda\nu)$. Hence u_3 is holomorphic in some small neighborhood of the segment joining 0 and $2\pi i/(\lambda\nu)$, which proves (A.1) and (A.2).

The condition (2.5) holds if $B_1(t) \neq 0$ and (q_2, p_2) is sufficiently small. The above argument is applicable to the equation

$$\frac{d^2w}{dz^2} = \sum_{n=1}^{N} b_n(z)w^n, \tag{6.11}$$

where $N \geq 2$ is an integer and $b_n(z)$'s are analytic in some domain $\Omega_1 \subset \mathbb{C}$.

7. Borel summability

7.1. Definition. Let Ω_1, U_0 and U_1 be the domain in $\mathbb{C}$. Consider the formal power series of h

$$v(q_1, v, r, h) = \sum_{n=1}^{\infty} v_n(q_1, v, r)h^n, \tag{7.1}$$

where $v_n(q_1, v, r)$ is holomorphic in the domain $(q_1, v, r) \in D_0 \equiv \Omega_1 \times U_1 \times U_0$. Define the formal Borel transform $\mathcal{B}$ by

$$\mathcal{B}(v)(q_1, v, r, y) := \sum_{n=1}^{\infty} \frac{v_n(q_1, v, r)}{(n-1)!} y^{n-1}, \tag{7.2}$$

where y is a dual variable of h.

Denote the set of nonnegative real numbers by $\mathbb{R}_+$. Let $dist(z, \mathbb{R}_+ e^{i\xi})$ be the distance from z to $\mathbb{R}_+ e^{i\xi}$. For $\theta > 0$ and a real number ξ, define

$$E(\xi, \theta) := \left\{ z \in \mathbb{C} \,\middle|\, dist(z, \mathbb{R}_+ e^{i\xi}) < 2^{-1}\theta \right\}. \tag{7.3}$$

Define the sector with the bisecting direction $\xi \in \mathbb{R}$ and the opening $\theta > 0$ by $\{z \in \mathbb{C} \setminus 0 \,|\, |\arg z - \xi| < \theta/2\}$.

We say that $v(q_1, v, r, h)$ is (fine) Borel summable in the direction ξ if there exists $\theta > 0$ such that $B(v)(q_1, v, r, y)$ converges when $(q_1, v, r) \in D_0$ and y is in some neighborhood of $y = 0$, and $B(v)(q_1, v, r, y)$ is analytically continued to $D_0 \times E(\xi, \theta)$ with exponential type of order 1 in $y \in E(\xi, \theta)$ for $(q_1, v, r) \in D_0$. Namely, there exist $K_0 > 0$ and $K_2 > 0$ such that

$$|B(v)(q_1, v, r, y)| \le K_0 e^{K_2|y|}, \quad y \in E(\xi, \theta), \quad (q_1, v, r) \in D_0.$$

For simplicity, we denote the analytic continuation with the same notation. The Borel sum of $v(q_1, v, r, h)$, $V(q_1, v, r, h)$ is given by the Laplace transform

$$V(q_1, v, r, h) := \int_0^{\infty e^{i\xi}} e^{-yh^{-1}} B(v)(q_1, v, r, y) dy. \tag{7.4}$$

7.2. Construction of formal solution. We assume (A.1)' and (A.1). Let $q_1 \in \Omega_1$ and $v \in U_1$. Let $L_0 > 0$ be given. Let r satisfy $|c_0 r| \le L_0$. Substitute u_1 into (3.13) and (3.14). Construct a formal solution of (3.13) - (3.14), (u_3, u_4) in the form

$$u_3 = \sum_{n=0}^{\infty} w_n(q_1, v, r)h^n, \quad u_4 = \sum_{n=0}^{\infty} \tilde{w}_n(q_1, v, r)h^n. \tag{7.5}$$

Here w_0 and $\tilde{w}_0$ are determined, respectively, by

$$\lambda w_0 + \mathcal{R}_3(u_1, w_0, \tilde{w}_0) = 0, \quad \lambda \tilde{w}_0 - \mathcal{R}_4(u_1, w_0, \tilde{w}_0) = 0. \tag{7.6}$$

w_0 (resp. $\tilde{w}_0$) is a holomorphic funtion of $(q_1, v, r) \in D_0$ by assumption. Then we determine w_n and $\tilde{w}_n$, inductively. Note that there is a division by $r^{\nu-1}$ in each step when we determine w_n.

We prove that w_n's ($n \ge 1$) are holomorphic functions of r^ν at the origin. Consider w_1. Multiply (3.13) with r. We calculate w_1 by dividing the next quantity with r^ν

$$r^\nu c_0 \frac{\partial w_0}{\partial \tilde{q}_1} + (1 - \lambda c_0 r^\nu) r \frac{\partial w_0}{\partial r}, \tag{7.7}$$

and by inverting the bounded quantity appearing from $\lambda + \nabla\mathcal{R}_3$ (resp. $\lambda - \nabla\mathcal{R}_4$). We note that (7.7) is equal to r^ν times some holomorphic function. Indeed, since w_0 is analytic in u_1, (3.15) implies that $r\frac{\partial w_0}{\partial r}$ is divisable by r^ν. It follows that w_1 is analytic in r^ν.

Next, we consider w_2. Similarly, we investigate $r\frac{\partial w_1}{\partial r}$. Since $r\frac{\partial}{\partial r}(1-\lambda c_0 r^\nu)^{-1}$ is divisable by r^ν, $r\frac{\partial w_1}{\partial r}$ has the same property. Hence w_2 is analytic in r^ν. Inductively, we see that w_n's are analytic in r^ν.

Concerning the estimate of w_n, we can easily show that there exist constants $K_1 > 0$ and $\rho > 0$ independent of n such that $|w_n| \le K_1 \rho^n n!$ by Nagumo's lemma.

7.3. Summability. We have

THEOREM 7.1. *Assume that the conditions (A.1)', (A.1) and (A.2) are satisfied. Then there exist $\delta_0 > 0$, a neighborhood of I_0, $\Omega_1 \subset \Omega_2$, a neighborhood U_1 of $v(I_0)$, $K_1 > 0$, and $L_0 > 0$ such that, if (2.8) holds, then (7.5) of (3.13) - (3.14) is Borel summable in the direction 0 for $(q_1, v, r) \in \Omega_1 \times U_1 \times (U_0 \cap \{r \mid |c_0 r| < L_0\})$.*

Theorem 7.1 is proved by the same argument as in [**Y2**]. We give the sketch of the proof in the appendix.

Appendix A. Global Borel summability

A.1. Basic properties. *Convolution.* For $0 < \theta < \pi$, set $\Omega := E(\pi, \theta)$. Define $H(\Omega)$ as the set of holomorphic functions in Ω. For $c > 0$, define $\mathcal{H}_c(\Omega)$ as the set of $h \in H(\Omega)$ such that there exists $K \ge 0$ for which

$$|h(z)| \le K e^{c|Re\, z|}(1+|z|)^{-2} \quad \text{for all } z \in \Omega. \tag{A.1}$$

Clearly, $\mathcal{H}_c(\Omega)$ is the Banach space with the norm $\|h\|_{\Omega,c} := \sup_{z\in\Omega} |h(z)|(1+|z|)^2 e^{-c|Re\, z|}$. Define the convolution $f * g$ of $f, g \in \mathcal{H}_c(\Omega)$ by $(f*g)(z) := \int_0^z f(z-t)g(t)dt = \int_0^z f(t)g(z-t)dt$.

Let D be a domain in $(q_1, v, r) \in \mathbb{C}^3$ and let $\theta > 0$. Set $\Omega := E(0, \theta)$. Let $H(D, \Omega)$ be the set of all holomorphic functions in $(q_1, v, r) \in D$, $y \in \Omega$. For $c > 0$ let $\mathcal{H}_c(D, \Omega)$ denote the set of those $f \equiv f(q_1, r, y) \in H(D, \Omega)$ such that there exists $K \ge 0$ satisfying

$$\sup_{(q_1,v,r)\in D} |f(q_1, v, r, y)| \le K e^{c|Re\, y|}(1+|y|)^{-2} \qquad \text{for all } \ y \in \Omega. \tag{A.2}$$

Define $u_3 =: w_0 + v_3$, $u_4 =: \tilde{w}_0 + v_4$, where $v_3 = O(h)$ and $v_4 = O(h)$. Set $v_3 = v$ (resp. $v_4 = \tilde{v}$) and $w = w_0 + v$ (resp. $\tilde{w} = \tilde{w}_0 + \tilde{v}$). Let $\hat{v} := \mathcal{B}(v)(q_1, v, r, y)$ be the formal Borel transform of v, where y is the dual variable of h. Substitute $u_3 = w_0 + v_3$ and $u_4 = \tilde{w}_0 + v_4$ into (3.13) and cancel the constant term, $h^0 = 1$. Divide the equation with h and apply the formal Borel transform to both sides of the equation. Define

$$\tilde{\mathcal{R}}_3 = \mathcal{R}_3(u_1, u_3, u_4) - \mathcal{R}_3(u_1, w_0, \tilde{w}_0) - h\left(c_0 \frac{\partial w_0}{\partial \tilde{q}_1} + r^{1-\nu}(1-\lambda c_0 r^\nu)\frac{\partial w_0}{\partial r}\right).$$

Because the Borel transform maps the multiplication h^{-1} to $\partial/\partial y$, we have

$$r^{\nu-1} c_0 \frac{\partial \hat{v}}{\partial \tilde{q}_1} + (1-\lambda c_0 r^\nu)\frac{\partial \hat{v}}{\partial r} - r^{\nu-1}\frac{\partial}{\partial y}\left(\lambda \hat{v} + \tilde{\mathcal{R}}_3^*(u_1, \hat{u}_3, \hat{u}_4)\right) = 0. \tag{A.3}$$

Here $\tilde{\mathcal{R}}_3^*$ is given by replacing the product in $\tilde{\mathcal{R}}_3$ with the convolution product. Similarly, if $\tilde{w} = u_4$, then we have the equation of $\hat{\tilde{v}}$, for some $\tilde{\mathcal{R}}_4^*$,

$$\text{(A.4)} \qquad r^{\nu-1}c_0\frac{\partial \hat{\tilde{v}}}{\partial \tilde{q}_1} + (1 - \lambda c_0 r^\nu)\frac{\partial \hat{\tilde{v}}}{\partial r} + r^{\nu-1}\frac{\partial}{\partial y}\left(\lambda\hat{\tilde{v}} - \tilde{\mathcal{R}}_4^*(u_1, \hat{u}_3, \hat{u}_4)\right) = 0.$$

There exists a neighborhood $y = 0$, W such that $\mathcal{B}(v)$ and $\mathcal{B}(\tilde{v})$ are analytic in $(q_1, v, r, y) \in \Omega_1 \times U_1 \times \Omega_0 \times W$ and that they give the unique analytic solution of the system (A.3) - (A.4). Assume $q_1^{(0)} = 0$, for simplicity. Theorem 7.1 follows from the next theorem.

THEOREM A.1. *Assume that the conditions (A.1)', (A.1) and (A.2) are satisfied. Let $c > 0$. Then, for every $L_0 > 0$ there exist a neighborhood of I_0, $\Omega_1 \subset \Omega_2$, a neighborhood of $v(I_0)$, $K_1 > 0$, $E(0, \theta) =: \Omega$ and δ_0 such that, if* (2.8) *holds, then* (A.3) - (A.4) *has a solution $(\hat{u}_3, \hat{u}_4)$ in $\mathcal{H}_c(D, \Omega) \times \mathcal{H}_c(D, \Omega)$, where $D = \Omega_1 \times U_1 \times (\Omega_0 \cap \{r \,|\, |c_0 r| < L_0\})$.*

A.2. Preparation of lemmas. Define $\mathcal{R} := (\mathcal{R}_3, \mathcal{R}_4)$. Let $\nabla\mathcal{R}$ be the Jacobi matrix with respect to (u_3, u_4). Denote by $\mathrm{diag}(\lambda, -\lambda)$ the diagonal matrix with diagonal components given, respectively, by $\lambda, -\lambda$. Let $\lambda(u_1)$ and $\tilde{\lambda}(u_1)$ be the first and second element of diagonal part of $\mathrm{diag}(\lambda, -\lambda) + \nabla\mathcal{R}(u_1, w_0(0), \tilde{w}_0(0))$, respectively. Consider

$$\text{(A.5)} \qquad \mathcal{L}\hat{w} \equiv r^{\nu-1}c_0\frac{\partial \hat{w}}{\partial \tilde{q}_1} + (1 - \lambda c_0 r^\nu)\frac{\partial \hat{w}}{\partial r} - r^{\nu-1}\lambda(u_1)\frac{\partial \hat{w}}{\partial y} = r^{\nu-1}\hat{g},$$

where $\hat{g} \in \mathcal{H}_c(D, \Omega)$. Similarly, we define $\tilde{\mathcal{L}}$ by replacing $\lambda(u_1)$ with $\tilde{\lambda}(u_1)$ in (A.5).

We consider the case $\tilde{q}_1 = q_1$, for the sake of simplicity. In order to solve (A.5) we use the method of characteristics. Consider

$$\text{(A.6)} \qquad \frac{dq_1}{r^{\nu-1}c_0} = \frac{dr}{1 - \lambda c_0 r^\nu} = \frac{dy}{-r^{\nu-1}\lambda(u_1)}.$$

By integrating (A.6) we have $q_1 = q_1^{(0)} - (\lambda\nu)^{-1}\log(1 - \lambda c_0 r^\nu)$, where $q_1^{(0)}$ is a constant. Substitute q_1 into (A.6) and solve (A.6) with respect to y. We have

$$\text{(A.7)} \quad y = y_0 - \Phi(r), \quad \Phi(r) := \frac{1}{c_0}\int_0^\sigma \lambda(s - q_1^{(0)})ds, \quad \sigma = -\frac{1}{\lambda\nu}\log(1 - \lambda c_0 r^\nu),$$

where $y_0 = y(0) \in \Omega$ is the initial value of $y = y(r)$ at $r = 0$. We have

LEMMA A.2. *Suppose that r_0 satisfies $1 - \lambda c_0 r_0^\nu \neq 0$. Assume that $\lambda(u_1)$ is analytic at $u_1 = -q_1^{(0)}$. Then there exists a curve γ_{r_0} on $\mathbb{C}$ passing through r_0 such that $\mathrm{Im}\,\Phi$ is constant on γ_{r_0}, where Φ is given by* (A.7).

LEMMA A.3. *Assume that $\lambda(u_1)$ is analytic at $u_1 = -q_1^{(0)}$. Suppose that r_0 satisfies $1 - \lambda c_0 r_0^\nu \neq 0$. Let γ_{r_0} be given by Lemma* A.2. *For $r \in \gamma_{r_0}$, define*

$$\text{(A.8)} \qquad P\hat{g} := \int_{r_0}^r \hat{g}(u_1(s), s, y_0 - \Phi(s))\frac{s^{\nu-1}}{1 - \lambda c_0 s^\nu}ds, \qquad \hat{g} \in \mathcal{H}_c(D, \Omega),$$

where the integral is taken along γ_{r_0}. Then $P\hat{g}$ satisfies (A.5). *Especially, $P\hat{g}$ is analytic in r in some neighborhood of the origin $r = 0$.*

LEMMA A.4. *Assume that $\lambda(u_1)$ is analytic at $u_1 = -q_1^{(0)}$. Suppose that r_0 satisfies $1 - \lambda c_0 r_0^\nu \neq 0$. Let γ_{r_0} be given by Lemma* A.2. *Let r_1 satisfy $1 - \lambda c_0 r_1^\nu = 0$. Assume that there exist a neighborhood W_{r_1} of r_1 and $M_1 > 0$ such that, if*

$\gamma_{r_0} \subset W_{r_1}$, then $M_1^{-1} \leq |r_1 - s||r_1 - t|^{-1} \leq M_1$ for every $s, t \in \gamma_{r_0}$. Then, there exists a constant $c_1 \geq 0$ such that, for every $\hat{g} \in \mathcal{H}(D, \Omega)$ we have

$$\|P\hat{g}\|_c \leq c_1\|\hat{g}\|_c, \quad \left\|\frac{\partial}{\partial y}(P\hat{g})\right\|_c \leq c_1\|\hat{g}\|_c. \tag{A.9}$$

Lemmas A.2, A.3 and A.4 are proved in [**Y2**]. Then the remaing argument of the proof of Theorem 7.1 is a standard iteration. The argument is similar to the one in [**Y2**].

Acknowledgments

The author thanks to Prof. Galina Filipuk, Prof. Alberto Lastra, and Prof. Slawomir Michalik for the invitation to the workshop. The author thanks the anonymous referee for useful comments.

References

[FH] G. Filipuk and R. G. Halburd, *Movable algebraic singularities of second-order ordinary differential equations*, J. Math. Phys. **50** (2009), no. 2, 023509, 18, DOI 10.1063/1.3068414. MR2494849

[KT] Takahiro Kawai and Yoshitsugu Takei, *Algebraic analysis of singular perturbation theory*, Translations of Mathematical Monographs, vol. 227, American Mathematical Society, Providence, RI, 2005. Translated from the 1998 Japanese original by Goro Kato; Iwanami Series in Modern Mathematics, DOI 10.1090/mmono/227. MR2182990

[Y1] Masafumi Yoshino, *Movable singularity of semi linear Heun equation and application to blowup phenomenon*, NoDEA Nonlinear Differential Equations Appl. **26** (2019), no. 1, Paper No. 8, 18, DOI 10.1007/s00030-019-0555-9. MR3908331

[Y2] Masafumi Yoshino, *Global Borel summability of some partial differential equation*, To be published in the Proceedings of "Formal and Analytic Solutions of Differential Equations", World Scientific Publishing Europe Ltd. DOI: 10.1142/q0335, ISBN: 978-1-80061-135-1.

Department of Mathematics, Hiroshima University, Hiroshima 739-8526, Japan
Current address: 2-29-3-404, Hachioji, Bessho, Tokyo, 192-0363, Japan
Email address: yoshinom@hiroshima-u.ac.jp

Contemporary Mathematics
Volume **782**, 2023
https://doi.org/10.1090/conm/782/15731

Some notes on moment partial differential equations. Application to fractional functional equations

Alberto Lastra, Sławomir Michalik, and Maria Suwińska

ABSTRACT. The present work provides a brief description on the state of the art regarding the theory of moment differential equations in the complex domain, and its applications to the study of fractional differential equations. All results are stated without any proof, giving appropriate references where the details can be found. We also describe how the theory can be particularized to fractional differential equations and the scope of the results achieved.

1. Introduction

The present work provides a brief description on the state of the art regarding the theory of moment differential equations in the complex domain, and its applications to the study of fractional differential equations. All results are stated without any proof, giving appropriate references where the details can be found.

The study of formal solutions of moment functional equations, their summability properties, and related results was firstly considered by W. Balser and M.Yoshino in 2010, in [**BY**]. In that work the authors define the formal moment differential operators on formal power series, generalizing the classical concept of formal differentiation of power series. In the last years several results have been achieved in the study of such functional equations, which are summarized and discussed in the present work.

2. Recent results on the study of moment partial differential equations

2.1. Gevrey case. As mentioned above, the study of formal solutions of moment functional equations, their summability properties and related results were firstly considered by W. Balser and M. Yoshino in 2010, in [**BY**]. In that work the authors define the formal moment differential operators on formal power series, generalizing the classical concept of formal differentiation of power series.

2020 *Mathematics Subject Classification.* Primary 35C10; Secondary 35G10.

Key words and phrases. Moment differential equations, summability, fractional differential equations.

The first author was partially supported by the project PID2019-105621GB-I00 of Ministerio de Ciencia e Innovación, Spain, and by Dirección General de Investigación e Innovación, Consejería de Educación e Investigación of Comunidad de Madrid (Spain), and Universidad de Alcalá under grant CM/JIN/2019-010, Proyectos de I+D para Jóvenes Investigadores de la Universidad de Alcalá 2019.

This concept is applied to study the formal series solutions of moment partial differential equations, in the form of an inhomogeneous moment partial differential equation with constant coefficients. The existence of a formal solution (Proposition 2, [**BY**]) and Gevrey estimates for their coefficients (Theorem 1, [**BY**]) give a first step in this theory.

Given a sequence of positive numbers $m = (m(p))_{p\geq 0}$, the operator of moment differentiation $\partial_{m,z}$ is defined for every element in $\mathbb{C}[[z]]$ by

$$\partial_{m,z}\left(\sum_{p\geq 0}\frac{a_p}{m(p)}z^p\right) = \sum_{p\geq 0}\frac{a_{p+1}}{m(p)}z^p.$$

This previous definition can be naturally extended to functions defined on a neighborhood of the origin by means of their representation in the form of an infinite series. However, this formal differentiation does not preserve convergence unless some regularity property is assumed for the sequence m. Usually, the sequence m is assumed to be a sequence of moments, i.e., a sequence defined by

$$m(p) = \int_0^\infty t^{p-1}e(t)dt, \tag{2.1}$$

for some kernel function for summability e. Here, the kernel function e of order k ($k > 1/2$) is a holomorphic function on the infinity sector in the direction 0 and opening π/k, such that e is exponentially flat of order k in this sector, $e(z)/z$ is integrable at the origin and $e(x) > 0$ for $x > 0$. The general moment summability described by W. Balser in Chapter 6.5 [**B**] generalizes the classical Borel-Laplace method by considering more general kernel functions and sequences than exponentials and Gevrey sequences to provide the sum of a formal power series. As W. Balser mentions in his book, such moment methods turn out to be equivalent to the classical ones in the sense that they sum the same formal power series to the same analytic functions.

As mentioned above, the first approach to the study of moment differential equations is that by W. Balser and M. Yoshino in [**BY**]. This first approach is based on the construction linked to a pair of kernel functions to summability, generalizing the classical Borel and Laplace operators which are associated with sequences $m = (m(p))_{p\geq 0}$ which are equivalent to some Gevrey sequence. To be precise, there exist $\alpha > 0$ and $C_1, C_2, A_1, A_2 > 0$ such that

$$C_1 A_1^p p!^\alpha \leq m(p) \leq C_2 A_2^p p!^\alpha$$

for every $p \in \{0, 1, \ldots\}$. We refer to such sequences as sequences of moments of Gevrey order α.

Let m_1 and m_2 be two sequences of moments of Gevrey order $k_1 > 0$ and $k_2 > 0$, resp. Let $p(t,z) \in \mathbb{C}[t,z]$ and $\hat{f}(t,z) \in \mathbb{C}[[t,z]]$. In Section 3 of [**BY**], the authors consider the following moment partial differential equation

$$P(\partial_{m_1,t}, \partial_{m_2,z})\hat{u}(t,z) = \hat{f}(t,z), \tag{2.2}$$

and study the set of formal solutions $\hat{u}(t,z)$ to (2.2) via a parametrization of the set of solutions (see Proposition 2, [**BY**]). The notion of formal solutions to a moment partial differential equation of the form (2.2) was clear. However, the study of analyticity of solutions on domains such as polysectors in $\mathbb{C}^2$ with $(0,0)$, when $\hat{f}$ is a holomorphic function on the same domain was quite unclear, and some light was put on it by considering Toeplitz operators which preserve summability properties and commute with products (see Section 4 in [**BY**]). Given a formal solution $\hat{u}(t,z)$ of (2.2), Gevrey estimates of a formal solution of (2.2) is characterized in terms of the Gevrey order of $\hat{f}$ (Theorem 1, [**BY**]).

In 2013, S. Michalik gives an answer to the problem of analyticity when considering problems of the form (2.2), where constant coefficients are being considered, in the work [**M**]. More precisely, S. Michalik considers the Cauchy problem consisting of a moment partial differential equation (2.2) and initial data given by

$$\partial^j_{m_1,t} u(0,z) = \varphi_j(z), \qquad j = 0,\dots,n-1, \tag{2.3}$$

with n being the degree of P with respect to its first variable, and φ_j are assumed to be holomorphic functions on some disc at the origin. The main results in this work state the relationship between analytic continuation of the initial data and convergence and analytic continuation of the formal solution. More precisely, multisummability of the formal solution of the main equation is characterized regarding the growth properties of the analytic continuation of φ_j, for $0 \le j \le n-1$. In a first step, S. Michalik considers an equation with $\hat{f} \equiv 0$, and splits the polynomial P in terms of the roots of the characteristic equation $P(\lambda,\xi) = 0$ with respect to its first variable. The formal solution can be written in terms of the formal solution of Cauchy problems associated with equations of the form

$$(\partial_{m_1,t} - \lambda_\alpha(\partial_{m_2,z}))^\beta \hat{u}_{\alpha,\beta} = 0, \tag{2.4}$$

for some α and $\beta \ge 1$, and where $(\lambda - \lambda_\alpha(\xi))^\beta$ divides the characteristic equation. After stating the Gevrey order of the formal solution of the main equation (see Theorem 2, [**M**]), and the conditions for its convergence near the origin, he is able in this latter case to characterize the exponential growth of the analytic solutions (with respect to t) of (2.4) on infinite sectors with appropriate bisecting direction, in terms of the exponential growth of the initial conditions (Theorem 3). Also, in the case that the formal solutions of (2.4) are not convergent, a characterization to summability along well-chosen directions of the formal solutions is provided in terms of the exponential growth of the initial conditions (Theorem 4). The final results (Theorem 5 and Theorem 6) put all the previous information together regarding equation (2.4) in order to state multisummability of the formal solution to (2.2), (2.3) along appropriate directions, provided exponential growth of the initial data. The inhomogeneous case ($\hat{f} \neq 0$) is also considered. The form of the formal solution (Theorem 7) together with the Gevrey estimates of the formal solution (Theorem 8) are stated. In this latter case, the Gevrey order of the solution is modified with respect to the homogeneous framework whenever the Gevrey order of $\hat{f}$ is large enough. We also refer to [**M2**] for further study in the homogeneous situation, while dropping the condition of the convergence of the initial data.

Further knowledge on the solutions to moment partial differential equations with constant coefficients is provided in [**MT**], where the authors study multisummable normalized formal solutions to (2.2), (2.3) and describe Stokes and anti-Stokes lines, jumps across Stokes lines and a maximal family of solutions. This is possible by taking into account the jumps across the Stokes lines in terms of hyperfunctions (see [**K**] for the details on this concept).

2.2. Some generalities. The construction of more general kernels via flat functions in Carleman ultraholomorphic classes which admit a nonzero proximate order is due to J. Sanz [**S**]. This novel development greatly enlarges the class of sequences m which can be considered as moments related to the kernel function e in (2.1). The sequences considered in the previous bounds satisfy certain conditions related to appropriate properties of the associated functional spaces, initially studied in the real setting: closeness under product and composition is guaranteed if the sequence is logarithmically convex (lc); the existence of functions within this class and with a prescribed Taylor expansion at the origin is characterized via the strong non-quasianalyticity condition (snc); it is also usual to consider sequences under the moderate growth condition (mg). Some more details on properties of the sequences and related spaces can be found in [**KMR**, **R**]. In these notes we will assume that the initial sequence is strongly regular, i.e., satisfies (lc), (snc) and (mg), starting from one. This concept was put forward by V. Thilliez [**T**]. Strongly regular sequences appear in the structure of the formal and analytic solutions of certain families of functional equations: Gevrey sequence of some positive order α, $(\Gamma(1+\alpha p))_{p\geq 0}$ is strongly related to fractional differentiation (see Section 3); 1+ level related to difference equations (see [**I**, **I2**], among others), etc.

It is worth mentioning that a branch of research is currently being performed in the direct study of classes of holomorphic functions defined in a sector of the Riemann surface of the logarithm whose derivatives satisfy certain bounds in terms of a given sequence, the so-called Carleman ultraholomorphic classes of functions. For more details on that topic we refer to the studies [**JS**, **JSS**, **JSS2**, **JSS3**] and the references therein. Also, a detailed introduction on the topic of summability analysis of formal power series can be found in [**S2**].

2.3. One example. As a motivating example, we illustrate the summability procedure of the generalization of Euler equation (see Example 1.1.4, [**L**]) in the context of moment differential equations, which resembles the classical procedure of summability of Gevrey formal power series solutions to ordinary differential equations. Let us consider a sequence of moments $m = (m(p))_{p\geq 0}$, which is associated with a kernel function for summability e. We consider the moment differential equation

$$z\partial_{m,z}(zy(z)) + y(z) = z. \tag{2.5}$$

Observe that in the special case, when we take the corresponding sequence of moments $(m(p))_{p\geq 0} = (\Gamma(1+p))_{p\geq 0}$ and the kernel function $e(z) = ze^{-z}$, we get the classical Euler equation $z\partial_z(zy(z)) + y(z) = z$.

It is straightforward to check that the formal power series

$$\hat{y}(z) = \frac{1}{m(1)} \sum_{p\geq 0} (-1)^p m(p+1) z^{p+1}$$

is a formal solution to (2.5). A generalized Borel-Laplace procedure for summability of formal power series can be applied. Indeed, one has that the generalized formal Borel operator $\hat{\mathcal{B}}_m$, defined for any formal power series by dividing each coefficient by the corresponding moment, can be applied to $\hat{y}$ to arrive at

$$\hat{\mathcal{B}}_m \hat{y}(z) = \frac{-1}{m(1)} \sum_{n=0}^{\infty} (-z)^{n+1} = \frac{1}{m(1)} \frac{z}{1+z}.$$

The properties of the kernel function e (see [**S**], for example) allow to define a Laplace-like transformation leading to a function

$$y(z) = \frac{1}{m(1)} \int_0^{\infty(d)} \frac{u}{1+u} e(u/z) \frac{du}{u},$$

valid for every direction $d \neq \pi$. The function y turns out to be a holomorphic function defined on some sector of opening depending on m. The formal power series $\hat{y}$ represents y asymptotically. Moreover, the opening of the sector in which this representation occurs guarantees uniqueness of the function y under this property (from a generalization of Watson's Lemma, see Corollary 4.12 in [**S**]). In addition to this, y is an analytic solution of (2.5).

2.4. Some remarks on moment integro-differential equations. The broader concept of generalized summability dealing with kernel functions associated with strongly regular sequences was applied in Section 7 [**S2**], where the previous techniques are adapted to study the growth properties of the solutions to (2.4) (Theorem 7.11, [**S2**]) and summability of the formal solutions to (2.2) along appropriate directions (Theorem 7.15). See also Section 5 in [**LMS**] for further details. In this direction, further results have been achieved on the development of a generalized multisummability theory of formal power series concerning different (comparable and nonequivalent) levels associated with ultraholomorphic classes in [**JKLS**].

We also remark on the existence of recent results regarding the appearance of functions in ultraholomorphic classes in the framework of the study of singularly perturbed linear partial differential equations in the complex domain of the form

$$L(t, z, \partial_t, \partial_z, \epsilon)u(t, z, \epsilon) = 0,$$

where L is a linear differential operator, polynomial in all its variables but in (z, ϵ), where ϵ is assumed to be a small complex perturbation parameter. The coefficients of L regarding such variables are assumed to be holomorphic functions near the origin in z, with coefficients being the sums of formal power series in the perturbation parameter ϵ, which are related to the generalized summability procedure described in [**LMS**], based on the construction of flat functions for generalized summability previously developed in [**S**].

The knowledge of the behavior of the solutions to linear moment partial differential equations with constant coefficients was enlarged in [**M3**], where sufficient conditions for the convergence, analytic continuation, and also for the summability of the formal solution are determined in terms of $\hat{f}(t, z)$ in (2.2). Summability is performed with respect to each of the variables and also in both variables. In [**MS**], a step forward is taken by considering moment partial differential equations of the form

$$P(t, \partial_{m_1,t}, \partial_{m_2,z})\hat{u}(t, z) = \hat{f}(t, z), \tag{2.6}$$

under initial data (2.3), where P is a polynomial in its second and third variables, with coefficients being holomorphic functions on some neighborhood of the origin (or formal power series of some Gevrey order). Newton polygon and formal norm techniques are applied in order to provide with Gevrey estimates for the formal solution of the problem which depend on the divergence of the coefficients, the forcing term $\hat{f}$, and/or the initial data. The results are described in terms of the generalized summability developed in [**B**], whereas in [**LMSu**] the generalized kernels for summability developed in [**S**] are being considered, after the choice of adequate moment formal norms and formal power series adapted to the problem. In this work, both the coefficients and the forcing term satisfy general bounds associated with strongly regular sequences, leading to a formal solution in powers of t with coefficients being holomorphic functions on a common polydisc, and satisfying bounds determined by the first positive slope of the Newton polygon associated with the equation, and also to the previous strongly regular sequence. The result is stated in terms of a finite number of variables z, and the two-dimensional case, i.e., z standing for just one component, allows to weaken the conditions of the strongly regular sequence. On the other hand, the multidimensional case, under the assumption of constant coefficients and certain technical restrictions on the sequence involved, allows to get to an analogous result as before. The estimates of formal solutions to moment partial differential equations with variable coefficients of the form

$$P(t, z, \partial_{m_1,t}, \partial_{m_2,z})\hat{u}(t, z) = \hat{f}(t, z), \tag{2.7}$$

is successfully studied in [**Su**] in the case of moment differentiation in the sense considered by Balser and Yoshino explained above. Regarding the problem (2.6), M. Suwińska is able to deal with coefficients not only depending on t variable, but also on the spatial variables z under certain assumption on the Newton polygon associated with the equation, and the Gevrey order of the formal solution to (2.7). The main tool in this generalization is the use of Nagumo-type norms in order to obtain Gevrey bounds for the formal solution in the shape of the previous works listed above (Theorem 1, [**Su**]). The work [**Re**] considers partial differential equations of a similar nature as those described above.

After the achievement of Gevrey and more general estimates for the formal solutions to the moment partial differential equations under consideration, it is a natural question to consider whether such formal solutions can be summed up following a summability technique such as a Borel-Laplace like procedure. More precisely, the next step taken in the theory was to determine summability properties for the equations in (2.6). For this purpose a deeper knowledge of the moment differential operator is needed, broadening the spaces of functions to which it can be applied successfully. An integral representation of moment differentiation (Theorem 3, [**LMSu2**]) is provided, determining exponential-like bounds for the subsequent moment derivatives of functions in ultraholomorphic classes. As a direct consequence, the space of summable formal power series along a direction with respect to some strongly regular sequence admitting a nonzero proximate order turns out to be closed under moment differentiation. Therefore, the definition of moment derivatives of the generalized sum of a formal power series can be defined as the generalized sum along that same direction of the formal moment derivative of the initial formal power series. As an application of this result, the summability properties of the formal solutions to the moment partial differential equation

$$(\partial_{m_1,t}^k - a(z)\partial_{m_2,z}^p)u(t, z) = \hat{f}(t, z),$$

under holomorphic Cauchy data near the origin are studied. Here $a(z)$ is also holomorphic on some neighborhood of the origin. A second result in that work (Theorem 4) states that summability of the formal solution with respect to t variable along certain direction is equivalent to summability of $\hat{f}$ (with respect to t) and the summability of the first terms of the formal solutions, written in powers of z. Further results making use of the previous property have been achieved in [**LMSu3**] when dealing with moment integro-differential equations of the form

$$(1-\sum_{i,q} a_{i,q}(z)\partial_{m_1,t}^{-i}\partial_{m_2,z}^{q})u(t,z)=\hat{f}(t,z),$$

with a finite sum, and holomorphic coefficients near the origin. The initial conditions associated with this moment functional equations are assumed to be holomorphic functions on some neighborhood of the origin. The natural definition of the operator $\partial_{m,z}^{-1}$ is determined by its action on formal power series by

$$\partial_{m,z}^{-1}\left(\sum_{p\geq 0}\frac{a_p}{m(p)}z^p\right)=\sum_{p\geq 0}\frac{a_p}{m(p+1)}z^{p+1}.$$

The consideration of appropriate norms defined on holomorphic functions near the origin, together with Newton polygon techniques, under certain properties linking the moment sequences m_1 and m_2 and under one slope in the Newton polygon one has that (Theorem 4.1, [**LMSu3**]) the formal solution is summable with respect to some positive sequence along certain direction $d\in\mathbb{R}$ whenever $\hat{f}$ and the first terms in the formal solution with respect to z are summable with respect to the same sequence along that same direction. The converse is only true (Theorem 4.2) under further assumptions on the sequence, which apply in the examples appearing in applications to fractional differential equations. The work is developed in terms of the approach in [**S**]. In this same proceedings volume an interested reader will find the version of this result in the framework of the kernel functions considered in [**B**]. We also refer to the work by P. Remy [**Re2**] in the study of integro-differential equations of a similar nature as those considered.

3. A focus on the fractional differential equations

The main application of moment functional equations might be that particularized to fractional differential equations and fractional integro-differential equations.

The moment sequence is then given by Gamma function $m=(\Gamma(1+\frac{p}{k}))_{p\geq 0}$ for some fixed $k>0$. The kernel functions associated with this sequence are given by $e(z)=\frac{1}{\alpha}z^{\frac{1}{\alpha}}\exp(-z^{\frac{1}{\alpha}})$ and $E(z)=\sum_{p\geq 0}\frac{z^p}{\Gamma(1+\alpha z)}$ with $\alpha=1/k$. Here, moment differentiation is quite related to Caputo $1/k-$fractional derivative $\partial_z^{1/k}$ as follows. For every $\alpha\in\mathbb{Q}_+$ the fractional derivative of order α is defined on formal power series in z^α by

$$\partial_z^{\alpha}\left(\sum_{p\geq 0}\frac{a_p}{\Gamma(1+\alpha p)}z^{\alpha p}\right)=\sum_{p\geq 0}\frac{a_{p+1}}{\Gamma(1+\alpha p)}z^{\alpha p}. \tag{3.1}$$

Recently, Caputo derivatives [**G**], asymptotic periodic solutions of evolution equations [**RWF**], or stability properties [**A**], or numerical studies [**Bh**,**F**] have been done regarding these kind of derivation operators. We also refer to the book [**Ba**].

This operator coincides with the classical derivative for $\alpha=1$. If $\alpha=1/k$, then the operator is known as Caputo fractional derivative for $1/k-$analytic functions (see [**KST**]). The aforementioned relation of the previous operator with moment differentiation is given by the equality

$$(\partial_m\hat{f})(z^{1/k})=\partial_z^{1/k}(\hat{f}(z^{1/k})),$$

with m being the moment sequence $m = (\Gamma(1+\frac{p}{k}))_{p\geq 0}$. The formal α-integral operator, naturally defined from (3.1) coincides with Riemann-Liouville fractional integral

$$I_{0+}^{\alpha} f(z) := \frac{1}{\Gamma(\alpha)} \int_0^z \frac{f(t)dt}{(z-t)^{1-\alpha}}. \tag{3.2}$$

We refer to [**GL, Ki, Ki2, O**] for further details on fractional calculus. In this case, the assumptions made on the sequences considered are satisfied and the equivalence mentioned in the last paragraph of the previous section holds. The equivalence result for summability regarding integro-differential equations holds in both, the Gevrey moment settings (see the contribution by M. Suwińska in this same volume) and also in the strongly regular settings. All the previous results mentioned in that section can be particularized in this situation.

References

[A] Ravi Agarwal, Snezhana Hristova, and Donal O'Regan, *Some stability properties related to initial time difference for Caputo fractional differential equations*, Fract. Calc. Appl. Anal. **21** (2018), no. 1, 72–93, DOI 10.1515/fca-2018-0005. MR3776054

[Ba] Dumitru Baleanu, Kai Diethelm, Enrico Scalas, and Juan J. Trujillo, *Fractional calculus*, Series on Complexity, Nonlinearity and Chaos, vol. 5, World Scientific Publishing Co. Pte. Ltd., Hackensack, NJ, 2017. Models and numerical methods; Second edition [of MR2894576]. MR3561379

[B] Werner Balser, *Formal power series and linear systems of meromorphic ordinary differential equations*, Universitext, Springer-Verlag, New York, 2000. MR1722871

[BY] Werner Balser and Masafumi Yoshino, *Gevrey order of formal power series solutions of inhomogeneous partial differential equations with constant coefficients*, Funkcial. Ekvac. **53** (2010), no. 3, 411–434, DOI 10.1619/fesi.53.411. MR2761091

[Bh] A. H. Bhrawy, M. A. Zaky, and R. A. Van Gorder, *A space-time Legendre spectral tau method for the two-sided space-time Caputo fractional diffusion-wave equation*, Numer. Algorithms **71** (2016), no. 1, 151–180, DOI 10.1007/s11075-015-9990-9. MR3439351

[F] M. A. Firoozjaee, H. Jafari, A. Lia, and D. Baleanu, *Numerical approach of Fokker-Planck equation with Caputo-Fabrizio fractional derivative using Ritz approximation*, J. Comput. Appl. Math. **339** (2018), 367–373, DOI 10.1016/j.cam.2017.05.022. MR3787701

[GL] A. O. Gel′fond and A. F. Leont′ev, *On a generalization of Fourier series* (Russian), Mat. Sbornik N.S. **29(71)** (1951), 477–500. MR0045812

[G] Mikhail I. Gomoyunov, *On representation formulas for solutions of linear differential equations with Caputo fractional derivatives*, Fract. Calc. Appl. Anal. **23** (2020), no. 4, 1141–1160, DOI 10.1515/fca-2020-0058. MR4151101

[I] G. K. Immink, *Exact asymptotics of nonlinear difference equations with levels* 1 *and* 1^+ (English, with English and French summaries), Ann. Fac. Sci. Toulouse Math. (6) **17** (2008), no. 2, 309–356. MR2487857

[I2] Geertrui Klara Immink, *Accelero-summation of the formal solutions of nonlinear difference equations* (English, with English and French summaries), Ann. Inst. Fourier (Grenoble) **61** (2011), no. 1, 1–51, DOI 10.5802/aif.2596. MR2828125

[JKLS] Javier Jiménez-Garrido, Shingo Kamimoto, Alberto Lastra, and Javier Sanz, *Multisummability in Carleman ultraholomorphic classes by means of nonzero proximate orders*, J. Math. Anal. Appl. **472** (2019), no. 1, 627–686, DOI 10.1016/j.jmaa.2018.11.043. MR3906392

[JS] Javier Jiménez-Garrido and Javier Sanz, *Strongly regular sequences and proximate orders*, J. Math. Anal. Appl. **438** (2016), no. 2, 920–945, DOI 10.1016/j.jmaa.2016.02.010. MR3466072

[JSS] Javier Jiménez-Garrido, Javier Sanz, and Gerhard Schindl, *Injectivity and surjectivity of the asymptotic Borel map in Carleman ultraholomorphic classes*, J. Math. Anal. Appl. **469** (2019), no. 1, 136–168, DOI 10.1016/j.jmaa.2018.09.011. MR3857515

[JSS2] Javier Jiménez-Garrido, Javier Sanz, and Gerhard Schindl, *Sectorial extensions, via Laplace transforms, in ultraholomorphic classes defined by weight functions*, Results Math. **74** (2019), no. 1, Paper No. 27, 44, DOI 10.1007/s00025-018-0951-1. MR3899178

[JSS3] Javier Jiménez-Garrido, Javier Sanz, and Gerhard Schindl, *Surjectivity of the asymptotic Borel map in Carleman-Roumieu ultraholomorphic classes defined by regular sequences*, Rev. R. Acad. Cienc. Exactas Fís. Nat. Ser. A Mat. RACSAM **115** (2021), no. 4, Paper No. 181, 18, DOI 10.1007/s13398-021-01119-y. MR4309827

[K] A. Kaneko, *Introduction to hyperfunctions*, Mathematics and its Applications (Japanese Series), vol. 3, Kluwer Academic Publishers Group, Dordrecht; SCIPRESS, Tokyo, 1988. Translated from the Japanese by Y. Yamamoto. MR1026013

[Ki] V. Kiryakova, *Gel'fond-Leont'ev integration operators of fractional (multi-)order generated by some special functions*, AIP Conf. Proc. 2018, 2048, 050016.

[Ki2] Virginia Kiryakova, *An open problem of Ljubomir Iliev related to the Mittag-Leffler function and fractional calculus operators*, Complex analysis and applications '13, Bulgarian Acad. Sci., Sofia, 2013, pp. 139–153. MR3312230

[KMR] Andreas Kriegl, Peter W. Michor, and Armin Rainer, *The convenient setting for non-quasianalytic Denjoy-Carleman differentiable mappings*, J. Funct. Anal. **256** (2009), no. 11, 3510–3544, DOI 10.1016/j.jfa.2009.03.003. MR2514051

[KST] Anatoly A. Kilbas, Hari M. Srivastava, and Juan J. Trujillo, *Theory and applications of fractional differential equations*, North-Holland Mathematics Studies, vol. 204, Elsevier Science B.V., Amsterdam, 2006. MR2218073

[LMS] Alberto Lastra, Stéphane Malek, and Javier Sanz, *Summability in general Carleman ultraholomorphic classes*, J. Math. Anal. Appl. **430** (2015), no. 2, 1175–1206, DOI 10.1016/j.jmaa.2015.05.046. MR3352004

[LMSu] Alberto Lastra, Sławomir Michalik, and Maria Suwińska, *Estimates of formal solutions for some generalized moment partial differential equations*, J. Math. Anal. Appl. **500** (2021), no. 1, Paper No. 125094, 18, DOI 10.1016/j.jmaa.2021.125094. MR4224842

[LMSu2] Alberto Lastra, Sławomir Michalik, and Maria Suwińska, *Summability of formal solutions for some generalized moment partial differential equations*, Results Math. **76** (2021), no. 1, Paper No. 22, 27, DOI 10.1007/s00025-020-01324-y. MR4198409

[LMSu3] Alberto Lastra, Sławomir Michalik, and Maria Suwińska, *Summability of formal solutions for a family of generalized moment integro-differential equations*, Fract. Calc. Appl. Anal. **24** (2021), no. 5, 1445–1476, DOI 10.1515/fca-2021-0061. MR4331893

[L] Michèle Loday-Richaud, *Divergent series, summability and resurgence. II*, Lecture Notes in Mathematics, vol. 2154, Springer, [Cham], 2016. Simple and multiple summability; With prefaces by Jean-Pierre Ramis, Éric Delabaere, Claude Mitschi and David Sauzin, DOI 10.1007/978-3-319-29075-1. MR3495546

[M] Sławomir Michalik, *Analytic solutions of moment partial differential equations with constant coefficients*, Funkcial. Ekvac. **56** (2013), no. 1, 19–50, DOI 10.1619/fesi.56.19. MR3099032

[M2] Sławomir Michalik, *Summability of formal solutions of linear partial differential equations with divergent initial data*, J. Math. Anal. Appl. **406** (2013), no. 1, 243–260, DOI 10.1016/j.jmaa.2013.04.062. MR3062417

[M3] Sławomir Michalik, *Analytic and summable solutions of inhomogeneous moment partial differential equations*, Funkcial. Ekvac. **60** (2017), no. 3, 325–351, DOI 10.1619/fesi.60.325. MR3701264

[MS] Sławomir Michalik and Maria Suwińska, *Gevrey estimates for certain moment partial differential equations*, Complex differential and difference equations, De Gruyter Proc. Math., De Gruyter, Berlin, [2020] ©2020, pp. 391–407. MR4401556

[MT] Sławomir Michalik and Bożena Tkacz, *The Stokes phenomenon for some moment partial differential equations*, J. Dyn. Control Syst. **25** (2019), no. 4, 573–598, DOI 10.1007/s10883-018-9424-9. MR3995954

[O] Nikola Obrechkoff, *On the summation of Taylor's series on the contour of the domain of summability*, Fract. Calc. Appl. Anal. **19** (2016), no. 5, 1316–1346, DOI 10.1515/fca-2016-0069. Translated from the Bulgarian original by Peter Rusev; Reprinted from *Nikola Obrechkoff, selected papers, Vol. II*, Prof. Marin Drinov Academic Publishing House, Sofia, 2006; With an editorial note by Virginia Kiryakova. MR3571014

[R] Armin Rainer, *Invariant functions in Denjoy-Carleman classes*, Ann. Global Anal. Geom. **35** (2009), no. 3, 249–266, DOI 10.1007/s10455-008-9135-7. MR2495975

[Re] Pascal Remy, *Gevrey order and summability of formal series solutions of some classes of inhomogeneous linear partial differential equations with variable coefficients*, J. Dyn. Control Syst. **22** (2016), no. 4, 693–711, DOI 10.1007/s10883-015-9301-8. MR3538953

[Re2] Pascal Remy, *Gevrey order and summability of formal series solutions of certain classes of inhomogeneous linear integro-differential equations with variable coefficients*, J. Dyn. Control Syst. **23** (2017), no. 4, 853–878, DOI 10.1007/s10883-017-9371-x. MR3688898

[RWF] Lulu Ren, JinRong Wang, and Michal Fe kan, *Asymptotically periodic solutions for Caputo type fractional evolution equations*, Fract. Calc. Appl. Anal. **21** (2018), no. 5, 1294–1312, DOI 10.1515/fca-2018-0068. MR3903612

[S] Javier Sanz, *Flat functions in Carleman ultraholomorphic classes via proximate orders*, J. Math. Anal. Appl. **415** (2014), no. 2, 623–643, DOI 10.1016/j.jmaa.2014.01.083. MR3178281

[S2] Javier Sanz, *Asymptotic analysis and summability of formal power series*, Analytic, algebraic and geometric aspects of differential equations, Trends Math., Birkhäuser/Springer, Cham, 2017, pp. 199–262, DOI 10.1007/978-3-319-52842-7_4. MR3706197

[Su] Maria Suwińska, *Gevrey estimates of formal solutions for certain moment partial differential equations with variable coefficients*, J. Dyn. Control Syst. **27** (2021), no. 2, 355–370, DOI 10.1007/s10883-020-09504-3. MR4231675

[T] Vincent Thilliez, *Division by flat ultradifferentiable functions and sectorial extensions*, Results Math. **44** (2003), no. 1-2, 169–188, DOI 10.1007/BF03322923. MR2011916

Universidad de Alcalá, Departamento de Física y Matemáticas, Alcalá de Henares, Madrid, Spain

Email address: alberto.lastra@uah.es

Faculty of Mathematics and Natural Sciences, College of Science, Cardinal Stefan Wyszyński University, Warszawa, Poland

Email address: s.michalik@uksw.edu.pl

Faculty of Mathematics and Natural Sciences, College of Science, Cardinal Stefan Wyszyński University, Warszawa, Poland

Email address: m.suwinska@op.pl

Selected Published Titles in This Series

782 **Galina Filipuk, Alberto Lastra, and Sławomir Michalik, Editors,** Recent Trends in Formal and Analytic Solutions of Diff. Equations, 2023

781 **Emanuel Indrei, Diego Marcon, and Levon Nurbekyan, Editors,** Geometric and Functional Inequalities and Recent Topics in Nonlinear PDEs, 2023

780 **Erik Koelink, Stefan Kolb, Nicolai Reshetikhin, and Bart Vlaar, Editors,** Hypergeometry, Integrability and Lie Theory, 2022

779 **Samuele Anni, Valentijn Karemaker, and Elisa Lorenzo García, Editors,** Arithmetic, Geometry, Cryptography, and Coding Theory 2021, 2022

778 **Carlos Galindo, Alejandro Melle Hernández, Julio José Moyano-Fernández, and Wilson A. Zúñiga-Galindo, Editors,** p-Adic Analysis, Arithmetic and Singularities, 2022

777 **Bang-Yen Chen, Nicholas D. Brubaker, Takashi Sakai, Bogdan D. Suceavă, Makiko Sumi Tanaka, Hiroshi Tamaru, and Mihaela B. Vajiac, Editors,** Differential Geometry and Global Analysis, 2022

776 **Aaron Wootton, S. Allen Broughton, and Jennifer Paulhus, Editors,** Automorphisms of Riemann Surfaces, Subgroups of Mapping Class Groups and Related Topics, 2022

775 **Fernando Galaz-García, Cecilia González-Tokman, and Juan Carlos Pardo Millán, Editors,** Mexican Mathematicians in the World, 2021

774 **Randall J. Swift, Alan Krinik, Jennifer M. Switkes, and Jason H. Park, Editors,** Stochastic Processes and Functional Analysis, 2021

773 **Nicholas R. Baeth, Thiago H. Freitas, Graham J. Leuschke, and Victor H. Jorge Pérez, Editors,** Commutative Algebra, 2021

772 **Anatoly M. Vershik, Victor M. Buchstaber, and Andrey V. Malyutin, Editors,** Topology, Geometry, and Dynamics, 2021

771 **Nicolás Andruskiewitsch, Gongxiang Liu, Susan Montgomery, and Yinhuo Zhang, Editors,** Hopf Algebras, Tensor Categories and Related Topics, 2021

770 **Stéphane Ballet, Gaetan Bisson, and Irene Bouw, Editors,** Arithmetic, Geometry, Cryptography and Coding Theory, 2021

769 **Kiyoshi Igusa, Alex Martsinkovsky, and Gordana Todorov, Editors,** Representations of Algebras, Geometry and Physics, 2021

768 **Dražen Adamović, Andrej Dujella, Antun Milas, and Pavle Pandžić, Editors,** Lie Groups, Number Theory, and Vertex Algebras, 2021

767 **Moshe Jarden and Tony Shaska, Editors,** Abelian Varieties and Number Theory, 2021

766 **Paola Comparin, Eduardo Esteves, Herbert Lange, Sebastián Reyes-Carocca, and Rubí E. Rodríguez, Editors,** Geometry at the Frontier, 2021

765 **Michael Aschbacher,** Quaternion Fusion Packets, 2021

764 **Gabriel Cunningham, Mark Mixer, and Egon Schulte, Editors,** Polytopes and Discrete Geometry, 2021

763 **Tyler J. Jarvis and Nathan Priddis, Editors,** Singularities, Mirror Symmetry, and the Gauged Linear Sigma Model, 2021

762 **Atsushi Ichino and Kartik Prasanna,** Periods of Quaternionic Shimura Varieties. I., 2021

761 **Ibrahim Assem, Christof Geiß, and Sonia Trepode, Editors,** Advances in Representation Theory of Algebras, 2021

760 **Olivier Collin, Stefan Friedl, Cameron Gordon, Stephan Tillmann, and Liam Watson, Editors,** Characters in Low-Dimensional Topology, 2020

For a complete list of titles in this series, visit the
AMS Bookstore at **www.ams.org/bookstore/conmseries/**.